THE NEW VIDEO ENCYCLOPEDIA

Garland Reference Library of the Humanities
(Vol. 1221)

CONTENTS

Preface	ix
Introduction	xi
Abbreviations	xiii
Sources	xvi
The New Video Encyclopedia	1

"The way we acquire information affects us more than the information itself."

MARSHALL MCLUHAN

PREFACE

The authors conceived this encyclopedia so that it will serve as a tool to help owners and users of video units to more fully understand and enjoy the wide array of equipment. Each new product or small improvement seems to generate more confusion about these components and their special features. It is hoped that the entries in this book will clarify the terminology tossed about by manufacturers, periodicals, reviewers and sales people.

All entries are listed alphabetically. Terms or products that are popularly addressed by their abbreviations—such as VCRs—generally are treated as basic entries while their full names are listed as cross-references. Another type of cross-reference pertains to seemingly esoteric terminology that may appear within an entry; we have capitalized these terms to signify a cross-reference where more information may be had on that topic. Some entries are followed by a string of related cross-references. These are listed alphabetically and not necessarily in order of importance. For reasons of simplicity, the authors have decided to use the term "video camera" to encompass both that piece of equipment and the camcorder. When information pertains only to the camcorder, then that specific term is used. Finally, since the book is basically about video, we have put more stress on video and less on audio terms.

ACKNOWLEDGMENTS

We wish to thank the Electronics Boutique of Wayne, New Jersey, for its help in ferreting out facts about video games; Irving Lin of Abest International, Jackson

Heights, New York, for his technical knowledge and for keeping our computers in working order; Kathy Brunetti for the loan of video material; and Canon, Chinon, Panasonic, Philips, Pioneer, Vivitar, Zenith and other manufacturers who supplied us with stills and specifications of their various products.

INTRODUCTION

Having completed the first edition of *The Video Encyclopedia* several years ago, I thought the contents would meet the needs of the most demanding video enthusiast for a reasonable period. However, after the book was printed, a parade of new products invaded the marketplace while several popular items faded into oblivion. The video field had changed drastically. Consumer camcorders replaced the awkward and heavy two-piece video cameras. Many useful features, originally built into high-priced videocassette recorders, soon found their way to low-cost VCRs. Manufacturers introduced Super- and ED-Beta and Super-VHS cameras and VCRs. Sony introduced electronic still video cameras. Laser videodisc players experienced a rebirth. Some models were able to handle a variety of disc formats. Major changes occurred in the audio and video portions of TVs, camcorders and VCRs. Digital videotape recorders and signal processing, time base correctors and editing consoles have changed the way professionals do their post-production work.

Focus on the rapid and dazzling technological advances in video during the last few years has overshadowed the many contributions of those early pioneers who struggled to bring this modern miracle into our homes. We who use all this equipment may not be fully aware of the international nature of video. English, German, French, Russian and Japanese inventors each made significant contributions. England's Sir William Crookes in 1879 experimented with cathode rays. Paul Nipkow invented a scanning device in 1884 that was capable of producing about 4,000 pixels (picture elements) per second. Germany's Karl Braun in 1897 brought the electron beam under control inside a cathode ray tube. H.R. Hertz, the German physicist, was the first to discover radio waves in the 1880s. Russian-born Vladimir Zworykin patented the electronic TV camera tube in 1923. John Logie Baird, a pioneer in British television, invented the first marketable home videodisc system in 1928. English scientist Philo Taylor Farnsworth, at 24 years of age, received a patent in 1930 for his electronic television system which was capable of transmitting visual images. He also invented an early camera tube in 1927. Boris Ritcheouloff, another Russian-born inventor and visionary, designed a video camera, receiver and video recorder in the 1920s. Japan's Dr. Norizaki Sawazaki invented the HELICAL SCAN principle in the 1960s, making possible the home VCR. These and other half-forgotten pioneers are responsible for the little bluish glow that emanates from millions of homes around the world each night.

The international roots of video should remind us of the potential worldwide commercial, social and political ramifications of this technology. Several countries, fully aware of the economic rewards, are presently vying to introduce a high definition TV format that will win international acceptance. Educational pundits are experimenting with practical ways to use TV in the classroom. Historians are busy analyzing the role TV has played in the recent social and political upheavals that have so profoundly marked the late 1980s. Numerous creative video documentarians, equipped with relatively low-cost, high-quality camcorders, are constantly prying into society's warts, including the problems of aging, the homeless, AIDS and runaways. Political analysts will continue to debate the influence of TV on Presidential elections. Obviously, the many potentials of video have only begun to be explored.

ABBREVIATIONS

ACTV	Advanced Compatible Television
CCD	Charge Coupled Device
ED-Beta	Extended Definition Beta
EP	Extended Play
HDTV	High Definition Television
IDTV	Improved Definition Television
LP	Long Play
NTSC	National Television Standards Committee
RF	Radio Frequency
SLP	Super Long Play
SP	Standard Play
S-VHS	Super-VHS
VHS-C	VHS-Compact
VCR	Videocassette Recorder
VDP	Videodisc Player
VTR	Videotape Recorder

SOURCES

Abramson, Albert. *The History of Television, 1880 to 1942.* Jefferson, NC: McFarland & Co., 1987.

Buchsbaum, Walter H. *Complete TV Servicing Handbook.* Englewood Cliffs, NJ: Prentice-Hall, 1982.

Gannon, Michael. *Workbench Guide to Semiconductor Circuits and Projects.* Englewood Cliffs, NJ: Prentice-Hall, 1982.

Genn, Robert C., Jr. *Workbench Guide to Electronic Troubleshooting.* West Nyack, NJ: Parker Publishing, 1977.

Graf, Rudolph F. *Electronics Learning Dictionary.* Indianapolis: Howard W. Sams & Co., 1981.

Grolle, Carl C. *Electronic Technician's Handbook of Time-Savers and Shortcuts.* West Nyack, NY: Parker Publishing, 1974.

———. *Grolle's Complete Guide to Electronic Troubleshooting.* West Nyack, NY: Parker Publishing, 1980.

Quinn, Gerald V. *Getting the Most Out of Your Video Gear.* Blue Ridge Summit, PA: Tab Books, 1986.

Rozman, Leo. *Electronic Troubleshooting With Simplified Circuit Analysis.* West Nyack, NJ: Parker Publishing, 1976.

A&E. See ARTS AND ENTERTAINMENT.

A-B Roll Editing. A generally professional editing procedure using two synchronized reels of the same program material.

A/B Switch. A device that permits any two units connected to it to pass their signals through, one at a time. The position of the switch, either A or B, determines which signal is to be emitted. The switch has two inputs and one output. It can accommodate such units as two VCRs, one VCR and a video game, etc. Since it does not affect the quality of the signals in any way, the A/B switcher is called a passive switcher. See SWITCHER.

AC Adapter. An accessory that connects a video camera to a home VCR by transforming the AC household current passing through the recorder into DC power required to operate the camera. AC adapters have either a 14-pin (Beta) connector or a 10-pin (VHS) configuration. In video, the AC adapter usually powers the PORTABLE VCR and the VIDEO CAMERA or charges the internal and any external battery. It is one of the four basic components of a portable VCR system, the others being the recorder, the camera and the TUNER/TIMER. The AC adapter is different from the ADAPTER used to link incompatible multi-pin connectors.

AC Interlock. A safety function, found on virtually all TV sets, that cuts off power when the back of the unit is opened. Experienced repair persons usually bypass the AC interlock by using a jumper cord designed for this purpose. Some TV receivers permit the service person to remove the interlock wire from the rear partition and connect it directly to the set while doing repair work.

ACC. See AUTOMATIC COLOR CONTROL.

Access Time. In video, the amount of time it takes to reach a selected point on a videotape or videodisc. Some of the more advanced VCRs offer stable program search at from 3 to 21 times normal speed in either direction. Other machines may include a feature

Accessory

called HALF-LOADING, which provides search speed in either the Rewind or Fast Forward mode at 120 times the SP speed.

Accessory. A device, gadget, component or unit that can be added to video equipment. Accessories vary from such simple items as lens filters and carrying cases that sell for a few dollars to highly sophisticated PROC AMPS (processing amplifiers) that can cost thousands of dollars. Some popular video accessories include AUDIO and VIDEO DISTRIBUTION AMPLIFIERS, COLOR PROCESSORS, EQUALIZERS, FADERS, IMAGE ENHANCERS, IMAGE STABILIZERS, IMAGE TRANSLATORS, RF AMPLIFIERS, RF CONVERTERS, RF MODULATORS, RF SWITCHERS, VIDEO FREQUENCY CONVERTERS, VIDEO GRAPHICS GENERATORS, VIDEO PRINTERS, VIDEO SYNTHESIZERS, etc.

Acoustics. In video, the reverberation of sound, or lack of it, in a room or an environment. Acoustics can affect the results of an audio recording. Some parts of a room have "dead" spots while others are more "lively." The built-in MICROPHONE of a camcorder operates better in dead areas; hiss and noise occur in live portions of a room.

Active Filter. In video, a single integrated circuit (IC) designed to reject noise and ripple that may otherwise be transmitted to the tuner of a TV receiver. Active filters replace SMOOTHING FILTERS.

Active Mixer. An audio accessory that corrects for signal losses caused by the circuitry of the mixer. Active mixers affect the signals by re-amplifying them. Some models contain equalizers which further alter the signals. Active mixers can also modify the signal by compressing it, giving it an echo effect or producing a reverberation quality. See MICROPHONE MIXER, PASSIVE MIXER.

Actuator. In satellite TV, a built-in device that controls the movement of the satellite dish or antenna so that it is in the proper position to receive a strong signal.

ACTV. See ADVANCED COMPATIBLE TELEVISION.

Adapter. A device that permits linking a multiple-pin connector of a video camera to an otherwise incompatible connector of a VCR. For example, Sony produces an adapter to fit its own camera's 14-pin K connector with a 10-pin portable VHS recorder. Toshiba also makes an adapter which allows Beta cameras (14-pin) to be used with VHS machines (10-pin connectors). Some independent companies manufacture adapters linking various models. An adapter employs microcircuitry to connect the trigger, pause, recording lights and playback operations. This type of adapter is different from the standard AC ADAPTER.

Adaptive Comb Filter. Special circuitry designed to produce high-quality freeze frame effects. These comb filters are usually found on such professional/industrial components as digital processors. See COMB FILTER.

Adaptive Range Coding (ARC). A video process that condenses the complete NTSC bandwidth into a digital signal that can then be recorded on tape. The technique requires the use of high-grade metal-particle tape with its more tightly packed magnetic field to hold the increased information.

Add-On Speakers. A stereo speaker system designed to increase the range, power and smoothness of the audio portion of TV receivers. Add-on speakers usually come equipped with built-in amplifiers and are connected to either the TV set or the stereo inputs of a VCR. Depending on the company, the power may range from 5 to 25 watts per channel. Because their design provides mutual compensation between speaker and amplifier, some of these high-fidelity speakers, with their superior bass response, are capable of simulating the effect of a movie theater in a living room.

Address Search. A VCR feature that permits the user to allocate a particular number to each index stop by marking it magnetically or electronically. An enhancement over the conventional index search, address search, using the keypad numbers on the remote control, can mark the beginning of as many as 99 scenes on a tape for later viewing. The address search system, also known as VASS or VHS Address Search System, is so advanced that it will automatically shift to reverse when the end of a tape is reached in the Forward mode during a search. Most VCRs use one or more search methods to find a specific scene or moment on a videotape. These machines automatically place an electronic mark on the tape each time the Record button is activated, thereby marking the beginning of every program recorded. Other features allow for specific scenes within a program to be marked. See CTL CODING, CUE MARK, ELECTRONIC PROGRAM INDEXING, INDEX SEARCH, SEARCH MODE.

Addressable Box. An attachment connected to TV sets to permit subscribers of PAY-PER-VIEW programming to receive special events for a fee. Addressable boxes allow the cable operators to deliver programs such as major sports events and films to those customers willing to pay for the service. See PAY-PER-VIEW SYSTEMS.

Advanced Compatible Television (ACTV). An experimental method of improving television broadcasting by adding an incremental signal to the existing one. The new signal produces an image of higher definition. The aspect ratio, or relationship between the height and width of the picture on the TV screen, will be similar to that of the wide-screen motion picture format. The proposed aspect ratio of ACTV has been acknowledged as 16:9, using modified NTSC equipment. (The conventional TV receiver has an aspect ratio of 4:3.) This wider image will encompass more of the original theatrical film than is currently seen on today's TV screens. However, ACTV, in its attempt to squeeze both signals into present broadcast channels, loses some image detail. The fi-

Advanced Editing

nal version of the system will produce 650 lines of horizontal resolution, 800 vertical lines and digital audio. Developed by the David Sarnoff Research Center at Princeton as an alternative to high definition television, ACTV will not make the millions of present TV receivers obsolete. Although the combined signal of ACTV can be received on current TV sets, a viewer will have to buy a new system to reap the benefits of the improved image. See ASPECT RATIO, HDTV, HIGH DEFINITION SYSTEM NORTH AMERICA, IMAGE ENHANCEMENT.

Advanced Editing. Special features usually built into a VCR to help in the editing of home video movies. Such facilities as ASSEMBLE EDITING, edit preview, digital image superimposer and the flying erase head aid in making glitch-free, professional-looking edits.

Advanced Video Entertainment System (AVES). A sophisticated video game system introduced by Atari in 1982. The AVES offered extended memory potential, more realistic graphics and higher resolution. The universal controllers, which included a 12-key panel, had a joy-stick which increased the directional movements from the usual four to eight. The AVES, equipped to accept future peripheral attachments, was incompatible with its other game console, the VCS (VIDEO COMPUTER SYSTEM). The AVES was also known as the Atari 2600.

AFC (Automatic Frequency Control). In video, special circuitry in a tuner or receiver designed to accurately retain the channel to which the component is tuned. The AFC prevents any drift in the selected channel.

AFM (Audio Frequency Modulation). See BETA HI-FI.

AFT. See AUTOMATIC FINE TUNING.

AFV (Audio-Follows-Video). An advanced feature of a professional/industrial editing console or switcher that permits the audio signal to follow the video edit operations, thereby facilitating audio crossfades to be produced under editor control. AFV offers a wide range of possibilities for professional editors. Scene transitions produced in the video mode can automatically activate fades between complex audio balances. Thus, video edits and scene transitions can contain more tightly synchronized crossfades. Also, AFV facilitates the addition of music, dialog and special audio effects to the multi-track master tape before a work print is produced.

AGC. See AUTOMATIC GAIN CONTROL.

Akai VT-350 Portable VCR. A unique portable videocassette recorder introduced in 1979 with its own non-standard tape format. Long discontinued, this unit from Akai had its own tape format slightly smaller than that of the Beta system and incompatible with any other machine. It also featured, unusual for its time, variable speed playback—slow motion to freeze frame. The unit, which had no tuner/timer, was part of a system that included a VC-300 video camera without lens or viewfinder (optionals) and an optional VT-300 monitor with a three-inch screen.

ALC (Automatic Level Control). See AUTOMATIC GAIN CONTROL.

Aliasing. In signal processing, the jagged or stairlike steps of a vertical line that appears on the screen. Aliasing may also be noticed with circular objects. Sometimes anti-aliasing electronic circuits are introduced to lessen the jagged effect. Aliasing is especially prevalent on many computer monitors.

Alignment. In video, the critical position or placement of video heads and the angle at which they make contact with the videotape. Misalignment often causes distortion, signal loss, video noise and snow. The term "alignment" also refers to television tuners and IF (intermediate frequency) amplifiers. The process is usually accomplished by using a sweep generator and oscilloscope. See TRACKING.

Alignment Tape. A special-purpose videotape containing audio and video reference signals that are used to correctly adjust the recording and playback heads of video recorders. Alignment tapes are produced by manufacturers and are not generally available to the public. They are for utilization in company and authorized service centers.

Alpert, Jon. VIDEO ARTIST, co-producer of documentaries, co-winner of the Grand Prize at the 1981 TOKYO VIDEO FESTIVAL. With his wife KEIKO TSUNO, Alpert started the Downtown Community TV Center in New York City in 1972. The husband-and-wife team has been responsible for award-winning documentaries. Their segment of a Public Broadcasting System show won the Grand Prize at the above festival. The documentary "Third Avenue: Only the Strong Survive" covers the daily struggle of an elderly Italian-American barber and his wife.

Alpha-Numeric Character Generator. See CHARACTER GENERATOR.

Alternate Channel Selectivity. The ability of a tuner to focus on one channel at a time while rejecting interference from adjacent channels on the dial. The ability of the tuner to suppress this signal overlap is measured in decibels (dB)—the higher the figure, the more effective the separation. An alternate channel selectivity rating in the vicinity of 80 dB is considered excellent. The term should not be confused with CAPTURE RATIO, which refers to two channels occupying the same frequency.

Ambience. Reflected light or sound that reaches the viewer or listener from a variety of directions. Light or sound waves bounce off the ceiling, walls and other areas of an area. The amount of AMBIENT LIGHT and AMBIENT NOISE (which can be measured) can affect the viewer's or listener's pleasure in terms of video and audio.

Ambient Light. Available light or the normal illumination of a room or environment. The term is used with PROJECTION TV systems, camcorders and TV sets. How these components function in ambient light is one method of measuring their effectiveness.

Ambient Light Filter. A special transparent device placed in front of a television screen designed to reduce the amount of ambient light that falls upon the screen. In addition, the filter helps to minimize the reflections of light from the glass facing of the picture tube.

Ambient Noise. Refers to unwanted sounds during a shooting session with a video camera. Ambient noise can be measured with a sound-level meter. An add-on microphone, used in conjunction with a video camera, may be employed to pick up distant conversations amid ambient noise.

American Museum of the Moving Image. A showplace that emphasizes the hardware of the TV and film industry, including costumes, sets and other paraphernalia. Located in Queens, New York, the museum has on exhibit a parade of equipment ranging from 19th century devices to the present Sony Walkman. Other highlights include interactive exhibits, video art displays, video screenings and a host of consumer products based on popular TV shows and personalities.

Amp Power. In audio, the number of watts per channel. The larger the number, the greater the power of an audio amplifier.

Ampex. The company that demonstrated the first videotape recorder in 1956. Developed by CHARLES GINSBURG, RAY DOLBY, Charles Anderson, Shelby Henderson, Alex Maxey and Fred Pfost, all engineers, the Ampex VTR recorded and played back only in black and white. Alexander M. Poniatoff founded the company. See VCR, VIDEOTAPE HISTORY.

Amplifier. See AUDIO/VIDEO AMPLIFIER, RF AMPLIFIER, VIDEO AMPLIFIER.

Analog. A physical method of representing information in a continuous form. Analog data yields an exact replication of the original information. Most conventional videocassette recorders, for example, record information using the analog process. Analog differs from digital, which duplicates information in a discrete, or discontinuous, form—as with more advanced VCRs. See DIGITAL, DIGITAL EFFECTS, DIGITAL SIGNAL PROCESSING.

Analog Encryption. An encoding procedure that operates within the standard 4.5 MHz bandwidth as applied to the video signal. Analog encryption can be applied to various transmission signals, including satellite, microwave and fiber optics. Some approaches to signal scrambling or encoding may result in degradation of the original video information when it is decoded. For professional purposes, analog encryption is less costly than its counterpart, DIGITAL ENCRYPTION. See ENCRYPTION.

Analog Signal Processing. The conventional method used by audio and video equipment manufacturers to reproduce a signal. A broadcast signal is produced in the shape of a series of waves, each wave height representing voltage, while the distance between peaks in these waves determines the frequency of that part of the signal. These components of the signal, along with others, are separated, amplified and fed into VCRs, TV receivers and other similar types of equipment for reproduction. Much of the original quality of the signal, however, is lost through this process, although some units are better able to rebuild the signal than others, thereby providing a better picture. A more sophisticated approach to reproducing a signal is by means of digital signal processing. See ANALOG-TO-DIGITAL CIRCUITRY, COMPOSITE VIDEO SIGNAL, DIGITAL SIGNAL PROCESSING, SIGNAL, VIDEO SIGNAL.

Analog-to-Digital Circuitry. Special, sophisticated circuits used by such units as time base correctors to convert analog video signals digitally, store them, and then reconvert them to analog form. The process assures that the video signal lines maintain their exact duration of 63.5 microseconds, thereby preventing time base error which results in "squiggly" pictures on screen. Some relatively expensive units, such as top-of-the-line laserdisc players, incorporate analog-to-digital circuitry in an effort to produce high-quality pictures. See TIME BASE CORRECTOR, TIME BASE ERROR.

Analog Tuning. A method of tuning a TV receiver, VCR, etc. Analog tuning permits setting the system to any channel within its frequency range. This tuner, because of its manual capability, either of the mechanical or electronic variety, differs from the FREQUENCY-SYNTHESIS TUNER, which is preset. See TUNER.

Anamorphic Lens. A specially designed camera lens that allows the user to make or view videotapes in wide-screen format. These anamorphic lenses for the consumer market operate on the same principle as those used by large movie companies.

Angle of View. The area or width of a subject or scene that a lens takes in or covers. The angle of view depends on the focal length of the lens. The smaller the number, the greater the angle of view. For example, a 12.5mm focal length has a wider angle than a 75mm lens. The different angles of view are in the descriptions of fixed focal length lenses: wide angle, normal and telephoto. The zoom lens ranges from a wide to a narrow angle of view. See FOCAL RANGE, LENS.

Animation. See FRAME-BY-FRAME RECORDING, INTERVAL TIMER, INTERVALOMETER, OPTICAL ANIMATION, PIPELINE ARCHITECTURE, PIXILATION, TIME LAPSE VIDEO.

Antenna. In video, that part of a transmitter or receiver facility that sends out waves into or accepts them from space. Also, a wire or set of metal rods constructed for the purpose of intercepting waves in the air and changing them into an electrical signal that is sent to a TV receiver. TV antennas are affected by various external factors,

Antenna Combiner

such as the location of the transmitters, the contours of the land and certain obstructions, and the physical condition of the antenna and connecting cables. Virtually all antennas utilize the dipole technique—two equal rods or arms, each as long as one-fourth the wavelength of the anticipated signal. The antenna lead-in is located at the center of the two arms. Since direction is important for maximum reception, most antennas have a combination of reflecting rods and directors (shorter rods). These provide additional directivity. Commercial TV antennas are usually designed for either local (15–20 miles), suburban or mid-range (20–30 miles), or fringe use. See LIGHTNING ARRESTER.

Antenna Combiner. A commercial accessory that joins the signals from several antennas, each of which is aimed at a different TV station. Antenna combiners are helpful, in some instances necessary, where TV transmitters are not located at a single source.

Antenna Coupler. An accessory that is used when more than one TV receiver is connected to a single antenna. The coupler, also known as an antenna splitter, helps to prevent both an impedance mismatch and interference between sets. Several commercial types are available. The resistance antenna splitter, which prevents some impedance mismatch and offers some isolation, unfortunately contributes to a reduction in signal strength. The transformer antenna splitter, on the other hand, reduces both impedance mismatch and insertion loss.

Antenna Rotator. A small motor mounted externally on the mast of an antenna and remotely controlled to adjust the antenna so that it captures the best possible signal from one of several TV transmitters or stations. In some areas where TV stations do not transmit their signal from a central location, the single dipole antenna is not effective. Either several antennas or a single antenna with a rotator must be installed for maximum reception.

Anti-aliasing Circuitry. An electronic circuit designed especially to lessen the effect of jagged edges or stairlike steps on diagonal lines or sections of circles or round objects that appear on screen. Many professional/industrial character generators offer anti-aliasing as one of their features. See ALIASING.

Anti-Copying Signal. See ANTI-PIRACY SIGNAL.

Anti-Piracy Signal. A commercial method of preventing pre-recorded videotapes from being "pirated" or duplicated illegally. One system places a special signal electronically on the tape; another modifies the vertical sync pulses, causing rolling or other forms of instability in the picture during the copying process. Supposedly, this signal has no effect during playback on a television set. Called by various names, the anti-piracy signal often is overridden by late model VCRs with advanced circuitry. Also, some companies sell accessories known as IMAGE STABILIZERS that correct the picture breakup the added signal may cause on some TV receivers. Earlier systems such as Copy-

Guard and Mag-Guard failed, for various reasons, to perform the task for commercial tape distributors. Macrovision, the latest and most effective anti-copying technique, modifies the vertical blanking interval from a video signal, adding other information, such as white pulses between video fields. This process, introduced in mid-1985, defeats the automatic gain control of a second VCR, thereby preventing that machine from copying the video information without some type of picture breakup, color noise or other interference. However, commercial rental cassettes that have been encoded with Macrovision result in a picture that may pulsate from dark to light—an effect that can appear annoying to some viewers.

APC. See AUTOMATIC PHASE CONTROL.

Aperture. The lens opening or iris diaphragm size that can be adjusted to control the amount of light entering the camera. The aperture size is expressed in f-stops, such as f/22, f/16, f/2.8. The smaller the number, the greater the amount of light that passes through the lens. An f/1.2 lens, therefore, provides more light than an f/2 lens. Also, the larger the lens aperture, the more costly the lens. Many camcorders feature AUTOMATIC IRIS CONTROL or automatic aperture. See LENS.

Aperture Grill. One of several techniques employed by TV receivers during picture reception to ensure that the correct electron beam hits the proper phosphor or dot on the picture tube. The aperture grill is also known as aperture mask. See FINE PITCH PICTURE TUBE, SHADOW MASK.

Aperture Reduction Ring. An accessory designed for some projection TV systems to make the image appear sharper by cutting down on the f-stop or APERTURE of the projecting lens. The disadvantage is that using a smaller aperture also decreases the amount of light transmitted to the screen.

APL (Average Picture Level). Refers to the normal luminance level of that part of a television line between blanking pulses. See FIELD BLANKING, LINE BLANKING, LINE SCAN, MUTING CIRCUIT.

ARC. See ADAPTIVE RANGE CODING.

Archiving. The videotaping of television shows, movies and other programs for storage and future playback. Archiving was one of the major points in the MCA/DISNEY VS. SONY LAWSUIT in which the Court of Appeals ruled that owners of VCRs would be gaining "economic control" of broadcast shows, making it difficult for the owners of these programs to fully exploit them. Some legal and video experts contend otherwise. They believe that most VCR owners do not indulge in archiving; instead, they tape a show to watch at a more convenient time, then tape over the material.

Arts & Entertainment (A&E). A CABLE TV advertiser-supported network specializing in cultural programs, documentaries, variety shows and chiefly foreign feature films. In 1990

Aspect Ratio

A&E, with its more than 42 million subscribers, was rated tenth among the leading cable networks.

Aspect Ratio. In video and film, the relationship between the height of an image and its width. In the early days of the film industry the screen ratio was standardized at three to four, also described as 1:1.33 or a 1.33 ratio. The TV industry simply adopted this ratio. In Europe a 1.66 aspect ratio is popular in films, while in the United States movies can be seen in ratios ranging from 1.85 to Cinemascope's 2.35. Since the average television screen cannot accommodate the entire image of films produced in these wide-screen ratios, many undergo a process known as SCANNING before they are televised.

Assemble Editing. A technique in video designed to permit the simultaneous electronic editing of audio and video information. Its counterpart, INSERT EDITING, replaces either the previous audio or video track or both with new information. Assemble editing has been automated with the introduction of edit controllers, a computerized system either built into some VCRs or added as an accessory. These controllers, which store editing information into their memory, can operate the functions of two VCRs to produce assemble edits. See CRASH EDITING, EDIT CONTROLLER, EDITING.

Atari. One-time leader in video game systems and producer of VCS (VIDEO COMPUTER SYSTEM) and AVES (ADVANCED VIDEO ENTERTAINMENT SYSTEM) game consoles. See VIDEO GAME, VIDEO GAME SYSTEM.

Attenuation Cable. A cord or wire with a mini plug at each end and designed to connect the output of a videocassette recorder to a low level microphone input. The attenuation cable is necessary since the recorder output is at speaker voltage level and requires low level routing. The cable can be utilized to add background music to a videotape from a cassette recorder by connecting it from the latter's "speaker out" to a VCR's "mic in." See ATTENUATOR, IMPEDANCE, IMPEDANCE ADAPTER.

ATV. Refers to any type of advanced television system not presently in general use or production. The most recent example of ATV is the high definition television system being demonstrated by more than 30 manufacturers at several consumer and professional shows and exhibitions.

Audimeter. An early device attached to home TV sets and designed to measure a family's viewing habits. The audimeter, placed in representative homes, was used by the A.C. Nielsen Company to measure the popularity of television shows. The rating information gathered from the accessories often determined the advertising rates of the shows and which shows would be renewed or canceled. In addition, the ratings revealed which channel or channels were watched the most.

Audio. The sound segment of a video tape, VCR, VDP or other component. Also, the input, output, cable wire, attachment or other feature, accessory or software referring to the sound portion of a system. For example, there are audio inputs, audio cables, audio mixers, etc. See AUDIO RESPONSE.

Audio Alarm. A feature, found on some camcorders and VCRs, that presents an audible signal to warn the user that certain functions have been activated. Especially useful with remote control, audio alarm is appreciated by camcorder owners. For instance, some models beep once when videotaping begins and twice when it ends. See TRIGGER ALARM.

Audio Bandwidth. In reference to videotape, the parameters or audio range of a tape. Although human hearing can respond to frequencies from approximately 15 or 20 Hz to 20 KHz, the audio portion of a videotape has a bandwidth that is much shorter, somewhere from 50 Hz to 10 KHz, depending on certain tolerance limits measured in decibels. This poorer response is caused by the small area of the tape allotted to the audio track and by the extremely slow speed at which the tape travels past the audio head. Higher-quality tapes extend these numbers on both ends of the bandwidth to produce less distortion, hiss, etc. However, the audio bandwidths of most tapes do not present any true limitations to many low-priced VCRs since these machines have an even shorter range than that of the videotape. See AUDIO RESPONSE.

Audio Cable Tester. A device designed to check cables for shorts, PHASING, continuity, etc. Used mostly by professionals, the cable tester is utilized with standard XLR3 pin-type cables and 3-conductor phone plugs.

Audio/Control Head. See AUDIO HEAD.

Audio Decoder. A costly accessory used in conjunction with VCR-equipped stereo sound to send the signals to various speakers for the purpose of creating a theatrical effect in the home. The audio decoder picks up the encoded stereo track on the videotape and interprets the appropriate paths for the signal, directing it to front, back and side speakers. The decoders are relatively expensive, ranging from about $600 to more than $9,000 for a close approximation of a theatrical sound system. See SURROUND SOUND.

Audio Distribution Amplifier. A device designed to improve the sound quality of videotapes. A typical model contains a special filter circuit which decreases buzz and other noise, a microphone input for mixing sound-on-sound or adding narration, bass and treble tone control, etc. Some models provide a bypass feature for comparing the affected and unaffected signal. The amplifier is often used to prevent GENERATION LOSS of the audio SIGNAL when duplicating tapes.

Audio Dub. A feature on a VCR that permits recording a new soundtrack over the existing one while maintaining the video portion of the program on the tape. Audio dubbing can be used to add a musical background from a stereo system or narration via a

Audio Dubbing Narration

microphone. Use of a microphone requires certain precautions: the volume of the TV set should be turned down to prevent feedback and avoid unwanted extraneous background noise. Some VCRs permit editing a second track to the existing audio track. See AUDIO DUBBING NARRATION, AUDIO DUBBING RECORDED MUSIC, SOUND-ON-SOUND.

Audio Dubbing Narration. The addition of narration to a videotape. The process requires the following steps. Connect an external microphone into the mic input of the VCR. If you decide to use the built-in mic of the camcorder, connect the camera to the VCR. Turn down the volume of the TV set to avoid feedback. Press Audio Dub, start the tape and begin the narration. See AUDIO DUB, SOUND-ON-SOUND.

Audio Dubbing Recorded Music. The addition of music to a previously recorded videotape. One simple procedure is to place the mic next to one of the speakers and switch the sound system to mono. Another, more desirable, method is to connect the amplifier or receiver of a stereo system or an audiotape recorder to the audio input of the VCR. See AUDIO DUB, SOUND-ON-SOUND.

Audio Equalizer. See EQUALIZER.

Audio Essay. A discussion of a specific film or program added to a commercial videodisc or videotape. Usually applied to classic works, the audio essay, which utilizes one of the stereo channels, presents an "expert" who takes the viewer on an oral and visual journey of the production. The historian or critic covers such items as biographical information pertaining to the performers or director, missing or added scenes, interviews and related still shots and trailers.

Audio Expander. A feature on an AUDIO PROCESSOR to improve the dynamic range of sound.

Audio Expansion Circuitry. A development found in higher-priced large-screen TV MONITOR/RECEIVERS to enhance the stereo effect of the system. By employing special circuits, these audio systems, with their sealed enclosures and multiple loudspeakers, produce superior sound to that normally provided by TV sets and other monitor/receivers.

Audio-Follows-Video. See AFV.

Audio Frequency Modulation. See BETA HI-FI.

Audio Head. In video, a stationary magnetic head capable of recording and playing back sound signals. After receiving the audio signal, the head pulses it onto the videotape during recording or takes it from the tape for reproduction during playback. The audio head is the third and last process that affects the videotape. The ERASE HEAD is the first, followed by the VIDEO HEADS. The audio head assembly is sometimes called the audio/control head and contains three heads. One performs the audio recording and playback, the second is designed for audio dubbing and the third is the control track head which transmits pulses onto the tape to con-

Audio Response

trol the start of each alternate field; i.e., to track the original recorded signal. See AUDIO DUB, CONTROL TRACK.

Audio Input. A jack, often located at the rear of a VCR, that accepts audio signals. It is usually used, along with the VIDEO INPUT, when copying or duplicating a videotape from another machine. A cable is connected from the audio-out of the playback machine to the audio-in of the recording VCR. Audio input is also utilized during audio dubbing. Beta machines take a mini-plug while VHS machines accept the standard RCA plug. See AUDIO DUB, COPYING, DIRECT AUDIO AND VIDEO INPUTS.

Audio-Mix Control. A stereo VCR feature that designates the amount of audio each channel feeds to the mono RF output. Table-model and portable stereo videocassette recorders produce dual channel sound by means of two individual audio tracks laid down on the top portion of the videotape. During the normal stereo playback, both tracks are utilized. The audio-mix control, however, permits an increase in either left- or right-channel sound by simply rotating the knob. See AUDIO TRACK, STEREO VCR.

Audio Output. A jack on a VCR used to redirect audio signals to other components. For example, to duplicate a videotape, a cable is connected from the audio-out of the playback machine to the audio-in of the recording unit. (Another cable connects the video input and output.) The audio output is also utilized in connecting the VCR's sound to a stereo or hi-fi system. Beta VCRs use mini-plugs while VHS machines require RCA phono jacks. See COPYING.

Audio Plug. The metal connector at either end of an audio cable that fits into component receptacles called jacks. Three basic types of audio plugs used in home video are mini-plugs, phono plugs and phone plugs. The mini-plug is a smaller version of and similar to the phone (from telephone) plug. Both have a shaft which protrudes from a metal sleeve. The phono (from phonograph) plug, often referred to as an RCA-type plug, also has a small shaft, but it is surrounded by a petal-shaped metal cup. See AUDIO INPUT, AUDIO OUTPUT, PLUG, PHONO JACK, etc.

Audio Processor. A device that can be utilized in video; e.g., between a VCR and a stereo system. The audio processor usually contains such features as inputs for microphones, VCRs and tape; a multiple-band EQUALIZER for improved video sound; a stereo delay simulator and an AUDIO EXPANDER to extend the dynamic range of the video sound signals.

Audio Response. In videotape, the ability of the tape to reproduce audio signals. Better-quality tapes, especially those listed as HG (high grade), produce less HISS, or an above-average signal-to-noise ratio. Tapes of poorer quality cause more audio distortion, hiss, etc. Audio response becomes more critical in the slower speed modes of both Beta and VHS machines. The average listener can respond to frequencies from approxi-

Audio Signal

mately 20 Hz to 20 KHz. Standard videotape, however, falls short of this range, and is somewhere between 50 Hz and 10 KHz. Distortion and poor response result beyond these parameters. Tapes that exceed this audio bandwidth range (within certain tolerance limitations measured in decibels) may be considered better than average, although most home video machines have a range narrower than that of most tapes. See AUDIO BANDWIDTH, HG.

Audio Signal. An electrical signal whose frequency falls within the audible range, the lowest measured at about 15 to 20 hertz and the highest at approximately 20,000 hertz or 20 KHz. See AUDIO BANDWIDTH.

Audio Signal-to-Noise Ratio. In videotape, a measurement that determines the loudness of an undistorted signal relative to tape noise. Audio signal-to-noise ratio is measured in decibels (dB). Videotapes tend to have an average ratio of 50 dB. The larger the number, the better the audio quality of the tape.

Audio/Video Amplifier. An accessory that adds sound processing to videotapes. The unit usually comes equipped with multiple audio/video inputs and outputs on its rear panel and digitally delayed audio modes that offer such special effects as stage, stadium, theater and matrix. The switchable amplifier may power several channels, depending upon the watts-per-channel used. Some models accommodate S-VHS and ED-Beta formats and feature a title generator and a video enhancer.

Audio/Video Dub. A video camera feature that permits the replacement of a current segment of audio and video information on tape with new material. When audio/video dub is activated, new information is inserted over both tracks. Most present cameras offer this feature while other models provide only AUDIO DUB.

Audio/Video Input. Basic RCA jacks found on VCRs and TV monitor/receivers. Stereo models provide one input for video and two for audio. Mono units offer only one audio and one video jack. See AUDIO INPUT, VIDEO INPUT.

Audio/Video Memory Function. A feature, found on some projection TV systems, that permits optimum control set-ups to be stored in memory for later recall.

Audio/Video Mixer. An editing accessory that allows switching back and forth between two video sources, such as two VCRs or a VCR and a camcorder. Some A/V mixers offer additional features such as a fader, wipe effects and special-effects generator. There are manual and electronic mixers. The electronic type may use computer software and infrared technology to "learn" the tape transport commands of the recording VCR. The user simply marks and names the scenes on the footage to be edited and instructs the mixer which scenes should appear in the final tape.

Audio/Video Mute. Special electronic circuits built into some AUDIO/VIDEO PROCESSORS that are designed to silence a TV set to circum-

AudioVideo International

Figure 1. An audio/video receiver, with multiple audio and video inputs, designed as a control center of home entertainment systems. (Courtesy Pioneer Electronics.)

vent annoying noise and static. The mute feature also darkens the screen when the tuner is between channels or a videotape ends. Muting is sometimes referred to as blanking. See also MUTING CIRCUIT.

Audio/Video Processor. A multifunction device for use with various video components. The processor usually provides inputs and outputs for audio and video switching, an audio and video distribution amplifier, a video stabilizer, an image enhancer, and an RF converter. Some sophisticated models may offer color tint control, color intensity control, split screen enhancer, audio/video mute, a bypass switch and a fade duration control. See all the above under individual entries.

Audio/Video Receiver. A separate unit designed to function as a control center of home entertainment systems. High-end A/V receiver systems usually accommodate several audio inputs (CD, phono, tape and line) and several video inputs (VCRs, videodisc player and television). Some units permit two-way dubbing, include S-video terminals and offer sound field memory which can store several surround sound settings as well as "memorize" 30 stations for instant recall.

Audio/Video Signal. See SIGNAL.

Audio/Video Switcher. See SWITCHER.

AudioVideo International. A popular monthly trade publication that covers such topics as sound processing, automobile sound systems and prerecorded video. Regularly scheduled departments report on world audio/video news, sales leaders, news from Washington, new products and the audio/video marketplace in general. For those interested in the more technical aspects of the field, the magazine offers an informative "tech talk" column. Also, *AudioVideo International* regularly presents awards for superior designs in video equipment. See VIDEO GRAND PRIX AWARDS.

Auto Channel Search. See AUTOMATIC CHANNEL SCAN.

Auto Eject. A VCR feature that releases the videocassette from its housing after it has been rewound. Some machines also provide power-off eject, which performs the same function when the VCR is turned off.

Autofocus. A process built into some camcorders in which an impulse of invisible light is emitted to the subject and returned to a pair of infrared sensors. This distance is then calculated by an integrated circuit. Finally, a drive motor adjusts the lens. Hitachi, Toshiba and Akai were among the first companies to feature autofocus in their cameras. Some of today's video cameras offer a more sophisticated—and more accurate—autofocus technique. Instead of relying on the not-too-precise infrared reflection to measure distances, the focal adjustment of these later cameras operate directly off the image-sensing elements. See PASSIVE PHASE DETECTION AUTOFOCUS.

Auto Frame. See AUTOMATIC FRAMING.

Auto Image Stabilizer. See LENS STABILIZATION.

Auto Lock Switch. A feature, found on some camcorders, designed to simplify and speed up the operation of the camera. When the user activates the auto lock switch, it simultaneously sets the autofocus system, white balance, shutter speed and backlight compensator.

Auto/Manual Aperture Control. A device that places the control of the f-stops or aperture openings into the hands of the user. Many camcorders feature AUTOMATIC IRIS CONTROL, a less desirable feature for some camera owners who prefer to make their own selections. Some users choose to open or close the lens one or two additional stops for special effects.

Auto/Manual Iris Control. See AUTO/MANUAL APERTURE CONTROL.

Auto Play. A VCR feature that automatically activates the Play mode of the machine when a videocassette is placed into the housing. The auto play function, which may also be found on videocassette players, operates only with cassettes that have had the safety tab removed. Auto play, which eliminates the need for pressing the Play button, is sometimes listed as auto playback function.

Auto Power On. A feature of more costly large-screen TV MONITOR/RECEIVERS that permits the viewer to set a built-in clock/timer for as much as 24 hours in advance. Also, a VCR feature that automatically turns on the power of the unit when a videocassette is inserted.

Auto Program. See AUTOMATIC CHANNEL SCAN.

Auto Repeat. A feature found on some videocassette recorders and players that permits the viewer to play back repeatedly an entire videotape. The Auto Repeat mode can usually be disengaged by pressing the button one more time. Auto Repeat differs from Repeat Play, a sophisticated feature found on more costly VCRs that plays a videotape up to an index signal, stops and rewinds to a previous index signal and continues to play that portion of the tape indefinitely until the Repeat Play mode is pressed again. See REPEAT PLAY.

Auto-Setup. A type of professional/industrial TV monitor that adjusts itself automatically, thereby eliminating the fine tuning previously required by technicians. Auto-setup monitors are especially useful in viewing the same image when interchanging videotapes from one facility to another. Another advantage is the capture of the same image when several auto-setup monitors, adjusted for the same color temperature, are arranged in a row. See MONITOR.

Auto Tracking. A feature, found on some videocassette recorders, that seeks out the most accurate playback tracking position for a given tape. Since machines differ in their video head placement, some tapes, especially those recorded at the slowest speed, often do not play properly on other VCRs from different manufacturers. Virtually all machines come equipped with a manual tracking control to adjust for these variations. Auto tracking handles these differences automatically. In addition, machines equipped with this feature usually provide manual tracking as well for those who prefer to make their own adjustments. Auto tracking differs from automatic tracking reset, which simply returns the machine to its tracking default after the viewer has manually changed the setting for a specific problem videocassette. See AUTOMATIC TRACKING RESET.

Auto-Winder. See REWINDER.

Automatic Backlight Compensator. See BACKLIGHT SWITCH.

Automatic Backspace Editing. A VCR feature that eliminates frame overlapping for glitch- and distortion-free transitions. A built-in microprocessor, after checking the signals of the control track, makes certain that a new recording starts at the end of the last frame each time a recording is begun from the Pause mode.

Automatic Channel Scan. A TV or VCR feature that automatically programs the TV or VCR tuner memory to lock in only active VHF and UHF channels. Usually operated from the remote control unit, the automatic channel scan, sometimes described as auto program or automatic channel search, searches up and down those channels active in a particular area and ignores the inactive ones that only bring in noise and static.

Automatic Chapter Search. A videodisc player feature that, when activated, takes the viewer to a particular selection on the disc. This chapter search feature is often found on the remote control unit of a player. See CHAPTER.

Automatic Color Circuitry

Automatic Color Circuitry. Electronic circuits built into some TV sets, TV monitors and monitor/receivers designed to retain factory-preset color levels. Automatic color circuitry locks in this balanced color arrangement regardless of discrepancies between channels and scenes. One disadvantage or criticism of this feature concerns the viewer's preference—the colors may appear too weak, too intense, too bluish, etc. However, the color circuitry usually comes with a switch that can be deactivated so that the colors can be adjusted manually. Also, a technician can modify the automatic color circuitry so that it operates more to the owner's liking.

Automatic Color Control (ACC). Special circuitry built into color TV receivers and monitors to automatically adjust the input signal, i.e., the chrominance bandpass amplifier, by controlling the bias. TV problems with color intensity, which cannot be immediately remedied by turning the color gain control, are usually the result of a defective ACC circuit. Sometimes listed as automatic color correction, ACC, which has been around for more than a decade, differs from automatic color compensation—a relatively recent development.

Automatic Color Compensation. An advanced feature, found only on some costly television sets, that monitors the three color guns or electron beams so that the colors retain their accuracy for the life of the cathode ray tube. Under normal conditions, tubes lose their color intensity as they age. With the addition of the special electronic circuitry, the TV set can compensate for this imbalance.

Automatic Color Tint Control. See COLOR TINT CONTROL.

Automatic Contrast Correction. A projection TV feature that helps to bring out almost imperceptible detail in overly bright or extremely dark sections of a screen image.

Automatic Degaussing Circuit. See DEGAUSS.

Automatic Digital Tracking. A VCR feature that automatically monitors its own playback. Special circuitry continually compares the RF signals on the videotape to reference signals in the circuit. If the two signal are not in sync, the special circuit emits a correcting signal to the capstan servo, which permits the video head to make adjustment for the best possible signal. See CAPSTAN SERVO, TRACKING, VIDEO HEAD.

Automatic Fade Control. A camcorder feature designed to provide fade-outs at the end of scenes and fade-ins at the openings. When the fade control is engaged during the middle of a scene, nothing happens until the end, when the fade-out occurs. If the control is pressed before starting the camera, the scene will open with a fade-in. Some cameras can be programmed to fade in and out on a scene.

Automatic Fine Tuning (AFT). A feature on such units as VCRs and TV receivers that permit the optimal setting for each channel to be locked in.

Automatic Image Stabilization

This eliminates the need to retune a channel each time it is selected for play. To get the best off-the-air recording with a VCR, select a channel, switch the AFT to Off and manually fine tune for the best picture. Then switch the AFT to On to lock in the channel.

Automatic Focus. See AUTOFOCUS.

Automatic Focus Compensation. A projection TV feature that adjusts for the disparity in projection differences between the lens and the center of the screen and the lens and the edges of the screen. See KEYSTONING.

Automatic Framing. A video camera zoom lens feature that keeps the size of the subject constant. Whether the subject moves toward or away from the camera, the automatic framing function maintains the original size of the image. Canon was the first company to offer the special zoom lens feature, also known as auto-framing, on some of its higher-priced 8mm camcorders.

Automatic Frequency Control. See AFC.

Automatic Gain Control. A camcorder feature designed to increase the signal only to the degree that the image gains an even intensity. AGC increases a blank signal to gray; in poorly lighted scenes it adds VIDEO NOISE to shadowed areas and produces less saturated colors. Usually in the form of a switch, the AGC when activated has one disadvantage: some deterioration occurs in the video image. In most VCRs a circuit controls the intensity of incoming audio and video signals so that they match predetermined output levels while taping off the air or recording with a camcorder. Automatic gain control is different from the SENSITIVITY SWITCH which affects the general amplification of the video signal. In audio, the AGC automatically boosts or attenuates audio signals to optimum levels. AGC is also known as automatic level control, a feature found on other instruments such as a COLOR VIDEO NOISE METER, where it serves to stabilize input levels.

Automatic Hue Control. A signal inserted into the vertical blanking interval to help a television set adjust the proper color. Found on only a few TV models, automatic hue control, which is placed on line 18 of the blanking interval, may cause a problem when prerecorded tapes encoded with the anti-piracy code Macrovision are played. The anti-piracy signal uses this same line for its white pulses which are placed here to defeat copying the tape information to another VCR. Automatic hue control is similar in function to AUTOMATIC COLOR CIRCUITRY.

Automatic Image Stabilization. As applied to video cameras, a method of achieving a steady recorded picture while the user is walking with the video camera. Normally, the results of such camera recording show up as images that are jumpy at best or unintelligible at worst. The use of servo mechanisms and rapidly responding compensating motors converts camera movement into relatively smooth pictures. All this is accomplished by

Automatic Iris Control

activating a special switch on cameras equipped with automatic image stabilization. See IMAGE STABILIZATION, IMAGE STABILIZER.

Automatic Iris Control. A camcorder feature designed to automatically operate the lens opening by "reading" the average light within a scene. If the automatic control cannot be overridden manually, then there is no way to correct for extreme light or dark backgrounds, etc. Some cameras are equipped with AUTO/MANUAL APERTURE CONTROL, a more desirable method, offering both flexibility and automation. See BACKLIGHT SWITCH.

Automatic Level Control. See AUTOMATIC GAIN CONTROL.

Automatic Light Control. In video, an electronic circuit which modifies any incoming light to a predetermined level. The ALC affects light the way the AUTOMATIC GAIN CONTROL affects audio.

Automatic Lock. A videodisc player feature that holds the optical assembly in place when the power is shut off. Similar to a "park" program for a computer disk drive, the automatic lock helps to prevent damage to the internal assembly whenever the machine has to be moved.

Automatic Phase Control. Special circuitry built into color TV receivers designed to synchronize the burst signal with the color oscillator. A TV set that displays hue changes as the receiver warms up reflects a problem related to the APC circuit, which compares the color reference burst with the 3.58 MHz oscillator output.

Automatic Program Delay. A professional/industrial unit designed to provide delays from a few minutes to several days, play back multiple feeds simultaneously and accommodate incoming feed record-only sessions. Delay actions are operated via time codes, are frame-accurate and are affixed to the studio reference clock. Some units can handle a schedule of up to 1,000 events which can be programmed for automatic operation.

Automatic Program Edit. A feature, found chiefly on top-of-the-line laserdisc combination players, that aids in the process of dubbing from disc to tape. Once the user enters the length of tape selected for recording, the player automatically calculates the number of tracks that can be recorded within that time range.

Automatic Programming. A feature on videocassette recorders designed for presetting a number of programs on different channels and at various times to record automatically. The most basic, and original, example of programming is scheduling a single event from one channel within a 24-hour period. With electronic tuners replacing the mechanical rotary models and other developments, VCRs can be programmed for eight events or more on as many channels to be recorded over a one-year period. Generally, the greater the range of multiple-event programming, the more expensive the machine. See ELECTRONIC TUNER, PROGRAMMABLE TUNER, TUNER.

Automatic Sag Compensation. Refers to a feature, built into some professional/industrial instruments such as COLOR VIDEO NOISE METERS, that helps to produce uniform input signals.

Automatic Scan Tracking. A feature, found on some video tape recorders, designed to provide distortion-free slow motion from FREEZE FRAME to Play mode. Originally only on professional/industrial VTRs, the function, like many other advances, eventually filtered down to home systems. See VISUAL SCAN.

Automatic Switchover. A feature, found on some high-priced professional/industrial video monitors, that enables the unit to accommodate either 110 or 120 volts without any manual adjustment. See MONITOR.

Automatic Timing. See PROGRAMMABLE TIMER.

Automatic Tracking Reset. An electronic feature, found on some VCRs, that returns the tracking control to its default or standard setting each time a videocassette is ejected or when the power is turned off. Tracking adjustment is sometimes necessary to play back a tape recorded on a different machine. Without automatic tracking reset, the VCR owner must reset the tracking control manually. Otherwise, the next recording will permanently be "off center," a familiar problem with forgetful VCR owners. Automatic tracking reset differs from auto tracking, which works completely automatically, requiring no adjustments. See AUTO TRACKING, DIGITAL TRACKING.

Automatic Transition Editing. A process that permits glitch-free editing by automatically winding videotape back a few frames when recording is stopped. When Record is resumed, ATE aligns the beginning of the new recording with the end of the previous one, thereby eliminating glitches and picture breakup. The problem with some types of automatic transition editing is that the last part of the previous scene is sometimes lost. Also, exact editing is almost impossible. JVC was one of the first manufacturers to offer this editing technique found today on many VCRs and portable models. Other VCR manufacturers use different approaches, all of which achieve similar results—almost glitch-free edits. Sony, for example, introduced its time-phase circuit. When the tape restarts, its movement is delayed electronically by special circuitry until the beginning of a FIELD rather than the middle of one. ATE is also known as Edit-Start, Edit-Start Control, Scene Transition Stabilizer, etc.

Automatic Turn-On. See AUTO POWER ON.

Automatic Variable Frequency Scanning. An advanced feature of sophisticated TV monitors that can receive a range of signals which permit the user to switch from video to computer graphics. Some models can scan from 15 to 36 KHz while others offer more limited ranges such as 31.5 and 35 KHz. Several monitors designed

Automatic White Balance

exclusively for computers call this feature "multisync."

Automatic White Balance. A feature on expensive camcorders to help simplify white balance control. By aiming the camera at a white card or similar object and pressing a button for a few seconds, a special circuit in the camera scrutinizes and compares the Red-Green-Blue channels and automatically corrects them. Sometimes automatic white balance is only one of two or three controls used on a camera for color adjustment. Other cameras have improved the automatic white balance adjustment by allowing it to be put on hold. By preventing the adjustment from changing automatically to match the shifting light conditions, the camera user can capture the dramatic changes in such scenes as sunsets without the camera compensating for these light changes. See WHITE BALANCE CONTROL.

Autosizing. See CHARACTER SIZING.

Autostereogram. A technique used for three-dimensional television without the use of special glasses required by the viewer. Large lenses are employed, and the viewer has a very narrow, fixed position from which to watch the 3-D picture. Autostereogram is not compatible with 2-D or current TV programming. See THREE-DIMENSIONAL TELEVISION.

Auxiliary Preset Button. A videocassette recorder feature found on some units and designed for setting in advance the output channel of a decoder that may be required for some cable TV systems.

Auxiliary Trigger. A video camera feature that provides an additional pause button. The auxiliary trigger can be useful in certain situations such as shooting in awkward or unusual positions. Not all cameras offer this sometimes-helpful feature.

AV Video. A monthly magazine aimed chiefly at professionals involved in "production and presentation technology," according to the subtitle on the cover. Columns discuss both video and audio techniques and applications while articles follow a similar pattern and include other topics as well, such as coverage of various equipment, slides and film, and contests. Several departments focus on useful tips, technical talk and product reviews. Perhaps the most noticeable feature of the periodical is its eye-catching color graphics.

Available Light. The amount of natural or artificial light that is present in an environment before a camcorder and accessories arrive. Some cameras are capable of recording quality pictures in available light. Light is measured in LUX or FOOT-CANDLE numbers. The lower the number, the greater the sensitivity of the camera. See FOOT-CANDLE, LUX, SENSITIVITY RANGE.

Azimuth. The angle of the recording head in relation to the tape path. To prevent crosstalk, or the confusion of the video heads in playing back the proper tracks which are crowded together, the head gap angle is tilted slightly away from the perpendicular. In the Beta format the tilt is seven degrees. Thus, each of the two heads lays down a different pattern on the tape. It is as if one recording head placed down a horizontal design within its diagonal track while the second head recorded a vertical pattern. When the tape is played back, each head can retrieve only the design or pattern it recorded, thereby eliminating crosstalk. The azimuth system provides a second advantage. The tracks that the two heads produce can be placed next to each other, eliminating the guard bands or spaces previously required between tracks. This permits storing more information on the tape. Sony first introduced the azimuth system in 1975. Some relatively recent Super-VHS camcorders have introduced a double-azimuth four-head system which reduces the size of noise bars during the search mode and provides noise-free still frames. See CROSSTALK, GUARD BAND, HELICAL SCAN.

Azimuth Error Correction. Special electronic circuitry designed to help rectify defects in prerecorded Dolby sound. Azimuth error correction can help with such problems as dialogue, targeted for the center channel, escaping to rear speakers.

B

Back Light. A photographic lighting technique that positions a light behind the subject to be recorded. The back light can be used to create or add depth to a scene. It can also be employed to create silhouette shots. See LIGHTING.

Back Stage. A weekly newspaper servicing the communications and entertainment industry. With main offices in the East, West, Midwest and Southeast, the publication provides national news about advertising agencies, teleproduction and post-production services, and their personnel. A separate section is devoted to stage, film and television casting, with news and reviews of theater, dance and cabaret. *Back Stage* will occasionally present a technical article on some phase of video, such as film-to-tape transfer, advances in video cameras or digital technology.

Background Music Jack. An audio feature on some camcorders that permits the user to connect an external sound source during the recording process. Adding background sound while the original recording is in progress eliminates the usual generation loss that is inherent in tape editing when music is recorded after a recording session.

Backing. The plastic base of videotape, called base film. Backing is designed to resist stretching and decomposition. A backcoat is usually placed on the opposite side of the base film to protect the plastic base, resist static and help the tape to pack evenly during play and rewind. In addition, backcoating increases the opacity of videotape. This is designed to prevent improper stopping of the tape on VCRs that use light to signal the end of a tape. The thin backcoat is measured in microns. A binder to hold a coating of magnetic oxide particles is placed on the face of the backing. See BINDER, COATING, OXIDE, VIDEOTAPE.

Backlight Switch. On some video cameras with AUTOMATIC IRIS CONTROL, a feature designed to provide one f-stop more light. In a scene containing a bright background with a dark subject, the automatic iris usually reads the darker part of the picture, causing the main subject to be underexposed. Activating the backlight switch compensates for this by adding more light. Different from a CONTRAST COMPENSATION SWITCH, the backlight switch only opens the lens wider. The backlight switch is sometimes listed as an automatic backlight compensator.

Backspacing. A feature on portable and home model VCRs designed to eliminate picture BREAKUP between scenes by backing up the tape when the Pause mode is engaged. Then when Record is pressed, the tape begins at the end of the previously recorded section. On some machines the tape is backed up over the last few frames of the previous scene. See AUTOMATIC TRANSITION EDITING.

Backward Compatibility. The capability of an improved or enhanced piece of hardware to accept software designed for an earlier model. For instance, S-VHS videocassette recorders, which require special tape to benefit from the improved features of the VCR, can play back standard videotapes. However, the conventionally recorded tapes will not reflect the higher quality of S-VHS VCR. See FORWARD COMPATIBILITY.

Baer, Ralph. Developed the first video game in 1971 while he was employed by Magnavox. See VIDEO GAME HISTORY.

Baird, John Logie. Pioneer in British television, inventor of the first marketable home VIDEODISC system in 1928. Using a TV system based on electro-mechanics (a light-sensitive cell and a mechanical revolving disc), he was able to send a television signal from London to New York in 1928. The British Broadcasting Company employed his system when television was introduced into England in 1929. Within a few years the process succumbed to an all-electronic TV system developed in the United States by such scientists as PHILO T. FARNSWORTH and VLADIMIR ZWORYKIN.

Balun. An adapter used for converting 300 ohms into 75 ohms. Usually supplied with a videocassette recorder to convert 300-ohm antenna wire, the balun is often needed to connect a video game, VCR or other component to a TV set, etc. One example of its function involves connecting a 300-ohm video game to a 75-ohm projection TV system. There are two types of baluns: a VHF-only and a UHF/VHF model. Sometimes called "balanced converter," the word "balun" is an acronym stemming from "BALanced" and "UNbalanced."

Band. In audio/video, a span or range of frequency signals. For instance, TV satellites work within two frequency ranges. Large dish antennas require the popular C-band whereas smaller dish antennas utilize the Ku-band. See BANDWIDTH.

Band Selection. A feature incorporated into electronic tuners and containing various positions, each of which permits the tuning of designated bands or channels. Electronic tuners on some TV receivers and VCRs usually provide 14 or more touch sensor tuner buttons. Each of

Band Separator

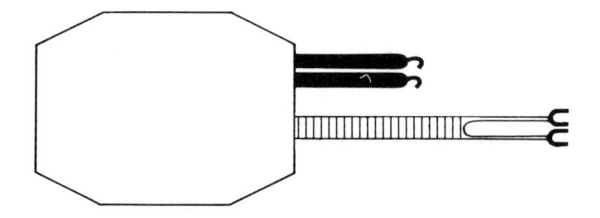

Figure 2. Band separator.

these positions can be adjusted for any one of three ranges: L (low band) for VHF channels 2–6; H for VHF channels 7–13, midband A-I and superband J-W; and U for UHF channels 14–83. However, some more recent and more sophisticated machines offer four ranges: VHF low (channels 2–6), VHF mid-high (channels 7–13 and cable A-I), UHF (14–83) and VHF super (for superband channels on cable). To bring in one of these channels, a channel button is pressed, the range selector is placed in position and then a small dial or knob called a tuning control is rotated until a satisfactory picture appears on the TV screen. Band selection is also known as channel range selector. See ELECTRONIC TUNER, MIDBAND CABLE TV, SUPERBAND CABLE TV, TOUCH SENSOR BUTTON.

Band Separator. An accessory that separates incoming UHF, VHF and FM antenna signals so that they can be directed to their respective terminals. Some band separators accept a 300-ohm (twin-lead) input while others take a 75-ohm input with a built-in matching transformer (75 to 300 ohms). Band separators provide a VHF as well as a UHF output. A band separator is different from a SIGNAL SPLITTER.

Bandpass Filter. In electronics, a filter that permits the passage of certain frequencies only. Bandpass filters, as well as some other types of filters, permit the boosting and rejecting of selected information at various designated frequencies. These filters are used in SIGNAL PROCESSORS to affect color and definition. See DEFEAT FILTER, FILTER.

Bandwidth. In video, the range of frequencies transmitted or received. The Bandwidth determines the amount of information that can be sent or picked up. The wider the bandwidth, the more detailed the image. Better-than-average video signals often fill a 4 MHz (megahertz) bandwidth. Average television sets cover less than this while COMPONENT TV systems very nearly approach this figure, thereby producing a picture with higher resolution. See AUDIO BANDWIDTH, AUDIO RESPONSE, MEGAHERTZ, RESOLUTION.

Bank Timer. A VCR timer-related feature that can store several sets of timer-recording instructions under different categories. On-screen menus help the viewer to code in timing information under such topics as news, cinema, cartoons and drama. These instructions are then entered into the

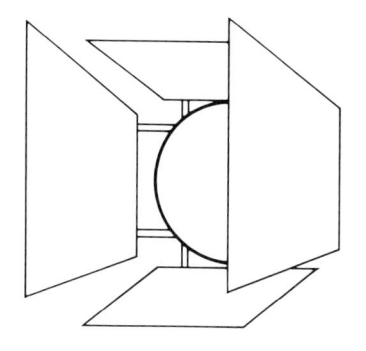

Figure 3. Barn doors.

timer section for upcoming recordings.

Bar Code Programming. A VCR feature that simplifies transmitting recording instructions to the clock/timer of the videocassette recorder. VCRs that come equipped with this programming function provide a pen-like device, called a bar code scanner, and a programming card containing a list of days, time segments and channel numbers. The owner, using the scanner, simply checks off the appropriate day, time and channel on the card for each program to be recorded. The information is then transferred to the VCR by infrared beam and can be displayed for confirmation on the TV screen.

Barn Doors. Adjustable metal flaps attached to the sides of a lamp unit. They control in part the amount of light directed onto a set.

Barrel Effect. Vertical edge distortion of a screen image. Better known as geometric distortion, the term has derived from the similarity of the distorted edge to the curved slats of a barrel. The effect tends to be more pronounced in rear projection TV systems. Barrel effect is sometimes measured in percentages, with line deviation judged by a straight-edge ruler. See GEOMETRIC LINEARITY.

Base Film. See BACKING.

Base Light. The general illumination of an area. The base light helps provide the camcorder with a lighting level above that which is needed to prevent electronic "noise." All video units, such as VCRs and cameras, produce video noise which affects the video signal. The base light, which is usually located over the subject, helps to overcome this. See LIGHTING.

Basic Cable. A term referring to the minimum services a subscriber of a cable TV system gets for the minimum monthly charge. These services usually include VHF and UHF channels, CNN (the CABLE NEWS NETWORK), religious and weather channels and other programming nationally distributed. Other services, like HOME BOX OFFICE, however, require additional monthly fees. See CABLE TV, TIER.

Battery. See LEAD ACID BATTERY, NICKEL CADMIUM BATTERY.

Beam Indexing. Refers to a signal generated by an electron beam that is deflected and fed back to a control device. Beam indexing is one of several methods of presenting images upon a screen. See DISPLAY.

Beck, Stephen. VIDEO ARTIST, electronics engineer. Working with a VIDEO SYNTHESIZER, he originated

Beeper Feedback

the concept of combining color, form, texture and motion to produce abstract kinetic VIDEO ART. His works are recognized world-wide, including Art De Moderne in Paris and the Museum of Modern Art and the Whitney Museum, both in New York City. His video art has been exhibited on national TV via Public Broadcasting System. In the early 1970s he worked on a PBS series call "Video Visionaries" along with performing live at concerts.

Beeper Feedback. See AUDIO ALARM, TRIGGER ALARM.

Beta Format. A system of home videotaping using a special two-hub plastic videocassette, 1/2-inch tape and recording speeds incompatible with other formats. Introduced for home use in 1975 by Sony, the Beta format uses a cassette smaller than that of its competitor, the VHS format. In its Beta II speed the cassette has a maximum play time of two hours with the L-500 tape, three hours with the L-750 and three hours and 20 minutes with the L-830. In Beta III the L-500 plays for three hours; the L-750 for four and a half hours and the L-830 for five hours. The Beta loading system originally pulled out about 24 inches of tape from the cassette once it was inserted and kept it threaded in all modes. Newer models released by Sony, particularly the compact home camcorders introduced in the fall of 1989, have a shorter, modified loading system. Although Sony was the originator of home VCRs, its Beta format fell behind VHS in popularity. Other companies, including Zenith, Toshiba, Marantz and Sanyo, had originally selected the Beta format for their VCRs and video cameras, but have since abandoned that format in favor of the more successful VHS. See BETA SPEED MODE, BETAMAX, SONY and WRITING SPEED.

Beta Hi-Fi. A full-frequency stereo process for VCRs developed by Sony in 1982. Conventional video stereo as found on VHS machines utilizes longitudinal sound tracks (tape passing across a stationary head). But this method produces poor sound quality because of two factors. The tape speed of video machines is very slow, only a fraction of that of audio recorders. Secondly, the small space of the tape allotted to the normal mono audio track has to be split in half to provide for the dual channels necessary for stereo. For these reasons, a noise reduction system such as Dolby B is required to improve some of the less-than-adequate sound. Sony, however, avoids these two shortcomings by using the video heads to place the FM-modulated audio signals onto the tape, superimposing the channels over the video signal. A greater dynamic range results, with a frequency response said to be from approximately 20–20,000 Hz. The portion of the tape otherwise assigned to the audio signal can still be used for a mono sound track (to keep the system compatible with other Beta machines). It can also be utilized as a third audio track for different functions, such as recording a foreign language.

Beta Speed Mode. A speed at which videotape plays or records in a Beta-type VCR. The three speeds are known as Beta I, II and III. Newer home machines no longer record in the discontinued Beta I mode, but some models are capable of playing back this speed. Beta II has evolved as the present standard and the one which is used for most prerecorded tapes. Beta III is used for maximum recording/playing time (five hours with an L-830 tape). Most Beta VCRs record and play back in both Beta II and III. Finally, Beta II is half the speed of Beta I, but Beta III is approximately two-thirds that of Beta II. Therefore, on an L-500 tape Beta I's playing time is one hour, Beta II's is two hours and Beta III's totals three hours. See TAPE SPEED, WRITING SPEED.

Betamax. Sony's trade name for its popular 1/2-inch Beta format videocassette recorder. The first Betamax, the SL-7200, was introduced in 1975. The term is occasionally used generically to refer to any VCR, just as Kodak has come to represent a simple camera.

Betamax Case, The. See MCA/DISNEY VS. SONY LAWSUIT.

Bezel. In video, the frame surrounding a video picture that has different proportions from that of the TV screen. For example, when some cable or broadcast stations present a wide-screen theatrical film in its correct aspect ratio (about 16:9), the top and bottom portions of the screen (4:3 aspect ratio) remain blank. Some local and network stations provide a decorative bezel to replace the normally black portions of the screen. Bezels may come in different proportions.

Bi-directional Microphone. A microphone that responds equally to sounds from two opposite directions, the front and back, as in a configuration of the number 8. The mic is designed to reject sound from the sides. See MICROPHONE, OMNIDIRECTIONAL MICROPHONE.

Binder. A chemical adhesive which holds the magnetic OXIDE particles to the BACKING or base of the VIDEOTAPE. The quality of the binder is important in that its composition determines the number of DROPOUTS that are likely to occur.

Black Box. A general term given to a variety of electronic devices because of their color and shape. They include IMAGE ENHANCERS, IMAGE STABILIZERS, VIDEO AMPLIFIERS and UP CONVERTERS which can be connected to a VCR or TV set.

Black Entertainment Television. A cable TV channel targeted for a black audience and offering entertainment, sports and films. BET also covers interviews and other news pertaining chiefly to its black audience. Founded in 1980, it presents its millions of viewers with a limited schedule: weekdays from 6:00 P.M. to 6:00 A.M. and Saturday from 12:00 Noon to Sunday 6:00 A.M.

Black Level. The lower portion of the VIDEO SIGNAL containing the sync, VERTICAL BLANKING INTERVAL lines and other control signals; i.e., the picture signal level that parallels the maximum limit of black peaks. See REFERENCE BLACK LEVEL.

Black Level Control

Black Level Control. A feature on some TV sets that controls the extent of black within picture areas. This may affect the contrast of the image, but it is not strictly a contrast control; it does not determine which portions of the image should turn black but rather the degree of blackness. The conventional contrast control, on the other hand, constricts or extends the range of contrast only. On broadcast-studio quality video cameras the black level control feature is called "pedestal" while the contrast control is known as GAIN.

Black Level Retention. The ability of a TV set, VCR or similar unit to reproduce black areas on a television screen. Although no TV receiver produces an absolute black, manufacturers have constantly experimented in this area to improve the overall image contrast. The range of color contrasts depends on the span between the darkest gray and purest white. The wider the range, the more noticeable the distinctions between hues. This variation in shades helps to give the appearance of depth or three dimensions to the image on the TV screen. The contributions over the years of several companies, including NEC, Proton, Zenith, Sony and Panasonic, have resulted in subtle improvements in black level retention. These enhancements are more prominent in the higher-priced TV MONITOR/RECEIVERS than in low-end TV sets. Black level retention is measured by percentage; e.g., 80 percent or better is rated satisfactory while 90 percent is considered good.

Black Stripe Projection Television. A process designed to increase the contrast of a picture in a projection TV system. It was introduced in 1981 by Sylvania and later incorporated by Magnavox into its large-screen TV sets. The black stripe projection method is considered a significant advance in projection TV systems since these models are continually compared to standard TVs for sharpness and brightness. The system is also known as black matrix lenticular screen. The black stripe refers to a slightly recessed black line which is imprinted onto the screen. This tends to add sharpness and contrast of the projected image.

Blacker-Than-Black. In video, the amplitude area of the composite video signal below the reference black level in the direction of the synchronizing pulses. See COMPOSITE VIDEO SIGNAL, REFERENCE BLACK LEVEL.

Blanking. In video, substituting during prescribed intervals a signal whose instantaneous amplitude makes invisible the return trace. This blanking process replaces the picture signal. See FIELD BLANKING, LINE BLANKING, MUTING CIRCUIT.

Blanking Level. That level in a composite picture signal which separates the scope of the composite signal holding the picture information from the part comprising synchronizing information. Blanking level is also referred to as pedestal level. See PEDESTAL LEVEL CONTROL.

Blay, André. Founder and originator of the home video prerecorded videocassette industry. In 1977 he purchased the rights to a package of 20th Century-Fox films for the purpose of home movie sales. Included in the 50 features were such classics as *The Grapes of Wrath*. Blay's company, Magnetic Video, transferred the films to Beta and VHS formats and retailed them for about $50 to $60 each, thereby introducing the first prerecorded programs for the home market.

Bleed Through. The result of one channel superimposed over another. With a VCR, this may occur if the wrong open channel is employed. A local station may be transmitting at that frequency. Switching from Channel 3 to 4 (both known as "open" channels, with a switch located on the back of most VCRs) or vice versa usually eliminates bleed through.

Blip. A white streak or speck that appears momentarily on a TV screen during playback of a videotape. Blips are caused by DROPOUTS, a characteristic of all tapes. These streaks occur occasionally, but if they seem to be excessive then the video heads need cleaning or the tape brand should be changed. See DROPOUT, VIDEO HEAD CLEANING, VIDEO NOISE.

Block Converter. See UP CONVERTER.

Block-Down Conversion. See DOWN CONVERTER.

Blue/Red Balance. See RED/BLUE BALANCE CONTROL.

Blooming. A fuzziness at the edges of bright objects or with subjects wearing white shirts in bright light, as seen on a TV screen. Blooming also occurs during slide-to-tape transfer when recording slides continuously. When a video camera records a dark, underexposed slide followed by a brighter one, the camera needs time to adjust between the two—causing an annoying overexposure or blooming effect. Blooming also occurs when the brightness control is turned too high. This leads to an increase in the size of the scanning spot on a cathode ray tube, or an out-of-focus image.

BNR. A noise reduction system introduced by Sony. The Beta Noise Reduction technique is similar to those systems already in use but not compatible. Developed by the Beta VCR camp to compete against the VHS system, BNR was not adopted by all Beta-format companies. Marantz, for example, decided to employ the Dolby C system for its Beta VCRs. See DOLBY, NOISE REDUCTION SYSTEM.

Boom Microphone. A sensitive, directional mic usually suspended above the camera or over the action by means of a mechanical support or boom. The function of a boom microphone is to pick up specific sounds or voices. See MICROPHONE.

Bounce Light. A light source or lighting technique used to soften shadows on a subject's face, etc. A bounce light may be the main light source, in which case it is aimed at the ceiling or a light-colored wall. The diffused light will "bounce" off the surface

Break-Out Box

and minimize harsh shadows on the subject. See LIGHTING.

Break-Out Box. An accessory designed to permit equipment with multiple-pin jacks to interconnect with separate, conventional audio and video plugs. The box often has two built-in jacks, one for accepting BNC connectors and the second for audio mini-plugs. In addition, it has an extended cable with a multi-pin connector (8- or 10-pin). Break-out boxes, chiefly used by professionals, are available in different configurations.

Breakup. Total picture distortion that lasts for only a second or two. This usually occurs between scenes when using a video camera or pressing the stop or pause function on an earlier model VCR and starting again. Most current camcorders and VCRS are equipped with electronic circuitry that helps to minimize these breakups or "glitches," as they are sometimes called. See AUTOMATIC TRANSITION EDITING.

Bridge. A brief audio or visual passage or sequence intended to connect segments of a program.

Brightness Control. On a video processor, a feature designed to adjust the video level. In this respect, it is used to improve scenes that are either too dark or too light, to fade in and out when eliminating commercials and to soften the effects of glitches and edits.

Broadcast. Transmission of radio or television intended for general public reception.

Broadcast Engineering. A slick, thick monthly magazine aimed at and mailed free to "corporate management, engineers/technicians and other station management personnel at commercial and educational radio and TV stations, teleproduction studios, recording studios, CATV and CCTV facilities and governmental agencies." General articles include a wide range of topics, such as managing broadcast operations, audio fidelity, the revolution in television and minimizing tape wear. Regular monthly columns cover satellite technology, news items, field reports, technical information on circuitry, business news, new products and an FCC update. The magazine provides a balanced blend of broadcasting topics and technical information.

Broadcast TV. Over-the-air transmission of television programs, in contrast to CABLE TV, MICROWAVE, etc. The major networks (ABC, CBS and NBC) and many local TV stations use broadcast TV for the transmission of their programs.

Broadcasting. The conveying of speech, music or visual information for public-service or commercial purposes to a comparatively large audience. Broadcasting differs from other methods of transmission, such as the two-way radio, which is aimed at a limited audience.

Broadcasting. A weekly magazine targeted at those working in broadcasting/telecasting and related fields. The magazine, which covers radio, television, cable and satellite TV, chiefly focuses on news items and up-to-date reports about legislation, businesses, personalities and competition in broadcasting. Individual departments (more appear here than in similar magazines) include information on business, journalism, law and regulation, radio, programming, stocks and syndicated markets. The publication, one of the oldest in the field, made its debut in the early 1930s.

Bulk Videotape Eraser. A device with an electronically generated neutral magnetic field that can clear a recording on tape by "scattering" the tape oxide particles. Erasers are available in various sizes and prices. Usually the wider the tape, the stronger the model necessary. Normally, home VCRs effectively erase videotape automatically before re-recording so that a bulk eraser in not required. The bulk videotape eraser, also known as a tape eraser or simply eraser, has several advantages: it works faster and more effectively and assures security by erasing the entire tape. VCRs only erase up to the point where the re-recording ends.

Burn. See IMAGE BURN.

Burst. A term used in video to refer to the color burst portion of a video camera signal. On a color vectorscope the burst portion of a video camera signal, for example, is seen as a heavy horizontal line along the 180° axis. If no line appears on this testing unit, it signifies that no color is being transmitted; hence, only a black-and-white picture will be produced on a TV screen. See COLOR BURST, COLOR VECTORSCOPE, COMPOSITE VIDEO SIGNAL, GRATICULE.

Burst Amp Control. A function usually found on a color processor and designed to manipulate the gain of the color subcarrier in relation to the luminance signal. The burst amp control mainly affects the color midrange, especially the flesh tones. Normally, the color deviations in the color subcarrier are corrected in VCRs and TV sets by the ACC (Automatic Color Control). However, in some components in which the ACC may not be effective, the burst control can help. See AUTOMATIC COLOR CONTROL, COLOR PROCESSOR, SUBCARRIER FREQUENCY.

Burst Phase Control. A feature found on a signal processor such as a color processor or processing amplifier and designed to adjust the tint of the color signal. When the burst phase control is rotated in one direction, red turns toward blue, blue changes toward green and green veers toward red. When the control is rotated in the opposite direction, the color shifts reverse their roles. The burst phase control differs from the chrome control which affects color intensity rather than tint. See CHROMA CONTROL, COLOR PROCESSOR, HUE, PROC AMP.

Bus. A row of buttons on a video MIX/EFFECTS SWITCHER that controls hundreds of special effects such as wipes, fades, etc.

Bushnel, Nolan. Inventor of the first successful electronic video game, PONG; founder of Atari. While working at Bally Manufacturing Company, he soon learned that his employers showed little interest in his video game technology. He therefore built a coin-operated version of Pong on his own. Shortly after, he founded the Atari company. See PONG.

B-Y Signal. In video, a blue primary signal for cathode ray picture tubes that is formed when the B-Y signal combines with a Y, or luminance, signal inside or outside the tube. The B-Y signal is one of the three color-difference signals that comprise color television.

Bypass Switch. An electronic circuit, found on some image enhancers and processors, designed to permit the user to circumvent all video processing for instant comparison between original and electronically enhanced signal information.

C

CAA (Constant Angular Acceleration). A laser video enhancement designed to eliminate crosstalk, or unwanted extraneous signals. CAA, a modification of Constant Linear Velocity (CLV), is accomplished by making rotational speed changes part of a sub-multiple of the horizontal sync frequency. As a result, rolling horizontal noise bars are kept out of the screen image.

Cable Converter. See UP CONVERTER.

Cable News Network. See CNN.

Cable Ready. A feature of many units such as VCRs and TV receivers permitting access to VHF channels 2–13, UHF channels 14–83, cable midband channels A through I and superband channels J through W. Some units are listed as 91-channel cable ready, 105-channel cable ready, etc. These numbers don't mean that all of these channels can be programmed; cable channels A-W are tuned in on the UHF positions, thereby tuning out certain UHF stations. Also, scrambled PAY-TV channels still need the decoder boxes to bring in a signal. More recent VCRs have electronic digital tuning. This feature allows the user to select any channel directly by entering numbers on a remote-control keypad. See CABLE TV, FREQUENCY-SYNTHESIS TUNER, MIDBAND CABLE TV, PAY TV, SUPERBAND CABLE TV.

Cable TV. A system capable of transmitting from 12 to 100 channels of video programming to subscribers by utilizing coaxial cable. CATV usually includes basic channels (local television at a minimum monthly charge), premium channels of films, sports and other services (again, at a monthly fee) and, in some areas, two-way communications. Originally conceived as a method of bringing improved reception of conventional TV to viewers outside the range of the few stations then broadcasting, CATV grew slowly from its humble start in the 1950s. It suddenly sprang to life in 1975 when Home Box Office introduced an entirely new market by offering first-run, uncut "blockbuster" movies. Just as CATV has cut deeply into conventional network televi-

Cable TV Adapter

sion's audience, competition from other video technologies has affected CATV's share of the market. In the late 1970s, for example, CATV controlled 95 percent of the number of TV subscribers, while Microwave (another system) controlled four percent and UHF-VHF a meager one percent. By the mid-1980s cable control dropped ten points to 85 percent. Microwave fell slightly to 3 percent while UHF-VHF climbed sharply to a 12 percent share of the market. There are approximately 5,000 cable systems in the United States of which about 35 are co-ops and municipally owned. By the beginning of 1990 54 percent of all television households were receiving CATV, compared with only 19 percent ten years earlier. In the early 1980s the cable industry was grossing over $3 billion per year from its 40 million subscribers; by 1990 the figure rose to $13 billion that consumers spent on both pay and basic TV. Forecasts for the 1990s predict the figure will rise to $20 billion. By 1990 there were 54 advertiser-supported networks and 11 pay-television stations.

Cable TV Adapter. An automatic switching accessory designed to permit the viewing and taping of various broadcast, cable TV and pay TV channels by utilizing certain programmable VCRs. Watching one program while recording another from pay TV or cable can present problems as well as complex connections. The components include a cable converter with a decoder for the pay TV channels, a VCR and a TV receiver. If the VCR is cable ready, it can be used to tune in various cable channels, but the pay TV decoder presents another obstacle. To overcome this, a cable TV adapter can be added as a fourth unit, providing greater flexibility. For instance, to record a pay channel while simultaneously watching a VHF channel, the VCR has to have one of the preset channels tuned to that used by the cable operator. Then the VCR/TV switch is set at TV, the receiver channel at the one to be viewed and the converter box positioned at the pay TV channel. The adapter offers other possible combinations such as recording a VHF channel while viewing a pay TV program. Without this device, a viewer must have either two decoder boxes, a patch box or an RF switcher to perform the same functions. The latter two accessories are preferable if more inputs are required, the patch box being less permanent than the switcher. See CABLE READY, CABLE TV CONVERTER, PATCH BOX, SWITCHER.

Cable TV Carrier System. An arbitrary method employed by cable systems in which the transmission of broadcast signals is "offset." This results in minimal screen interference from neighboring channels. Although typical TV broadcast channels start and finish with MHz numbers that are even (e.g., 66–72 MHz for channel 4), different carrier systems use varying offset frequencies. Two basic types are the HRC (Harmonically Related Carrier) and the ICC (Incremental Coherent Carrier) systems. With HRC, the cable operators offset all channel frequencies, channels 5 and 6 by 1.25 MHz and the remainder by .75 MHz. Cable companies utilizing the ICC system offset the frequencies of channels 5 and 6 by 2 MHz. Offsetting, however, can create problems. The cable converter may have to be retained, whereas with normal channels it can be eliminated since CABLE-READY tuners can lock in on these stations. Finally, another disadvantage of offset frequencies is that these channels become difficult to tune in with any degree of accuracy.

Cable TV Converter. A device used with CABLE TV to carry multiple channels to subscribers who could receive them on an unused channel. Based on the principle that many channels could be carried in exactly the same electronic space as one, cable operators began by offering 12 channels through their converter box. (The Focus-12 device was first used in New York's borough of Manhattan.) The number of channels capable of being carried by the converter grew and the Gamut-26 was introduced. Because of the shielding of the converter, the channels are transmitted to the box by cable, not over the air.

Cable TV Cooperative. A CABLE TV station owned and operated by subscribers. There are approximately three dozen cable cooperatives of the 5,000 cable systems in the United States. Most of these co-ops are in the Midwest. Cable co-ops differ from municipally owned cable stations which are owned by the local governments. See PUBLIC ACCESS TV.

Cable TV Networks and Their Subscribers.

Advertiser Supported

ESPN	54+ million
CNN	53+
TBS	52+
USA	50+
Nickelodeon	49+
MTV	49+
The Family Channel	48
TNN	48+
Lifetime	47
Arts & Entertainment	42+

Pay Television Networks

Home Box Office	17 million
Cinemax	6+
Showtime	6+
The Disney Channel	4+
The Movie Channel	2+

Caddy. A protective plastic jacket, sleeve or holder which stored the now defunct CED videodisc. The caddy was inserted into the videodisc player which automatically removed the disc and prepared it for play. The Capacitance Electronic Disc, developed by RCA, contained delicate microscopic grooves which were protected by the caddy. The CED system, which employed a stylus that physically tracked the disc grooves, differed from the LaserVision system. The latter uses a laser beam which "reads" the information of a grooveless videodisc that is virtually indestructible and therefore needs no protective caddy. By the mid-1980s the public virtually rejected both systems, selecting videotape as the medium of its choice. However, the laserdisc made a comeback by the late 1980s. See CED VIDEODISC SYSTEM, VIDEODISC, VIDEODISC PLAYER.

Camcorder. A combined video camera/VCR unit only slightly larger than a typical 8mm movie camera. Older home video cameras had to be connected to separate portable or table

Camcorder History

model videocassette recorders which contained the necessary tape deck and transport systems, usually in the 1/2-inch format. Camcorders, on the other hand, are self-contained single units that employ 1/2-inch videotape or smaller cassettes. Manufacturers have endowed the camcorder with great versatility. Many cameras offer automatic adjustments for focus, exposure and color balance, allowing the user to concentrate only on the subject matter. Some models have manual controls designed to override many of the automatic features—a boon to users who wish to experiment with special effects in lighting and other areas. Other models can accept video and audio signals from external sources (with the aid of an adapter). Camcorders come is a variety of formats—VHS, S-VHS, VHS-C, S-VHS-C, 8mm, Hi8. The conventional full-sized VHS format accounts for about 65 percent of sales, with 8mm at about 20 and VHS-C at about 15 percent. Although camcorders have not come down in price as VCRs have, they are present in about 10 percent of American homes.

Figure 4. A camcorder that features the S-VHS format and a 10:1 power zoom lens. (Courtesy NEC Technologies, Inc.)

Camcorder History. Sony in 1980 showed off its VideoMovie, a one-piece camera/VCR combination that used 5/16-inch tape. JVC introduced its camcorder prototype in 1982. Made up of a portable VCR and camera, the two components connected as one unit with a shoulder brace. By 1984 JVC and Zenith brought out their VHS-C models of camcorders. Sony introduced the 8mm camcorder format in April 1985, offering a compact and light camera with excellent sound. That same year Panasonic introduced the first full-size VHS camcorder, called Omni-Movie. The Super-VHS camcorder made its debut by way of JVC in 1987, followed the next year by Canon's 8mm Hi8 (High-band) camcorder. Finally, Sony unveiled its ED-Beta, complete with two CCD chips and a price tag of more than $7,000, in 1988.

The VHS-C format equaled Sony's 8mm model in picture quality and compactness. But it failed to measure up in the areas of audio quality and recording time, which was limited to 20 minutes with its mini-cassette. By 1986, one million camcorders were sold in the United States. In contrast, more than two million units were sold in 1989 alone. While camera enthusiasts in the Far East and Europe preferred the 8mm format, those in the U.S., who chiefly owned VHS videocassette recorders, remained loyal and chose the VHS-C format. The mini-cassettes, they reasoned, could be played on their present equipment with the addition of a simple adapter. By 1988, the VHS format received a boost with the introduction of Super-VHS, which improved the picture quality over that of 8mm. However, the sound still remained less than desirable. Sony, refusing to take a back seat, countered with Hi8, which offered 400 horizontal lines, or superior image detail to that of S-VHS, and improved color. VHS manufacturers in 1989 parried with yet another enhancement in their Super-VHS-C models that introduced high-fidelity audio. By the early 1990s, the camcorder became the third most successful audio/video electronics product, topped only by the VCR and the CD player.

Cameo. A lighting technique used in TV production in which only the foreground is lighted. The background remains intentionally dark or unlit. See LIGHTING.

Camera Jack. An input on VCRs which accepts video camera connections. Manufacturers use 10-pin, 14-pin (Type K) or 5-pin Din jacks. However, two similar jacks may not necessarily be compatible. For example, a 10-pin jack on a specific camera may fit a 10-pin VCR, but not all functions assigned to each pin will operate as intended. Adapters are available for connecting jacks of different pin numbers and configurations. See MULTIPLE-PIN CONNECTOR, PIN.

Camera Mounting Head. An accessory video camera support designed for moving and manipulating the video camera efficiently. Mounting heads, originally simple devices based on friction and gears that were designed for the film industry, have taken on new engineering techniques, resulting in complex fluid and cam heads for use with professional video cameras and camcorders.

Camera Search. A video camera feature that allows the user to review existing scenes while the camera remains in Record mode.

Camera-to-VCR Adapter. See ADAPTER.

Camera Tube. The video camera component that collects the light focused on it which is admitted by the lens and then generate the electrical impulses that are eventually translated into images by a television cathode ray tub. There are three basic types of camera tubes, the VIDICON, the SATICON and the TRINICON. It is these that make up the majority of tubes found in home video cameras. A metal oxide semiconductor chip (MOS) can replace the conventional camera tube, although the cameras using this process are more costly. The camera tube is also known as an image pickup tube. Each tube has several internal elements. The vidicon, for example consists of (starting from the front) a faceplate, a target plate or target area, a mesh screen, deflection and focusing coils and, at the end, a cathode source. Other types of tubes are the PLUMBICON, NEWVICON, etc. The camera tube suffered its greatest setback in the spring of 1990 when several manufacturers of professional video cameras presented no tube-type models at the National Association of Broadcasters show. See MOS.

Capacitance Electronic Disc. See CED VIDEODISC SYSTEM.

Capacitor

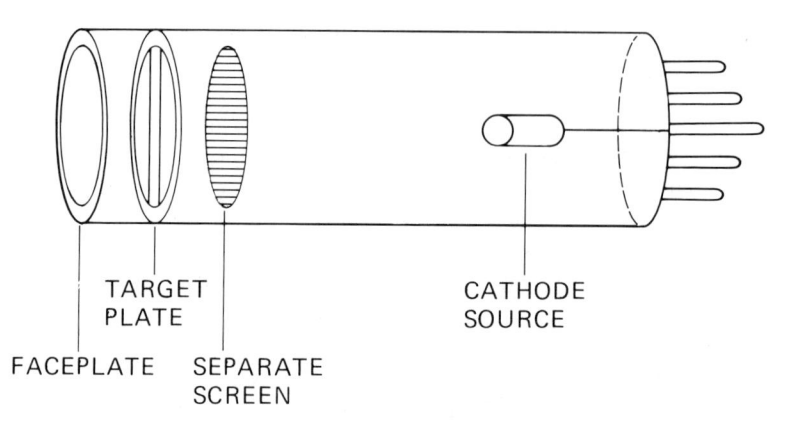

Figure 5. Camera tube.

Capacitor. An electronic component which accepts an electrical charge that it can dispense at a consistent and predictable rate. In relation to a VHD videodisc system, a capacitor is formed by the combination of the disc and a tiny piece of metal, known as an electrode, which is part of the stylus.

Capstan. A motor-driven vertical shaft designed to control the speed of the videotape as it proceeds through the VCR from the supply to the take-up reel. To do this, the capstan itself is controlled by a sensor. A PRESSURE ROLLER holds the tape firmly against the capstan.

Capstan Servo. An electronic circut built into editing VTRs to provide smooth edits. The servo is a speed control system which locks into the sync of another recorder to insure that both video signals are interlocked for GLITCH-free edits. Without this feature, picture BREAKUP appears between scenes. The capstan servo may also be found on VCRs where it operates in conjunction with the tracking control during playback. It adjusts the running speed of the tape either slightly faster or slower than normal.

Caption Center, The. See CLOSED-CAPTIONED DECODER.

Caption Generator. See CHARACTER GENERATOR.

Capture Ratio. The ability of a tuner to restrict or reject a second, weaker signal in proximity to the main signal. This permits the strong signal to be received without interference from secondary signals traveling on the same frequency. Capture ratio is measured in decibels (dB); the lower the number, the more effective the tuner in terms of "capturing" the strong signal. Moderate-priced tuners may have a capture ratio of approximately 3 dB while more costly units approach a figure of 1 dB. The term, which applies more to audio than video, should not be confused with ALTERNATE CHANNEL SELECTIVITY which refers to adjacent stations.

Cardoid Microphone. A mic that picks up only what it is pointed at. It has a pick-up pattern designed with one "strong" side, while the other side tends to diminish in strength. Similar to a heart-shaped polar response, the cardoid mic pattern offers a stronger front response than that at the rear. See MICROPHONE.

Carrier. In video, a constant wave that can be adjusted by changing frequency, pulse or amplitude.

Carrier Color Signal. A monochrome signal combined with a chrominance subcarrier and additional sidebands, all of which carry color information in a color television system.

Carrier Deviation. See FM DEVIATION.

Carrier Frequency. A certain wavelength of a special frequency on which a signal is registered for transmission in a clear way to a receiver. The audio or video signal is then isolated from its carrier frequency, amplified and reproduced. See WHITE CARRIER PEAK.

Carrier System. See CABLE TV CARRIER SYSTEM.

Cartridge. A plastic container holding a single reel, closed loop of videotape. In contrast, a videocassette contains two reels. Cartridges are presently found only on industrial machines.

Cartrivision. A now-defunct videotape recording system developed by Avco and distributed by Sears. Cartrivision featured almost two hours of playing time from one cartridge using 1/2-inch tape. The system used three heads, recording only one of every three fields of video on tape moving at 3.8 inches per second.

Cassette Eject. The method used to remove a videocassette from the VCR housing. Older machines simply allowed the cassette chamber to spring up with a loud thud. Today's more sophisticated top-loading methods include the use of gears, air-dampened pistons, coil springs and pneumatic fan blades—all much quieter and gentler.

Cassette Indicator. An icon that appears in the display window of some VCRs to indicate whether the machine contains a videocassette and the direction of the tape movement. A series of LEDs connects the two circles that make up the icon and light up sequentially to describe in which direction the tape is moving. The indicator is not illuminated when the machine is shut down.

Cassette Shell. The container that holds the videocassette. The shell usually has several vital parts to help move the tape accurately from the supply to the take-up reel. These include a flange, hub and clamp, leaf spring, tape pad, guide pin, guide roller, tape guide and tape guide rib. A window is provided on the top to view the two tape reels. See VIDEOCASSETTE.

Cathode Ray Tube. The vacuum picture tube of a TV monitor or receiver. The CRT receives the electrical im-

Cathode Ray Tube Display

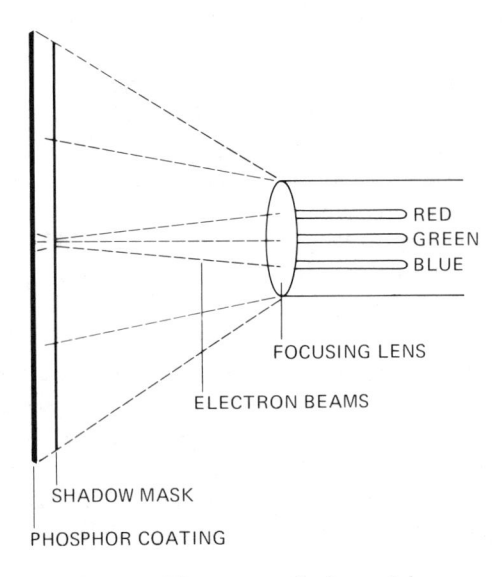

Figure 6. Three-gun cathode ray tube.

pulses that are generated by the video camera tube and converts them into a series of glowing brightness values or images on the TV screen. One end of the CRT contains a cathode and heater element from which electron beams travel to either a phosphor-coated TV tube or to an oxide coating which produces a voltage (a camera tube). There are three types of cathode ray tube displays: the familiar TV RECEIVER, the MONITOR and the MONITOR/RECEIVER. The electronic viewfinder of a video camera is actually a miniature monitor. See CAMERA TUBE, ELECTRONIC VIEWFINDER.

Cathode Ray Tube Display. That part of a system (a TV receiver, monitor or monitor/receiver) in which controlled electron beams provide data in visual form.

CATV. Originally an abbreviation for Community Antenna Television. CATV grew from a select service to small communities who were isolated from the reception of conventional TV programs. Local companies supplied transmission to these remote areas—for a slight fee. Since the telephone lines could not carry the "broadband" TV required, cables were used instead. Thus started cable TV. See CABLE TV.

CATV Adapter. See CABLE TV ADAPTER.

CAV. One of the two formats of the LaserVision (LV) videodisc system. A Constant Angular Velocity (CAV) disc plays for 30 minutes per side and is capable of many special effects not available to the CLV, one-hour-per-side format. The CAV disc turns at a constant speed of 1800 rpm or 30 revolutions per second, matching television's standard of 30 frames per second. This permits such special effects as FREEZE FRAME, SINGLE

FRAME ADVANCE, SLOW MOTION, VISUAL SCAN, etc.

C-Band. The MICROWAVE frequency range of a satellite TV signal. These signals usually fall between 5.9 and 6.4 GHz when transmitted to a satellite and range from 3.7 to 4.2 when returned to earth. See DOWNLINK, SATELLITE SIGNAL, UPLINK.

CBS Cable. A cable service offering various cultural programs such as documentaries, ballet, jazz, classical music, etc. Owned by CBS, Inc., the service, which began in 1981, programs its events to its more than three million subscribers.

CCD. See CHARGE COUPLED DEVICE.

CCD Color Comb Filter. Advanced electronic circuitry, usually found only on a TV monitor/receiver, designed to improve image detail and definition without color fringing or video noise appearing on screen. The color comb filter makes use of a Charged Couple Device to gain greater luminance/chrominance (brightness and color) separation. Separating these two elements of the signal—while keeping the full luminance bandwidth—results in less distortion in the final picture.

CCTV. See CLOSED-CIRCUIT TELEVISION.

CD+G. A CD disc, similar in appearance to the conventional disc, that contains visual information in the form of still images or graphics. When a specially equipped CD graphics player is connected to a TV set, these discs can provide pictures, musical information or text. The resolution is similar to that produced by computer graphics. CD+G discs, which first appeared in 1988, can be played on standard CD players, but the added visual benefits will not be available. Because of its limited technology, the CD+G disc produces only still images, differing sharply from the CD-Video disc or CD-I format, which can turn out "moving pictures." However, this relatively recent format allows the disc to retain its full 80 minutes of playing time. Another advantage of CD+G discs is that they will work with different TV formats, such as PAL or SECAM. Since the discs are encoded using the binary system (combinations of "0" and "1"), each system translates this code to fit its own TV standard before it appears on screen. The music portion of the disc, of course, is not affected. See CONTINUOUS MOTION, GRAPHICS DECODER.

CD Graphics Player. A CD unit, resembling the conventional CD player, but equipped to play back still images, in addition to the usual 80 minutes of music, recorded on CD+G (Graphics) discs. Such graphics players open a whole new world of entertainment and information to the user. Once the unit is connected to a TV set, the owner can view song lyrics, biographical data about performers or the music, photographs and drawings—items usually appearing as liner notes. Normal CD players do not have the necessary features, such as outputs for video or S-video, to play back these images; neither do they offer a graphics output jack for an add-on de-

CD-I (Interactive)

coder that can access these still images. See GRAPHICS DECODER.

CD-I (Interactive). A Compact Disc containing text, graphics, still images and high-quality sound, all designed to work with a computer. CD-I discs usually provide educational and recreational programs that the user can interact with. The CD player needs a special interface port for hooking up to a computer. Some players have such a port built in. CD-I, in other words, blends audio reproduction capabilities with video and graphics reproduction abilities, thereby providing the user with a variety of ways to interact with these features. For example, the user viewing a disc relating to an art gallery can concentrate on only one painting—or call up the biographical material on a particular artist. See CONTINUOUS MOTION.

CD ROM. A memory-storage compact disc designed only for reading and used chiefly with computers. ROM (Read Only Memory) discs take advantage of CD (compact disc) technology.

CD Video. A 5-, 8- or 12-inch disc, similar to that of the CD, containing audio and video information. A recently developed concept (introduced in 1988), the 3-inch CD Video originally offered a 20-minute audio track of popular music with about five minutes of video. The disc, selling for about $10, can be played on a CD Video player. The advantages of this format include the superb CD sound and the excellent image resolution, both of which surpass any videotape format. CD video almost immediately expanded in disc size and types of programs to include classic films, rock concerts, documentaries and operas. See AUDIO ESSAY, CONTINUOUS MOTION, VIDEODISC, VIDEODISC PLAYER.

CD Video Player. A multi-purpose unit that can play back a variety of laserdisc formats. A typical CD video player can accommodate 3-inch and 5-inch CDs (20 minutes or more of music), 5-inch CD videos, or CDVs (music followed by video), 8" and 12" CD-LDs, or laserdiscs (for feature films and other programs). Some models offer multiple-track programming for both audio and video, headphone jack and separate level control. In addition, other models feature digital audio output terminals and track, timing and frame number functions, freeze frame and slow motion and an S-video output to feed separated color and brightness signals to a similarly equipped TV monitor/receiver.

CED Videodisc System. A discontinued videodisc format that played back discs containing both sound and pictures on a standard TV receiver. The CED (Capacitance Electronic Disc) player was simpler than the LaserVision system. Like conventional audio disc recorders, the CED player used a grooved disc whose information was "read" by a stylus. The player offered visual scan, fast random access and other features. The stylus operated on the disc, which traveled at 450 rpm, from the outside in. Not as sophisticated as its competitor, the LV, the CED player sold for much less. Both systems failed to capture the imagination of the consumer by the mid-

Figure 7. A CD video player that can accommodate 8- and 12-inch Laserdiscs and 3- and 5-inch compact discs and compact disc videos. This model features dual-side play, digital time base correction and remote control. (Courtesy Pioneer Electronics.)

1980s. CED, introduced by RCA and other companies in 1982, virtually disappeared from the market in 1984, although parts and repairs were available. The moribund Laserdisc format experienced a moderate revival in the late 1980s. See LV VIDEODISC SYSTEM, VIDEODISC PLAYER.

Ceramic Microphone. See PIEZOELECTRIC MICROPHONE.

C-Format Videotape Recorder. An industrial machine with one-inch tape format used by professionals. Introduced in the 1970s, the C-format VTR is gradually replacing the two-inch machines which have been in use since the development of VTRs in 1956. The C-format, with its two quality audio channels and the addition of Dolby noise reduction, has brought forth a potential for improved sound. C-format VTRs are manufactured by several major companies.

Channel. A broadcasting band or frequency, ordinarily six MHz wide, that a TV or VCR tuner can be set to receive. The Federal Communications Commission appoints each band or frequency to individual television stations. Also, an independent signal path occupying a portion of the spectrum designated for operating a particular carrier including its sidebars required to transmit information. Channels include VHF, UHF and cable TV. Stereo has two independent channels.

Channel Capacity. Refers to the number of channels a CABLE TV system can handle simultaneously. Most cable systems at the present time can carry from 12 to 35 channels. Future

Channel F Video Game System

systems, some under construction at the present time, will be capable of handling over 100 channels. See CABLE TV.

Channel F Video Game System. One of the first video games to introduce programmability with a large selection of game cartridges. Introduced by Fairchild in 1976, Channel F had a short-lived career after selling a few hundred thousand systems. The company decided to abandon the game a short time after its inception. Then, in 1982, Zircon took over the Channel F inventory and brought out a revised version—Channel F II.

Channel Index. An advanced TV receiver or VCR feature that displays every channel in the tuner memory. Channel index, or channel search as it is sometimes called, accomplishes this by showing a consecutive sequence of images, in the form of freeze frames, along the side and bottom of the screen while the main picture appears on screen in a larger format. Advances in digital technology have made this feature, along with others, possible.

Channel Labeling. A TV feature that permits the viewer to enter the identification letters of an area network into a channel/time display. For instance, ABC, CNN or HBO can be listed on screen along with the channel number and time each time that display mode is pressed. Some TV receivers allow up to five networks to be customized in this manner.

Channel Lock. A television feature designed to lock out the viewing of a program or sequence of programs on a specific channel. This may be particularly appealing to parents who don't want their children to see certain programs. Some lock-out techniques involve the use of a conventional lock and key.

Channel Modulator. See SATELLITE TV MODULATOR.

Channel One. A private enterprise that broadcasts news and commercials into classrooms by way of conventional television. Owned by Whittle Communications, the project is designed to present news events at the students' age and learning level on a daily basis. The service is offered free to any school system that wishes to take advantage of the project. In return, the company receives revenues from advertisers whose commercials are oriented toward the students—captive audience. Introduced on an experimental basis in 1989, the service signed up 2,500 schools by the end of that year for its 12-minute news program. It is estimated that Channel One is seen by almost two million teen-agers and generates $200 million in advertising spots. The service is not without its critics, who contend that the school should not become a conduit for selling merchandise. Its only competitor is the Turner Broadcasting System with its CNN Newsroom, which claims to be installed in 7,000 schools. See ELECTRONIC CLASSROOM, INTERACTIVE VIDEO.

Charge Coupled Device (CCD)

Channel Range Selector. See BAND SELECTION.

Channel Search. See CHANNEL INDEX.

Chapter. In videodisc terminology, one of several arbitrary portions of a program into which a disc can be divided electronically. These chapters—for example, musical numbers or short film subjects—can readily be located by special features on the videodisc player. Some VDPs can be programmed to play back a sequence or more than a dozen chapters on one side of a disc. Each chapter is made up of many frames, also easily accessible.

Chapter Search. See AUTOMATIC CHAPTER SEARCH.

Character Generator. A professional/industrial device designed to produce electronic letters and numerals by way of a specially designed keyboard. It is basically used for titling and graphics as well as for special KEY and fade effects. Broadcast-type units sell for thousands of dollars and offer such advanced features as anti-aliasing, resolution of 1506 x 483 pixels, up to 65,000 colors per page and 70 standard fonts including italics, drop shadows, outlines, autosizing, embossing and bevels. Some home video cameras and top-of-the-line VCRs have built-in character generators. Generally, this accessory does not have the complexity and sophistication of the SPECIAL EFFECTS GENERATOR (SEG). See SUPERIMPOSITION, VIDEO EFFECTS TITLER.

Figure 8. A character generator, which produces electronic letters and numerals, is used for titling and graphics effects. (Courtesy MFJ Enterprises, Inc.)

Character Sizing. Refers to a generally standard feature of an industrial/professional CHARACTER GENERATOR designed to create electronically or digitally various sizes and types of letters, numbers and symbols. These characters can often be produced with such enhancements as italics and with drop shadows, bevels, etc.

Charge Coupled Device (CCD). A type of image sensor that uses a microprocessing chip to replace a video camera picture tube. This picture-sensing or image pickup device can be employed in conventional video cameras and flat-wall TV. A combination of charge coupled devices make up a single metal oxide semiconductor chip which forms the replacement for the image pickup tube. The first public demonstration of a solid-state imaging device occurred in 1967. However, its resolution failed to meet broadcasting standards. A considerable improvement occurred in the 1970s with the introduction of the CCD. The charge coupled device, which permitted photo-generated charges to be moved to MOS (metal oxide semiconductor) capacitors, stored these charges. They

47

could then be read as a string of image scanning lines or image fields. Other important developments in the history of the CCD took place in 1979, 1983 and 1984, each relating to professional camera use by Bosch, NEC and RCA, respectively. By 1990, several major manufacturers of professional video cameras for the first time featured only CCD cameras at the National Association of Broadcasters show. CCD technology provides video cameras, especially professional models, with high resolution, excellent sensitivity and improved signal-to-noise ratio. Some cameras come equipped with a frame-interline transfer (FIT) CCD chip while others feature a lower-priced interline transfer (IT) chip. See CAMERA TUBE.

Christian Broadcasting Network. A CABLE TV channel offering Christian news, sports, children's and family entertainment, free call-in prayer and counseling service. Programming emanates from over 60 sources. CBN, owned and operated by Christian Broadcasting Network, began transmitting in 1977 and has more than 20 million subscribers.

Chroma. A term used to describe the general value of a color—or the measure of color purity or color intensity—without referring to its brightness. Chroma, also known as chrominance, is a composite of HUE and SATURATION. Black, gray and white contain no chroma. See CHROMINANCE SIGNAL.

Chroma Amplitude Modulation. See CHROMA SIGNAL-TO-NOISE.

Chroma Control. On color processors, process amplifiers and similar units, a feature designed to modulate the color intensity of the video signal. When the chroma control is turned up, the intensity of the colors is increased and they become more vivid; when the knob is turned down, the intensity is decreased and the colors develop a pastel-like quality. The chroma control, which is similar to the color control of a TV set, differs from the burst phase control (also located on these signal processors) which adjusts color tint rather than intensity. See BURST PHASE CONTROL, COLOR PROCESSOR, PROC AMP.

Chroma Decoder. A video processor designed to decipher a composite video signal into its red, green and blue components. Used chiefly by professionals, the decoder is a chroma demodulator compatible with RGB monitors, video projectors, chroma keyers, etc. See COMPOSITE VIDEO SIGNAL.

Chroma Key. See CHROMAKEYING, KEY.

Chroma Noise. See CHROMINANCE NOISE.

Chroma Noise Reduction. See HQ CIRCUITRY.

Chroma Phase Modulation Noise. See CHROMA SIGNAL-TO-NOISE

Chroma Signal-to-Noise. The measurement of the amount of interference influencing either the color saturation or the hue within the picture.

There are two types of chroma noise. Chroma amplitude modulation applies to color saturation and appears as minor changes in color strength, especially in large blocks of a particular color. The effect within the color takes on a mottled pattern. The second type, chroma phase modulation noise, is manifested by traces of different colors from those of the original. Both types of chroma noise subtract from the fine detail and purity within the color signal which is reproduced on the TV screen. Chroma signal-to-noise is measured in decibels (dB); the higher the number, the sharper and more detailed the color picture. See COLOR RESPONSE.

Chromakeying. A video process which creates a particular illusion by superimposing a subject (from a subject/key camera) over a different background (background camera). To produce this effect, the subject usually stands or is placed in front of a blue background. The cameras are then connected to a SPECIAL EFFECTS GENERATOR with a chromakeyer which automatically switches the subject/key camera to the background camera each time the camera beam "sees" blue. There are different types of chromakeying, such as upstream and downstream chromakeying, each using various combinations of cameras and backgrounds.

Chrominance/Luminance Delay Inequality. In video, an effect caused by the color and black-and-white signals being recorded separately on the tape, resulting in colored edging or fringing around objects. See COLOR-UNDER RECORDING, FRINGING.

Chrominance Noise. Refers to a particular kind of video interference that affects color signals in the form of temporary traces of color aberrations. Chrominance noise differs from luminance noise which affects both black-and-white and color signals. See LUMINANCE NOISE.

Chrominance Reversal. See IMAGE REVERSAL.

Chrominance Signal. That part of a video signal containing color information. The video signal, actually a color video signal, comprises both black-and-white and chrominance signals, the latter consisting of the color information in electronic form. The chrominance signal, composed of hue and saturation elements but excluding luminance and brightness, takes the form of chrominance subcarrier sidebands that are appended to a TV signal to carry color information. Chrominance and luminance components are recorded separately, the former as an amplitude modulation and the latter on a frequency modulation (FM) carrier. See COLOR SYNC, LUMINANCE SIGNAL.

Chromium Dioxide. One of the coatings available in the composition of VIDEOTAPE. It is magnetically sensitive and used mostly on Beta tapes. The VHS format generally utilizes FERRIC OXIDE. See BACKING, BINDER, COATING.

Chromostereoscopy. Part of a three-dimensional television system employed in Japan but developed in California. Chromostereoscopy uses a technique in which red-colored ob-

Cinemax

jects appear closer than blue-colored ones, thereby giving the impression of depth. The process is compatible with 2-D or conventional TV broadcasting. But for a full, 3-D effect, the viewer needs special glasses. See THREE-DIMENSIONAL TELEVISION.

Cinemax. A PAY TV service offered to subscribers for a monthly fee. Cinemax, similar to Home Box Office, offers current Hollywood films, classics and foreign features 24 hours a day. Owned by Time, Inc., the channel has several million subscribers. It began operations in 1980.

Circuit. In video, refers to the interconnecting of a number of electronic devices in one or more closed paths to carry out an assigned task. Also, an electronic channel between two or more points that can produce a number of paths.

Clean Edit. A transition between scenes that is GLITCH-free or without picture BREAKUP. Most later model camcorders provide clean edits. See AUTOMATIC TRANSITION EDITING, EDITING.

Cleaner. See VIDEO HEAD CLEANER.

Cleaning Solvent. The special cleaning fluid used on video and audio heads and other elements along the tape path of a VCR. The most popular and recommended liquid solvent is TRICHLOROTRIFLUOROETHANE or Freon. This solvent is more effective than alcohol which evaporates more slowly and may leave a residue.

Clock/Calendar. A video camera feature that permits the user to stamp the time or date, or both, over a video image being shot. The information is superimposed over the recorded image for further reference. Many cameras, regardless of their price range, offer this useful feature.

Closed-Captioned Decoder. A special accessory box primarily intended for the hard-of-hearing. It decodes an otherwise invisible signal and presents captions at the bottom portion of the screen, revealing what the performers are saying. Captioned subtitles are inserted on one of the lines within the VERTICAL BLANKING INTERVAL. The captions can only be viewed with a decoder. A viewer can record these captions on tape while recording the program by simply connecting the decoder box between the antenna and the VCR. More than 375 hours of closed-captioned shows are broadcast each week by both networks and cable systems. The first TV season in which the three major networks presented their primetime lineup closed-captioned occurred in the fall of 1989. This contrasts to only 10 hours weekly in 1980. Another function of the decoder serves senior citizens and others interested in learning English as a second language. Several major organizations produce captions for television, including the National Captioning Institute located in Washington, D.C., and The Caption Center, a part of Boston's public TV station WBGH. See VERTICAL INTERVAL REFERENCE SIGNAL, BLACK LEVEL.

Closed-Circuit Television. A non-broadcast TV system. CCTV is used for such purposes as monitoring and security (banks, stores). The closed system usually includes a CCTV video camera, a videocassette recorder, a CCTV monitor/receiver, etc.

Close-Up Lens. A special video camera adapter lens designed for extremely close work, such as recording coins, stamps, insects, etc. Many recent camcorders provide macro settings to accomplish this function. When using either a supplementary lens or the macro feature built into the camera, the operator will notice that the DEPTH OF FIELD is severely restricted. See MACRO LENS.

CLV (Constant Linear Velocity). One of the two formats of the LaserVision (LV) videodisc system. CLV discs can play for up to one hour per side but lack the many special effects features offered by the CAV format of 30 minutes per side. Since the larger outer tracks of the disc hold more information than the smaller diameter inner tracks, the playing time is extended. However, the 30 revolutions per second of the CAV speed, which matches TV's 30 frames per second, is altered, thereby eliminating such features as freeze frame, single frame advance and slow motion. However, some advanced models of laserdisc players incorporate digital memory to handle smooth freeze frames, slow motion and other special effects. See CAV, LV VIDEODISC SYSTEM, VIDEODISC PLAYER.

C-Mount. A standard lens mount used on many video cameras and camcorders to permit compatibility and interchangeability with other lenses by diverse manufacturers. A typical use would be temporarily fitting a wide angle or telephoto lens to the camera in place of the normal one. Since almost all camcorders come equipped with a zoom lens, which offers a wide range of focal lengths, the average home videographer has little need to change lenses.

CNBC. An advertiser-supported CABLE TV network specializing in business programs in the daytime and a diversified schedule of talk shows, health programs and other consumer-related shows in the evening. Owned by the National Broadcasting Company and General Electric, CNBC began operations in April 1989 with 13 million subscribers. Cable operators have been less than enthusiastic about promoting the station because of its undefined programming focus and its affiliation with the giant TV network.

CNN (CABLE NEWS NETWORK). The first 24-hour news channel available to subscribers of cable TV systems. Based in Atlanta, Georgia, CNN offers its service at a nominal fee to cable operators in exchange for its connection to their local news information. Modest in its staff size compared with network news channels, the CABLE TV advertiser-supported network has often presented major news-breaking stories first because of its continuous format. It also provides interviews, commentaries by celebrities and special features. The channel, owned by the Turner Broadcasting System, had more than 53 million viewers as of 1990. See TED TURNER.

Coating. The magnetic oxide particles on videotape which are formed into diagonal fields of information by video heads. Beta tapes are generally coated with chromium dioxide (original Sonys were designed for optimum efficiency with chrome tapes). VHS tapes generally are composed of ferric oxide. Beta tapes, when combined with cobalt, provide better resistance to dropout or flaking of particles but are weaker in the area of high frequencies. VHS tapes offer better frequency response, but the oxide composition may present a problem with excessive snow or distortion, especially with low-grade tapes. See CHROMIUM DIOXIDE, DROPOUT, FERRIC OXIDE, OXIDE.

Coaxial. A special cable designed to carry one or more TV channel signals with minimum power loss and high video frequency transmission. Rated at 75 ohms, this cable offers a thicker surrounding of the center conductor and rejects undesirable interference. In video, a few types of coaxial cable are available, but the RG 59/U is the most popular. It has a thick coating around the center conductor for strength and protection from interference. It is designed for cameras, monitors, VCRs, etc.

Coaxial Connector. See F CONNECTOR.

Coercivity. In videotape, the ability of the particles that compose the magnetic medium to be magnetized. Therefore, the higher the coercivity, the better the quality of the tape. See OERSTEDS OF COERCIVITY, VIDEOTAPE.

Colecovision. A revised video game system introduced in 1982 by Coleco. The system had realistic and above average color graphics, eight-direction control sticks, a push-button keyboard and a special controller for changing the speed as well as the action during the game. An optional adapter permitted playing the Atari VCS game cartridges.

Color. In video, color formed by the three primary colors (red, green, blue) and their composites. Mixing green and red produces yellow or orange; blue and red produces purple. Combining all three primary colors results in white. The general value of a color is called CHROMA, which is a combination of HUE (a color shade) and SATURATION (intensity of a color). See CHROMA, COLOR TEMPERATURE, HUE, SATURATION.

Color Adjustment Circuitry. An advanced television feature that allows the user to modify the color temperature of white while retaining natural skin tones of the subject. The special circuits adjust for cooler or warmer white color temperatures.

Color Bandpass Amplifier. Special circuitry of a color TV receiver that takes signals of particular frequencies through a stage devised to uniformly amplify them.

Color Bar Chart. See VIDEO TEST CHART.

Color Bar Generating Signal. In video cameras, a signal produced as a set of vertical color bars. This can be used as a test pattern when compared to a color bar chart such as the MUNSEL COLOR CHIP CHART. By using this chart in conjunction with a COLOR VECTORSCOPE, it is simple to determine whether the video camera components are in good working order. See VIDEO TEST CHART.

Color Bar Signal. A test pattern used by television stations, TV set manufacturers and others as a standby and check signal. The color bar signal gives a graphic picture of the condition of the equipment. The signal consists of the standard six broadcasting color bars—yellow, cyan, green, magenta, red and blue. It is the pattern normally seen on a TV channel before the day's programming begins. The color bar signal can also be reproduced by some video cameras, accessories such as a TV SIGNAL GENERATOR and a VIDEO TEST CHART known as the MUNSEL COLOR CHIP CHART. See COLOR BAR GENERATING SIGNAL, COLOR BARS.

Color Bars. A standard reference designed by the TV industry through the Society of Motion Picture and Television Engineers (SMPTE) for the coordination of levels and phasing. The bars, which are used by engineers to adjust broadcasting equipment, are electronically generated. A 1,000 Hz tone accompanies them for audio reference.

Color Burst. A group of high frequency pulses at the beginning of each horizontal line scan to determine the phase or relative timing of the color signal. Color burst, sometimes referred to as reference burst, consists of about nine cycles of the chroma subcarrier. A TIME BASE CORRECTOR, among its many functions, modifies the color burst and intensity of colors of video signals that are weak or have degenerated.

Color Constancy. In video, the continuity of COLOR INTENSITY either in a single frame or over a short range of time.

Color Contamination. Refers to the appearance of small traces of color that spill over to black-and-white portions of a video image. The amount of color contamination that a video camera produces can be determined by the use of a black-and-white pattern and is measured in IREs. A smaller numerical reading represents a purer or better picture.

Color Control Unit. An accessory connected between a video camera and recorder which adjusts or corrects color temperature for both indoors and outdoors. The CCU, used if no filter or electronic system or correction comes with the camera, offers two advantages. It is a more flexible method of adjusting color temperature and, because it is a separate unit, the camera remains lighter in weight. However, there are also a few inconveniences. The CCU is another piece of equipment that has to be toted around; there are more cables to worry about; and color corrections have to be made off the camera.

Color Conversion Filter. A colored glass disc fitted over the front of a lens of a video camera to change the COLOR TEMPERATURE. Since the camera tube is normally balance for

Color Correction Filter

proper exposure with tungsten (indoor) light, some type of conversion is required when the camera is used outdoors. One method is using a special orange-colored filter. However, manufacturers have devised more sophisticated ways of correcting color temperature: built-in, behind-the-lens optical filters, electronic switching, etc. Also known as color correction filter. See COLOR CONTROL UNIT, COLOR TEMPERATURE SWITCH.

Color Correction Filter. See COLOR CONVERSION FILTER.

Color Corrector. In VCRs, a feature that helps correct any defects in videotape. The special circuitry, along with dropout compensators, noise reduction and picture enhancement circuits, helps to compensate for any imperfections in tape manufacture and design. In a camcorder, an optional accessory designed to allow the user to adjust white balance, correct for color and add color wipe transitions between scenes. The unit may have additional features, such as the capability for audio mixing and the transfer of film negative to positive video pictures.

Color Cycling. A special process that permits dramatic color changes of screen images ranging from diagrams to music-video effects. Color cycling is often performed by professional videographers using such highly technical equipment as computers with special computer-generated animation programs that allow for color ranges and user-definable timing.

Color Demodulator. See DEMODULATOR.

Color Edging. See FRINGING.

Color Enhancement Light. A special light, found on some camcorders, designed to illuminate an area several yards in front of the camera. The small 10-watt light, first introduced in 1989 by Panasonic on one of its camcorder models, usually provides enough light to capture the natural colors of a particular scene.

Color Flash. A color dot rushing across a monochrome telecast as seen on a color TV receiver. These red, green and blue dots—or color flashes—occur when random noise, after being picked up by the decoder and matrix, reaches the picture tube. See OXIDATION.

Color Fringing. See FRINGING.

Color Intensity. In video, the saturation of a color. The original color signal and the concentration of the electron beam determine color intensity or saturation. If a certain color has high saturation, that color is usually considered bright; if the color consists of low saturation, it is said to be dull. More recent TV sets have introduced special circuitry to improve saturation. These models prevent the automatic color circuitry from oversaturating individual scenes so that the color of objects are not emphasized to the detriment of the entire picture. Saturation differs from HUE, which refers to the shade of a color.

Color Shift

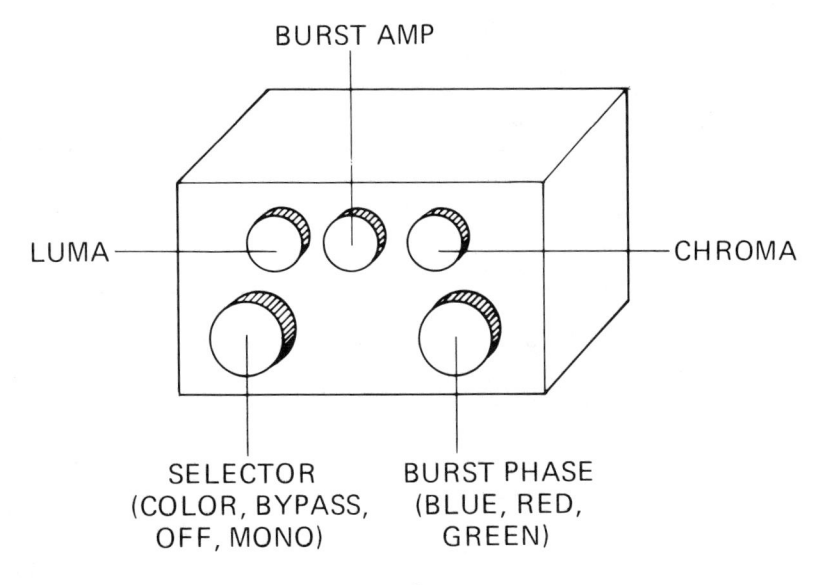

Figure 9. Color processor.

Color Intensity Control. A feature, found on some video processors, designed to adjust color level. This increase or decrease in color intensity helps to produce rich, natural color; boost weak, faded colors; and reduce overly bright colors. Especially useful in copying videotapes, color intensity control allows boosting color before the recording process to help prevent generation loss.

Color Phase. Another term for tint. See HUE.

Color Phase Correction. That which produces the correct color hues. Color phase refers to the color signal in terms of its timing relationship.

Color Processor. A device designed to enhance a picture by individually controlling brightness, color tint and intensity, and the skin tones. Among its many uses, it can color-correct duplicate tapes while recording or during playback. The color processor is similar to a PROC AMP but without its number of corrective steps or special effects.

Color Response. In videotape, the ability of the tape to reproduce color signals. Various tapes respond differently to color signals. Those of poorer quality display signs of color SMEAR. Color response is also known as chroma signal-to-noise. Both terms refer to the ability of the tape to accurately reproduce color.

Color Saturation. See CHROMA SIGNAL-TO-NOISE, COLOR INTENSITY.

Color Shift. In video, the extent to which colors hold their hue after being recorded in a particular format.

Color Signal

Color shift may occur within a single format. For example, a top-of-the-line VHS machine often displays less color shift than one of the economy models. In addition, the recording speed affects color shifting; standard play (SP) normally provides better color rendition than does extended play (EP), or the six-hour recording mode. Also, viewers who participated in comparison tests of different formats (LaserVision, 8mm, Super-Beta, ED-Beta, VHS, S-VHS, Hi8), conducted by a leading video magazine, selected LaserVision videodiscs as having the least color shift.

Color Signal. See CHROMA SIGNAL-TO-NOISE, CHROMINANCE SIGNAL.

Color Slide Theater. A television system developed by Sylvania in 1968 to show slides on a TV screen. A color console TV had a built-in Kodak Carousel slide projector which projected slides onto the TV screen. The console also contained an audio tape recorder for presenting an accompanying narrative while the slides were advanced by remote control. The model remained on the market for only one year.

Color-Striping. An anti-copying process designed to prevent VCR users from copying information from videodiscs or pay-per-view cable broadcasts. This technique is added to the pressings of videodiscs so that a videotape which has copied the material will play back an image filled with rotating color bands. Color-striping, still in the experimental stage, has interested cable broadcasters who seem concerned about having unauthorized parties duplicating their programs.

Color Sync. The reference and control signal that is required to record and play back color—designated by the figure 3.58 MHz. This number becomes the reference point to which VCRs lock in for color.

Color Temperature. A method of rating the color quality of various light sources. Color temperature is measured in degrees Kelvin (expressed as °K). Fluorescent light as well as sunlight is listed generally at 5400°K while most TV lights have a color temperature of 3200°K. The higher the Kelvin rating, the more bluish the picture; the lower the number, the more reddish the image. Color filters, electronic controls on the video camera and separate COLOR CONTROL UNITS are some methods of correcting color temperature, which is different from WHITE BALANCE. Some more recent PROJECTION TV systems have control panels that permit switching among color temperatures. This is useful in rear projection work in which a subject is to be videotaped in front of the screen. The color temperature must be matched to that of the studio lighting.

Color Temperature Switch. A control on many video cameras which adjusts for different lighting conditions such as bright sun, indoor light, etc. The more sophisticated cameras may have a switch with several settings: incandescent indoor lighting (3200 °K), fluorescent lighting (4500°), sunlight (5200°K) and cloudy bright (6000°K). Color temperature control, which corrects the camera for various kinds of light, is different from WHITE BALANCE control, which helps to set or establish colors, but both may be used together to adjust for natural color rendition. See COLOR CONTROL UNIT.

Color Tint Control. A function on most television receivers designed to adjust color when a channel is changed. Different channels transmit colors that are not always consistent. Some of the reasons for this diversity include the frequent adjusting of the transmitter for color balance, the individual channel's deliberate decision to enhance its color in the hopes of attracting more viewers, and the peculiar whims of station engineers whose personal visual and esthetic tastes affect the color that is telecast to home TV sets. As a result, the TV viewer often has to adjust the color tint control when he or she switches channels. Some more sophisticated TV receivers, such as Bang and Olufsen's models, provide a memory chip which stores the proper tint adjustment for each channel, thereby automatically making corrections as the view changes channels.

Color TV. A conventional monochrome television receiver with the addition of special circuitry. A special screen is utilized to receive phosphors which produce the three primary colors (red, green, blue). These colors can be blended to create other colors. The first color television set appeared in 1951. It was a crude arrangement which consisted of a mechanical color wheel designed to tint black & white pictures. It was not until 1954 that the electronic color system that most of us are more familiar with first appeared by way of RCA.

Color-Under Recording. A process employed by all home video recorders to record color and black-and-white information separately. The color portion of a video signal is converted to a lower frequency while the black-and-white part of the video image is left unchanged. These two video signals are rejoined during playback. But since the union is never exact, FRINGING, erratic color or VIDEO NOISE may result from what is known as CHROMINANCE/LUMINANCE DELAY INEQUALITY.

Color Vectorscope. A testing instrument used to measure the color purity, frequency response and other signals of video components such as TV receivers, VCRs and video cameras. The vectorscope, a variation of the OSCILLOSCOPE, provides a circular time base of extreme stability. It can test RF and video signals as well as check the time delay between two signals. When used in the field, the unit can help with multiple camera setups by providing precise phase matching (genlock) adjustments.

Color Video Noise Meter. A professional/industrial instrument designed to measure luminance noise, chroma AM noise and chroma PM noise for either NTSC or PAL signals. These electronic meters usually provide other features, such as automatic level control, automatic sag compensation, character displays that indicate present operations and warning messages, special circuitry for automatic

Color Video Printer

testing, and automatic memory. See AGC, AUTOMATIC SAG COMPENSATION, GENERAL PURPOSE INTERFACE BUSS, NTSC, PAL.

Color Video Printer. See VIDEO PRINTER.

Coloration. See MICROPHONE.

Colorimetry. In video, the procedure of measuring color and analyzing the results. Colorimetric characteristics include such elements as wavelength and primary-color content.

Colorization. A process that converts black-and-white movies to color. The technique involves the use of computers that assign pre-selected colors to individual areas or shapes of the original monochrome film. These colors are then carried throughout the scene or sequence for the purpose of consistency. Those film makers connected with the original work and other critics of the process have voiced their protest against colorization on the grounds that it tampers with the original creative talents that went into the making of the film.

Comb Filter. A technique designed to provide optimum picture detail without color interference on a television screen. Brought out in 1978 by Magnavox, the comb filter contains a 63.5 microsecond horizontal delay line together with special circuits for color and brightness. This circuitry either augments or diminishes chroma or luminance. The addition of a comb filter increases the number of viewable horizontal lines, reduces color fringing and provides a purer black-and-white picture. Comb filters separate the overall video signal information into its black-and-white and color components. In addition, these filters eliminate a variety of visual disturbances such as luminous dots skimming across the TV screen. Most comb filters utilize either the glass delay line technology or a CCD (charge-coupled device). The first is the more common and less costly method while the latter provides higher resolution. See CHROMA, DIGITAL FIELD COMB FILTER, FRINGING, HORIZONTAL LINES, LUMINANCE.

Combi Player. A general term used by journalists and columnists to describe a unit that can play various sizes of audio and video discs. See CD VIDEO.

Commercial Eliminator. A device designed to eliminate commercials during unattended recording on a VCR. One type works only during black-and-white programs and films, cutting out color commercials by a special color burst signal that activates the VCR. A second type operates with recordings of black-and-white and color programs. This unit disconnects an internal circuit when the program fades to black and activates another circuit when the audio goes silent. Because these devices, also known as commercial cutters or killers, require that the VCR be in Pause mode, they don't work on many new machines which cannot operate in Pause when the automatic timer is engaged. Commercial eliminators of either type are not 100 percent accurate; often part of a program is edited out along with the unwanted commercials. By the late

1980s, they all but disappeared from the video marketplace.

Commercial Killer. See COMMERCIAL ELIMINATOR.

Communications Satellite. An orbital device which can receive a signal transmitted from a source on earth and can, in turn, send it back to large geographical areas. Satellites orbit at 22,300 miles over the equator, at which point they follow the rotation of the globe. In this way they appear fixed in the sky. This is known as GEOSTATIONARY. TV services like HOME BOX OFFICE, SHOWTIME and others utilize satellite services. See SATELLITE TV.

Compact Videocassette. See CVC.

Compansion. In audio, a compression-and-expansion technique used in some noise reduction systems to improve sound quality. These systems expand audio signals during recording and condense them during playback. This results in weak segments of the signal being boosted during recording and decreased to normal when they are played back, thereby minimizing background noise. See DOLBY NOISE REDUCTION SYSTEM, NOISE REDUCTION SYSTEM.

Compatibility. See CAMERA/RECORDER COMPATIBILITY.

Component. A self-contained unit of an operating system, or a part of equipment that performs a function required for the operation of the entire system. A closed-circuit television system, for example, is made up of several components, such as a video camera, a TV monitor and a videocassette recorder.

Component Digital Editing. A sophisticated editing process used in relation with professional/industrial equipment such as digital or production switchers. This advanced editing technique usually employs other features, such as compositing in real time and color correction functions. See DIGITAL SWITCHER.

Component Television. The use of individual units such as a TV MONITOR, tuner, speakers and other items which make up a television system otherwise combined in one box—the TV set or receiver. Proponents of component TV speak of advantages like better video and improved sound as well as flexibility in component selections. The center of such a system is the MONITOR which, although it lacks a tuner, has direct audio and video inputs, higher resolution, more circuitry, etc. The component tuner is capable of better electronic isolation, thereby eliminating more noise, distortion and signal loss than its counterpart built into the conventional TV set. Component tuners also feature frequency-synthesized tuning, various audio/video inputs and outputs, wireless remote control and other sophisticated features. Sony introduced the first component TV in 1980 in Japan and later to the United States where it failed to gain popular support. It provided a choice between two monitors with dual channel 10-watt amplifiers, external speakers and other options. Component TV is capa-

Component Tuner

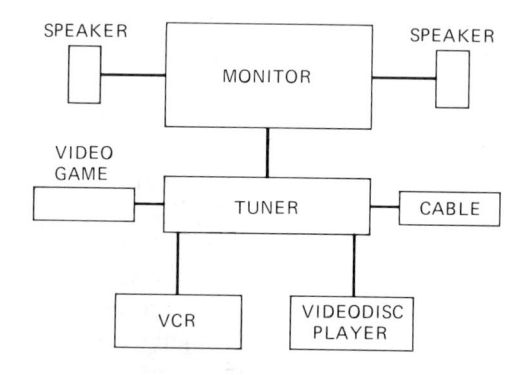

Figure 10. Component television.

ble of covering a wider BANDWIDTH (close to 4 MHz), thereby providing higher image resolution. Instead of a resolution of 200–240 lines provided by conventional TV sets, component TV can attain figures of 350 lines or better. Discriminating American videophiles have opted for TV monitor/receivers instead of component TV. See FREQUENCY-SYNTHESIS TUNER, MONITOR, RESOLUTION, VIDEO NOISE.

Component Tuner. An individual unit which, along with a TV monitor, audio amplifier, speakers and other parts, make up a television system or component video. The tuner portion of this system provides a video signal with less loss, noise and distortion than is usually present in conventional TV sets. This is the result of quality components within the tuner, better isolation of the parts and generally higher manufacturing standards. Component tuners offer several features, including, among others, wireless remote control, frequency-synthesis tuning and direct inputs and outputs for connections to monitors, VCRs, cameras and speakers. See COMPONENT TELEVISION, FREQUENCY SYNTHESIS TUNER, ISOLATION, TUNER, TV MONITOR.

Composite Color Signal. A signal made up of the color picture signal, all blanking and sync signals, luminance and chrominance signals, vertical and horizontal sync pulses, vertical and horizontal blanking pulses and, finally, the color burst signal.

Composite Video Signal. A signal containing various elements which control different aspects of the image. This signal includes horizontal and vertical blanking pulses, horizontal and vertical sync pulses, horizontal and vertical drive signals and the video image. There are several types of video signals. See HORIZONTAL SYNC, SYNC PULSE, VIDEO SIGNAL.

Compositing System. A sophisticated professional/industrial device designed to produce shadows, transparencies, reflections, blue foreground objects and other effects. Used chiefly in post-production, the compositing system can correct corner darkening and unnatural shadows in a multi-layered image. Some models provide a built-in time-code reader, capabilities for hundreds of events to be programmed for real-time on-line compositing and a menu-driven remote control.

Compression/Expansion Noise Reduction. See Noise Reduction System.

Computerized Editing. A system of editing that permits the numbering of each frame of video as it is being shot on location with a video camera. This reference then makes it possible to edit the tape immediately. Computerized editing, or computer-assisted editing as it is sometimes called, is available only with industrial VCRs. See FRAME.

Condenser Microphone. A wide-range mic usually built into a video camera or camcorder. It is designed basically to pick up all the sound in the shooting area and is characterized by wide frequency range and low distortion. The mic contains circuitry that uses a condenser and requires batteries.

Connector. The metal fitting or connection at the end of a wire or cable. A connector can be male or female, thread-type or slip-on. The barrel-shaped female connector has external threads and is usually located at the rear of the component while the male part is on the cable. There are various types of connectors. The BNC is most often utilized for professional hookups. The PL-259 is double the size of its look-alike, the F-fitting, and is sometimes used with video cameras and some industrial equipment as well as for transmitting video-only signals. The F-connector, the most popular, is designed basically for RF signals which transmit both audio and video signals. All the above connectors use COAXIAL cable. The twin lead cable usually does not need connectors. Audio/video inputs and outputs require other types of connectors. Generally, VHS machines use RCA phono type while Beta VCRs employ mini-plugs for audio and RCA phono plugs for video. Another type is the MULTIPLE-PIN CONNECTOR used between a video camera and a VCR.

Constant Angular Velocity. See CAV.

Constant Linear Velocity. See CLV.

Consumer Electronics Show. A seasonal event originating in the 1960s and consisting of independent dealers, buying groups, domestic and foreign manufacturers, government officials, importers, chain and department store buyers, trade guests and members of the press. New products and innovations are highlighted and displayed.

Continuous Motion. In video, a smooth, moving image as that which is produced on videotape or some videodisc formats such as CD-I or CD-V. Other disc formats, such as CD+G, produce graphics, with some new images displayed on screen in less than one second. However, the total effect resembles STROBE DISPLAY— a sequence of images—rather than the more familiar continuous motion that

Contour Control

viewers know from television and films.

Contour Control. Refers to the capabilities of a projection TV system or similar unit to adjust the sharpness at the edges of a screen image in relation to its contrast. See HORIZONTAL IMAGE DELINEATION.

Contrast. In video, the variation in luminance (brightness) between the darkest blacks and the purest whites. Also, the variation in density between the highlights and shadows of an image. Contrast is measured by a ten-step gray scale.

Contrast Compensation Switch. On some video cameras with AUTOMATIC IRIS CONTROL, a feature designed to provide more or less light by opening or closing the lens approximately one f-stop. In scenes with lighter or darker backgrounds than the subject, the automatic iris may not give the desired lighting, thereby resulting in over- or underexposure. The contrast compensation switch corrects for this by either opening or closing the lens. This feature is different from and more flexible than the BACKLIGHT SWITCH which can only open the lens.

Contrast Ratio. The relationship between the whitest portion of an image and the blackest portion. Because television has a high contrast ratio, it is difficult to perceive the subtle gradation of gray.

Control Track. The signal on the lower edge of videotape that regulates the synchronizing pulses which are necessary during Record and Playback modes. If this portion of the tape is wrinkled, broken or stretched, the damaged control track will cause an unstable picture. The control track can also be affected by storing cassettes horizontally for long periods of time or by improper winding. See AUDIO HEAD, CONTROL TRACK PULSE, SCATTERWIND.

Control Track Pulse. An electronic control signal placed on the bottom portion of the videotape during recording. When the tape is played back, the control track pulses guide the video heads in reproducing accurately the original information. The pulses, which make up the control track, usually govern the speed of the VCR and designate the start of every second video FIELD recorded on the tape. See AUDIO HEAD, CONTROL TRACK.

Convergence. The precise alignment of the three primary colors (red, green, blue) in the form of three electron beams on the TV screen. Convergence affects both accuracy of color and sharpness of detail. Poor convergence at any point of the screen results in colors running over the outlines of subjects and objects, sometimes blurring the entire image. The quality of the DEFLECTION YOKE (a complex set of coils encircling the neck of a picture tube) affects convergence. Convergence is judged by a special test pattern. In electronics, there are two types of convergence: static (for center white dots) and dynamic (for white dots or lines over entire screen). See CONVERGENCE ALIGNMENT, CONVERGENCE CONTROL, CON-

Copying

VERGENCE TEST SIGNAL GENERATOR.

Convergence Alignment. A process which lines up or overlaps the three primary colors of television (red, green, blue) and brings them into registration to form a perfect image. See CONVERGENCE, etc.

Convergence Control. An adjustment or set of adjustments that bring together the three primary colors (red, green, blue) into one focal point. In projection TV using a three-tube system, alignment is needed to merge the three separate images into registration without color fringing. Controls for two colors are adjusted both horizontally and vertically while the third primary color (usually green) serves as a reference for the other two. See FRINGING, REGISTRATION.

Convergence Test Signal Generator. A device used by service personnel which utilizes a signal consisting of a white dot or grid pattern to adjust television receivers for color convergence. Proper convergence occurs when the white dot or pattern is perfectly white.

Conversion Filter. See COLOR CONVERSION FILTER.

Converter. See CABLE TV CONVERTER, SINGLE CONVERSION BLOCK CONVERTER, UP CONVERTER.

Coppola, Francis Ford. Film director; early proponent of PREVISUALIZATION, videotaping of artists' views and scenes of a film to help form a rough version of the finished work; and the first director to utilize ELECTRONIC CINEMA during the making of his film *One From the Heart* (1982). Although more popularly known for his films, which include *The Godfather* and *Apocalypse Now*, Coppola remains in the forefront of video and electronic experimentation as they relate to filmmaking.

Copy Guard. See ANTI-PIRACY SIGNAL.

Copy-Protection Remover. A device or "black box" designed to eliminate certain copy-protection signals from videotape. The crystal-controlled unit, often using digital filters, works with all VCRs to prevent rolling, picture breakup and other interferences. Because the copy-protection remover, also known as a digital video stabilizer, utilizes digital filters, it differs from the image stabilizer, both of which use automatic vertical blanking level to achieve the same effect. Although these devices allow a commercial prerecorded tape to be duplicated, their manufacturers usually warn the user that copying copyrighted material is illegal. See IMAGE STABILIZER.

Copying. In video, the term refers to making a duplicate copy of the audio and video material of a tape. The process can include "dubbing," as it is often called, from a 1/2-inch machine to an industrial 3/4-inch, one-inch or two-inch VTR. Or it can involve a Beta format and a VHS machine or vice versa. Thee are two basic methods employed, RF copying or direct. In the first, the open channel output is connected to the VHF input of the machine doing the recording. In the second method, cables are connected from the audio and video outputs of the first machine to the audio and video inputs of the recording VCR. This method is much preferred since its direct connections produce a better copy. However, special cable connections may be needed since Beta and VHS use different phono plugs, especially in the audio input and output. If the original tape is a master, the new copy is called a first generation tape; if your tape is a prerecorded one, then the copy you have just made is called a second generation tape. Each generation removed from the original or master adds to the degradation of the picture quality. Duplicating copyrighted material is not legal.

Copyright Infringement. See TITLE 17.

Coring Noise Reduction. Part of a video noise reduction system designed to help eliminate or reduce video noise. This is accomplished by copying a signal and comparing it with the original. This approach recognizes unwanted interference and eliminates it. See DIGITAL VIDEO NOISE REDUCTION SYSTEM, VIDEO NOISE.

Corner Insert. A special video effect in which a picture from a second video source is placed in one of the corners of a TV picture already in progress. The electronic effect is accomplished by interrupting the horizontal and vertical scanning of the first image in a preselected portion of the picture and introducing the image from the second source. See PIP.

Corporate Video Decisions. A monthly magazine catering chiefly to those employed in the video industry. From its first issue appearing in the fall of 1988, the magazine has been offering corporate professionals information about the business side of video. News articles and feature stories focus in on how general businesses use video to market their products; the purchase, sale and mergers of video companies; and related events around the globe. One regular feature of the periodical is its "Screening Room," which reviews the latest business videotape releases. Other departments include technical trends and new products. Perhaps its most unusual feature—for a video magazine—is its periodic discussions of controversial topics such as drug and alcohol addiction at the corporate level.

Correlator Filter. Used in certain noise reduction systems to separate and cut out noise from overtones. The filter permits the passage of overtones which it "reads" based on the original tone. When no tone is present, the filter blocks the passage of any extraneous sound (noise) from getting through. The results of the correlator filter are similar to those of the DYNAMIC NOISE FILTER, but the methods of achieving them are different. See FILTER.

Coupler. An accessory which accepts two sources or units and permits them to be fed through one output. In other words, a two-way splitter can be used

Crosshatching

in reverse to form a coupler. The splitter/coupler should be the signal splitter type, not the VHF/UHF version. See SIGNAL SPLITTER.

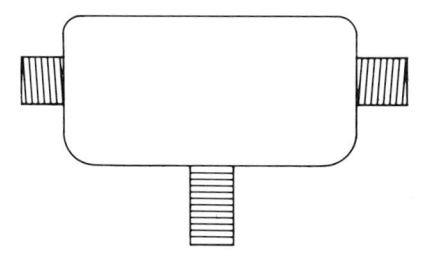

Figure 11. Coupler/splitter (two-way).

Crash Editing. A simple, basic editing technique of adding one segment after another of recorded programming from one tape onto another. The method utilizes the Pause control (instead of Stop), then Record, to help minimize picture BREAKUP between scenes. Most recent video equipment compensates for pausing and recording, thereby virtually eliminating the annoying breakup that used to show up between scenes edited on older VCRs. As in all editing, two home video recorders, audio and video input and output connections and a TV set used as a monitor provide the bare essentials. Crash editing is also called ASSEMBLE or assembly EDITING. See EDITING, INSERT EDITING.

Crawl. Lettering that moves across the bottom of the TV screen from right to left. The crawl can refer to news of national importance, local election results, dramatic changes in local weather conditions, announcements of delayed televised programs, etc. Crawls are usually superimposed during a scheduled television show, network or local, so that the current program need not be interrupted.

Cross-Channel Fade. See CROSS-FADE.

Cross-Fade. A feature on some signal processors, switcher/faders and other devices which permits the fading out and fading in simultaneously of two pictures. The true cross-fade provides a superimposition of the two images half-way through the fades. This special effect is used for smooth transitions between scenes. Some accessories feature a variation of the cross-fade in which the first image fades to black before the second picture begins to appear. Cross-fade is also known as cross-channel fade or fader. See FADE THROUGH BLACK, FADER, SIGNAL PROCESSOR, SUPERIMPOSITION, SWITCHER/FADER.

Cross-Hatch Signal. A grid-like pattern produced on a TV screen designed to check horizontal and vertical linearity as well as to align convergence. The cross-hatch signal pattern consists of 14 by 17 lines (NTSC standard) or 14 horizontal by 19 vertical lines to match the PAL 625-line standard. Some units, such as the TV SIGNAL GENERATOR, produce a cross-hatch signal which can be internally switched to a dot pattern, the dots appearing at the crossing points of the lines. See VIDEO TEST CHART.

Crosshatching. In projection TV, the converging of colored lines on a screen to bring the three color tubes into alignment. In three-tube systems (each tube or "gun" representing a

65

Crosstalk

primary color) it is essential that all three converge exactly. Projection TV systems, therefore, provide each color tube with a test pattern of intersecting lines to facilitate crosshatching. See CONVERGENCE CONTROL, PROJECTION TV.

Crosstalk. An effect created by crowding together the diagonal tracks on a tape, thereby confusing the heads during playback. Packing the signals may get more information into a given space on the videotape, but it eliminates the GUARD BANDS or spaces between tracks. To correct crosstalk, the AZIMUTH technique was introduced by Sony in 1975. In videodisc playback, crosstalk refers to the rolling horizontal noise bars and other extraneous signals that "spill over" from adjacent pits to affect the main picture area. More advanced solid-state laser pickups have virtually eliminated this problem. See AZIMUTH, CAA (Constant Angular Acceleration), GUARD BAND.

CRT. See CATHODE RAY TUBE.

CRT Coating. A method designed to increase the brightness and contrast of a television picture tube image. The technique, called internal angular reflection coating, accomplished these gains by improving the focus of the path of light from the cathode ray tube to the projection lens.

Crystal-Controlled 2:1 Interlace. A process using improved circuitry in video cameras to correct the placement accuracy of odd- and even-numbered scan lines. With earlier, less expensive video cameras, the random interlace of these scan lines posed a problem in the final image these units produced. Crystal-controlled 2:1 interlace ensures that the odd-numbered scan lines are placed between the even-numbered lines.

Crystal Microphone. See PIEZO-ELECTRIC MICROPHONE.

C-Span. A cable TV channel begun in 1979 which presents live sessions of the U.S. House of Representatives via satellite. The channel also offers local government access, programs related to special issues and community council meetings. Owned by Cable Satellite Public Affairs, it programs its service to its millions of subscribers Monday to Friday.

CTL Coding. A videocassette recorder feature that magnetically marks the beginning of individual recordings for future reference. CTL coding differs in some ways from conventional indexing. The unique digital coding function, very similar to address search, permits the user to add or erase codes at any point on the tape. In addition, by specifying a particular code, the user can locate in any direction a particular program or segment. The user simply enters a number on the remote control and activates the shuttle search mode. When the marked number is located, the VCR will automatically play back the selected segment. See ADDRESS SEARCH, CUE MARK, ELECTRONIC PROGRAM INDEXING.

Cucaloris. A small sheet, usually of metal, with a pattern stamped out of it. When placed in front of a light source, it casts its pattern onto a background.

Cue and Review. A very general term which describes different features on different VCRs. Sony has used it for its visual scan; on its SL 3000 portable, for example, Sony applied Cue for Fast Forward without a picture. According to Akai, Cue has meant high speed search with a picture. JVC has used the term Cue with its program indexing feature. Finally, there is Cue/Review, another concept when applied to video cameras. See VISUAL SCAN.

Cue Mark. A signal or code placed on a videotape or film to warn the VCR or the movie operator where a section begins or when to change reels. In video, cue marks are placed on tape electronically when a special control is activated. Each time the VCR goes into Record, an electronic signal is encoded on the tape. The machine can then quickly locate those sections in either Fast Forward or Rewind. Cue mark is also known as index mark. See ELECTRONIC PROGRAM INDEXING.

Cue/Review. A feature on virtually all video cameras with an electronic viewfinder which permits replaying almost instantly a portion of recorded tape. Cue/review may go under different names, depending on the manufacturer. The term is also used with VCRs and is synonymous with "search" or the individual speed of the search in each mode, such as 6x in standard play, etc. See SEARCH MODE.

CVC Format. A 1/4-inch VCR format incompatible with other video formats. No longer being produced, the format was introduced by Technicolor in 1980. The Compact Video Cassette (CVC) recorder, which weighed about seven pounds, used 1/4-inch videotape in a cassette approximately the size of an audio cassette. The maximum record/play time was about 60 minutes. CVC was unlike UCM, another 1/4-inch system which featured a one-unit combination of camera/recorder. These two systems were also incompatible. See UCM.

CX. Introduced by CBS Records, a noise reduction system which compresses and expands a program, thereby extending the dynamic range. However, like its competitors dbx and DOLBY, it must be played back through a decoder to benefit from the extension. The advantage that CX offers is that a program played back without the decoder will still be intelligible. Although CX also operates on the compression/expansion principle, it compresses and expands only loud signals. CX was used with the first opera videodiscs, produced in stereo by Pioneer Artists in 1982. See NOISE REDUCTION SYSTEM.

D

Daisy Chain. Refers to a single playback VCR, used as a "master" in a duplicating process, feeding more than one machine. The recording units are called "slave" units. Many home VCR owners have gotten together in this manner to duplicate videotapes. However, signal loss is bound to occur if too many machines are connected. A VIDEO DISTRIBUTION AMPLIFIER may be employed to increase the signal level.

Day/Event. Refers to a VCR's programming capabilities. "Day" pertains to the length of time a machine can record in advance; e.g., 14 days, one year, etc. "Event" points out the number of programs the VCR can record in that time period; e.g., 4 events over 21 days.

dB. See DECIBEL.

DBS. See DIRECT BROADCAST SATELLITE SERVICE.

dbx. A noise reduction system designed to eliminate unwanted sound from a program while still maintaining that program's full audio range. By compressing a program before it is recorded and expanding it during playback, dbx restores or recaptures the original dynamic range. A linear compression/expansion system similar to DOLBY, CX and others, dbx covers the entire frequency range, yielding approximately 30 dB (decibels) improvement in signal/noise over the whole band at mid frequencies. It also provides better than 40 dB of noise reduction. See NOISE REDUCTION SYSTEM.

DC Light. A video camera feature that draws power from the internal battery of the unit for the purposes of improving image detail. In addition, the feature is used to enhance color. Only a small number of cameras offer DC light.

DC Restoration. The capability of the television monitor, receiver or camera to respond to alterations in brightness as viewed by a video camera. The better the DC restoration, the greater the picture detail in night scenes or low-lit shots. DC restoration makes use of a special circuit that returns the direct current signal to the TV or camera outputs.

Decibel. A measure of sound strength or volume, generally subjective. Decibel is usually expressed as dB. Signal-to-noise ratio is one area measured in decibels. See AUDIO SIGNAL-TO-NOISE RATIO, VIDEO SIGNAL-TO-NOISE RATIO.

Deck. See VCR DECK.

Decoder Box. A device supplied by a pay TV system designed to unscramble signals so that subscribers, who usually pay a monthly fee, can receive a clear picture of a particular channel. Pay TV companies scramble their signals so that their channels appear on screen as unintelligible to non-subscribers. See CABLE TV, PAY TV.

Dedicated Chip. A small piece of silicon imprinted with logic circuits. Used in early video games such as PONG, these chips permitted the games to provide such basic activities as paddle movement. The games, however, were non-programmable since the printed chip was an inherent part of the game console. See VIDEO GAME, VIDEO GAME SYSTEM.

Dedicated Design. A term applied to an accessory or software which attaches to or fits equipment only of the same manufacturer. For example, because of a particular thread design, a SUPPLEMENTARY LENS of one manufacturer may only fit that company's camera-equipped standard lens. The problem is even more prev-

alent with MULTI-PIN CONNECTORS and ADAPTERS.

Defeat Filter. In electronics, a filter which cuts any frequencies in the band it rejects. Defeat filters, as well as such others as BANDPASS FILTERS, permit the boosting and suppressing of selected material at a choice of frequencies. These filters are used in SIGNAL PROCESSORS to affect color and DEFINITION. See BANDPASS FILTER, FILTER.

Definition. The degree or extent of detail or sharpness in a TV image. Definition depends on various factors. With a camera, sharpness is affected by the quality of the lens, the lighting, etc. With a VCR, it is influenced by the quality of the tape, the selection of the speed mode, the condition of the video heads, etc.

Deflection Yoke. The wire wrapped around the base of a CATHODE RAY TUBE. The deflection yoke transmits voltage from horizontal and vertical sweep to special circuitry which affects the electron beam. The horizontal directs the beam in a diagonal direction from the end of one line (on the right as seen by the viewer) to the start of the next line scan (left). The vertical sweep moves the electron beam down one stage. The quality of the deflection yoke affects both GEOMETRIC LINEARITY and CONVERGENCE. See CATHODE RAY TUBE, FIELD BLANKING, LINE BLANKING, SEQUENTIAL SCANNING.

Degradation. The deterioration of the video information on a second tape that has been copied from another, or

Degauss

master, tape. Each generation removed from the original or master adds to the degradation of the picture quality. Other factors besides duplicating tapes can cause degradation. Some methods of signal scrambling or encoding, for example, may result in degradation of the original video information when it is decoded. On the other hand, digital VCRs, as opposed to conventional analog machines, maintain the integrity of the original information. Using this process, signals converted into numbers not only are safe from distortion and noise, but can produce unlimited copies without degradation. Since the digital signal remains permanent, there is no loss in detail with successive generations of recordings. Professionals have isolated five other basic types of degradation that affect NTSC picture quality—NOISE, INTERMODULATION, MICROREFLECTIONS, ENVELOPE DELAY and PHASE NOISE.

Degauss. To remove the excess magnetism from a TV screen. Degaussing, which is done automatically when the TV set is turned on, is essential to prevent interference from a variety of electric household motors in the vicinity. Such stray magnetic fields can affect the purity and convergence action of color picture tubes. To prevent such problems, a special degaussing coil is mounted around the television picture tube or its screen. Today's TV manufacturers employ several different methods to perform automatic degaussing. See GAUSSIAN FILTER.

Delay Line Aperture Control. An electronic method of improving contrast by artificially increasing the strength of the original video signal. Similar to another technique known as HORIZONTAL IMAGE DELINEATION, delay line aperture control strengthens the beginning and end of the signal as it repeatedly passes from black to white. See VELOCITY-MODULATED SCANNING.

Demodulator. In video, an instrument that restores the basic characteristics of a wave that has previously been modulated. Since 1975, color TV manufacturers have applied IC (integrated circuitry) to demodulators and other electronic functions. Some more sophisticated professional units can receive, process and provide dual-buffered composite baseband video outputs for each of 139 channels, including UHF, VHF and Cable.

Demonstration Mode. A feature on some VCRs designed to help the user understand the various programming functions. The demonstration mode leads the user through these functions and the contents of the several on-screen menus. Time shifting, or setting up the VCR to record one or more programs at a later time or day, has been one of the more difficult operations for many VCR owners who simply end up using their machines only for playing back prerecorded tapes. Manufacturers, realizing this, have come up with several methods for making recording easier. These include OTR (one-touch recording for taping a show presently being telecast), on-screen programming (which guides the viewer step by step) and, finally, demonstration mode.

Depth of Field. Refers to the range of focus of a camera lens, or, in other words, that which is in focus behind and in front of a subject. Depth of field changes with the lens opening (the smaller the opening, the greater the depth of field), the distance between subject and camera (the closer the subject, the shorter the depth of field) and the type of lens (wide angle lenses provide greater depth of field than telephotos).

Depth of Modulation Chart. A method of measuring the sharpness of a video camera. The chart consists of a predetermined pattern of short vertical lines spaced according to "bursts" that are calibrated in megahertz. The bursts range from 0.5 MHz to 4 MHz. Depth of modulation, or modulation transfer function, refers to determining the "quantity" of resolution or sharpness of an image. Normally, the degree of sharpness is highly subjective. Depth of modulation charts seek to remove this aspect, replacing subjectivity with mathematical calibrations that result in percentages. For example, a pattern can now be considered properly registered if the depth of modulation totals 80 percent. Depth of modulation problems, such as dark objects appearing light or light objects becoming darkened, may be the result of a faulty camera or lens.

Detail Enhancement Circuitry. An image processor feature designed to add contrast and produce sharper outlines to images. A special electronic circuit splits the incoming video signal and sends it along two paths, one ignoring the split signal and the other modifying the black-to-white transitions so that certain objects appear sharper. The two signals are then joined to produce a better TV image. See IMAGE PROCESSOR, SIGNAL PROCESSOR.

Detail Enhancer. Refers to electronic circuitry built into a VCR to help deliver a clear screen image during playback. This electronic feature is usually part of the HQ (high quality) circuitry of many videocassette recorders. Sometimes described as a "detail switch," this feature may have two settings—one for use with conventional videocassettes and another for HG (High Grade) tapes for producing sharper images. See HQ CIRCUITRY.

Detailer. See IMAGE ENHANCER.

Deviation. See FM DEVIATION.

Dew Indicator. A warning light that goes on when there is too much moisture in the atmosphere. This feature appears on some VHS machines but rarely on Beta VCRs. Simple remedies consist of shutting off the machine for a while, using a hair blower near the VCR to minimize the moisture, putting on an air conditioner, etc. Some machines have a dew control which shuts off the VCR when moisture is excessive. See HUMIDITY ELIMINATOR UNIT.

Diagonal Audio Recording. Another method of recording sound on videotape, as opposed to linear audio and stereo. Diagonal recording places the audio track along with the diagonal video track for better quality. In contrast to linear stereo, or linear track

Diaphragm

stereo as it is sometimes called, the superior Beta or VHS Hi-Fi technique uses diagonal audio recording. See LINEAR AUDIO, LINEAR STEREO.

Diaphragm. The internal segment of a microphone that is vibrated by sound waves entering the mic. These vibrations are changed into voltage fluctuations, creating audio signals. See MICROPHONE.

Dichroic Daylight Conversion Filter. See COLOR CONVERSION FILTER.

Differential Phase and Gain. A fault in a video camera resulting in color shift on a TV screen. Differential phase and gain becomes noticeable during scene changes. If the gain is insufficient or excessive, the colors will be distorted in hue and intensity (either too soft or too bright). Equipment demonstrating this anomaly should be brought to a service center for proper adjustment. In other units, such as a process amplifier, also known as a PROC AMP, differential gain refers to the degree of alteration in the chrominance level when a change occurs in luminance. Differential phase applies to the degree of change in hue or tint when the luminance level is altered.

Digital. In video, refers to the use of encoded numbers ("1" and "0") in transmitting data or in performing "on-off" operations. TV receivers, VCRs, video cameras or similar units equipped with digital circuitry include a special chip for storing frames in a digital memory. See DIGITAL EFFECTS.

Digital Clock. The component of a VCR or a TUNER/TIMER serving a dual function: time-keeping and time-setting for unattended recording. Usually different controls set the clock or timer.

Digital Color Art. A VCR feature that permits the viewer to place a tint over the video image. Provided the videocassette recorder is equipped with the necessary digital circuitry, this digital playback function offers the viewer a choice of several colors from which to choose.

Digital Compression. An experimental process designed to permit full-motion video to be picked up and stored in a computer for possible future display. Digital compression will help professional broadcasters, who use the NTSC standard, record and play back in real time (30 frames per second), a procedure not presently available.

Digital Effects. A series of special image enhancements resulting from digital circuitry built into some VCRs, television receivers and video cameras. For example, a VCR equipped with this circuitry can produce steady still pictures without the usual streaks and other visual interferences that often accompany this function. These stills can be gotten from a live broadcast as well as from a videotape. In addition, the audio portion of the program continues while the still remains on screen. Other features of digital circuits include video noise reduction, picture-in-picture (PIP), slow motion (without video interferences) and a quick succession of still pictures known as strobe display. Strobe display, which differs from the conventional continuous slow motion function, produces images without interference and offers a wider range of slow motion. Digital circuitry also allows the viewer to watch two pictures at the same time. The second image usually appears as a corner insert. Some VCRs apply digital techniques to improve the overall image quality of poorly recorded tapes. Advanced circuitry accomplishes these feats by digitally "capturing" video frames from the moving videotape and parading them on the TV screen. Unlike conventional VCRs, which use analog signal processing, machines with digital circuitry also reduce the graininess of multi-generation tapes. See DIGITAL VCR, DIGITAL VIDEO EFFECTS SYSTEM, GENERATION.

Digital Effects Video Palette. A console-type device used for electronic drawing. The console, which has a 256-color capacity, consists of an 11x11-inch "drawing board," an electronic stylus, a computer to store information and a monitor to display the artists' work. Success of the initial unit and technology have contributed to produce advanced models. Consoles have functions like Broken (to create small broken lines), Stream (to make a flowing line), Cycle (to produce a range of colors), Point (to draw connecting lines), etc. See VIDEO PALETTE.

Digital Electronics. An experimental field being developed for video theaters to rival movie houses. Digital electronics offers special effects and the promise of economic advantages. Currently, it can produce transitions, montages, mattes, mosaics, dissolves and a string of other tricks competitive with conventional filmmaking. And it can present these—as well as tape editing—at less cost.

Digital Encryption. An electronically advanced method of encoding or scrambling video information for security purposes. Digital encryption which rephrases original video information line by line is generally more costly a procedure than is its counterpart, ANALOG ENCRYPTION, but results in a higher quality video image after the information is restored or decoded. The digital system requires a larger bandwidth than does the analog technique, which uses the standard 4.5 MHz bandwidth. See ENCRYPTION.

Digital Field Comb Filter. A feature, usually found in advanced television receivers, that greatly reduces the NTSC interference effects often accompanying conventional TV sets. These special comb filters reportedly surpass the performance of standard line filters by delivering as many as 480 lines of horizontal resolution from such wide band sources as CD-V/laserdiscs and S-VHS tapes. See COMB FILTER.

Digital Gain-Up. A feature, found on some video cameras, that enhances image contrast through the use of special electronic circuitry. Digital gain-up is particularly important in improving pictures shot under extremely low light conditions. Some camcorders advertise a digital gain-up circuitry response of 1 LUX.

Digital Image Superimposer

Digital Image Superimposer. A video camera feature designed to electronically store high contrast images. These images are usually used for a variety of purposes, such as superimposing titles, graphics or captions. When the user activates the button of this feature, the digital superimposer memorizes an image. Some cameras provide a built-in superimposer with a four-page, eight-color memory. The digital image superimposer, also known as a digital superimposer or word register, is more sophisticated than a character generator. The latter also creates titles but more slowly and with limited typefaces.

Digital Memory. An electronic system that stores, transmits and processes different types of information by way of electrical pulses that represent numbers. The information may be sound, video images or written characters. Digital memory is particularly effective with video signals, which are waveforms usually recorded as "analog" shapes. The analog system captures electronic interference or video noise along with the desired image. Digital memory, by converting the video signals into a set of binary numbers, virtually eliminates the unwanted noise and thus presents a highly improved image and such glitch-free special effects as slow motion and freeze frame. Some VCRs with digital memory, introduced in 1986, double the image information provided by the tape. This results in a clearer picture on the TV screen without the addition of video noise. The benefits of digital memory have crossed over to videodisc players as well. These units currently offer excellent freeze frames and variable-speed strobe effects on CLV laserdiscs. Video cameras with digital memory can produce still frames of live action, store pictures in memory for later use, etc. See ANALOG, DIGITAL EFFECTS, STROBE DISPLAY, VIDEO SIGNAL.

Digital Multi-Audio. A feature built into a few VCRs that assigns auxiliary audio tracks to the videotape in lieu of video information. With Hi8 VCRs, for example, digital multi-audio permits up to 24 hours of audio recording.

Digital Multi-Effects System. A professional/industrial system designed chiefly for television stations that is capable of manipulating flat video images into three-dimensional images in real time. In addition, the system, which sells for about $350,000, can produce such effects as sparkle, degrade (image breakup) and centipede (elongation of an object or subject). Although some of these special effects can be accomplished on other pieces of equipment, the more costly digital multi-effects systems can reduce the time it takes to produce these effects from hours to minutes.

Digital Noise Reduction. See DIGITAL VIDEO NOISE REDUCTION.

Digital Oversampling. See OVERSAMPLING.

Digital Paint Art. A feature, found only on digital VCRs, that produces a rough gradation to a screen image to simulate the effect of an oil painting.

Digital Still-Frame Memory

Digital paint art usually offers several levels of contrast.

Digital Photography. See STILL VIDEO.

Digital Picture Memory. An advanced electronic feature, found on some videodisc players and videocassette recorders, which permits the audio portion of a program to continue while a picture "freezes" on the screen.

Digital Scan. A feature on specially equipped VCRs designed to allow the viewer to see a picture in fast forward or rewind mode. Only some VCRs with digital effects include this function.

Digital Servo Transport Circuitry. A VCR system that controls both the capstan and head-drum motors to provide better track positioning. Some motors may produce phase and frequency errors. These defects in destabilization are sensed and corrected by a quartz crystal oscillator. Image quality depends upon accurate tracking which, in turn, relies upon the precision and rigidity of the motors. Imprecise tracking may result in audio wow-and-flutter or picture jitter or both. See JITTER.

Digital Signal. An electronic signal of a two-state or binary structure, performing such functions as on/off, positive/negative or high/low. Digital signals, because of their inherent binary form, are said to be discontinuous, as opposed to analog signals, which are described as continuous. See ANALOG SIGNAL.

Digital Signal Processing. A sophisticated method of reproducing a broadcast signal by converting its components into a set of binary numbers or a digital code. In the more conventional analog signal processing system, the signal, in its components of wave peaks (voltages) and distances between these peaks (frequencies), is dissected and reconstructed before it is fed to a TV set or VCR. However, much of the signal quality is lost in this procedure. Digital signal processing, in contrast, encodes the signal with the aid of computer circuitry as near to its origin as possible (the videotape, broadcast studio, videodisc, etc.). This minimizes distortion and deterioration. The digital information is then transformed to an analog signal of high quality, ready for playback by a TV set. One major drawback of the digital process at the present time is that it requires a larger bandwidth than its counterpart, the analog process. See ANALOG SIGNAL PROCESSING, BANDWIDTH, COMPOSITE VIDEO SIGNAL, SIGNAL, VIDEO SIGNAL.

Digital Signal Sound Processor. An accessory designed to offer a simplified version of surround sound audio effects. Usually composed of dual speakers, an amplifier and digital delay circuits, the sound processor comes equipped with jacks for hooking it up to VCRs or TV receivers. The unit can be used without the complex wiring often necessary for surround sound.

Digital Still-Frame Memory. An electronic feature, found on some VCRs and video cameras, that can play back

Digital Stereo

single frames with perfect steadiness. Videocassette recorders and cameras with digital circuitry usually provide unstable freeze frames containing electronic interference or video noise in the form of lines, streaks or snow. Digitally produced special effects such as still-frame eliminates these interferences. In addition, still-frames can be gotten from live broadcasts as well as videotape. See DIGITAL EFFECTS, DIGITAL MEMORY, DIGITAL PICTURE MEMORY.

Digital Stereo. In VCRs, the enhancement of audio by means of special digital and error-correction circuitry that virtually eliminates interference. Discrete right and left digital audio signal processing helps to produce a dynamic range of over 90 dB. Digital stereo has proven effective with video cameras as well. This is especially true with digital audio dubbing, which adds high-fidelity sound to special effects and music.

Digital Superimposer. See DIGITAL IMAGE SUPERIMPOSER.

Digital S-VHS. An experimental audio enhancement process that records audio signals in a lower level of a videotape than conventional video signals. The digital system is compatible with existing S-VHS recorders and S-VHS-C camcorders. Developed by JVC, which calls the process D-MPX (depth-multiplexed signal AC bias recording system), digital S-VHS promises benefits for both consumer and professional applications. The system provides sampling frequency of 48 KHz in two-channel mode and 32 KHz in four-channel mode.

Digital Switcher. A professional/industrial unit that speeds up component digital editing by providing additional sophisticated capabilities to the production printer, including multi-layer compositing in real time and color correction functions. In addition, digital switchers can accommodate several chroma keyers simultaneously to isolate different areas during the composite procedure. See PRODUCTION SWITCHER, SWITCHER.

Digital Sync. A television feature that eliminates the need for a manual vertical hold control. Digital sync automatically produces vertical hold modifications by sensing variations in video sync and locking into the correct reference for vertical adjustment.

Digital Television. A TV system designed to outperform all present methods as well as those, such as high definition TV, already proposed as a future standard. Suggested by such influential institutions as Massachusetts Institute of Technology, Columbia University and Zenith, proponents of digital TV see the system as the wave of the future—integrating with digital communications which is expected to take the lead in the next few decades. They also view digital TV as a way for the U.S. to recapture a large share of the future international electronics information industry. Digital TV differs from its counterpart, analog TV, by translating information into numbers, which remain unvaried and, therefore, undistorted during transmission or recording. This results in clear, sharp pictures with no loss of detail. The first use of digital circuitry to appear in television systems showed up in 1985 in TV monitors.

Digital Television Standards Converter. A professional electronic accessory which uses digital technology to convert different broadcasting and recording standards of other countries. Ordinarily incompatible standards such as Secam, PAL and PAL M are connected to the television converter which processes the signal for LINE and FIELD conversion. A memory output compensates for distortion usually caused by such conversion. The converter also operates with satellite broadcasts of foreign programs. Other functions of the device include its use as an editor, mixer and frame synchronizer to integrate local shows into network programs.

Digital-to-Analog Converter. Advanced circuitry, usually found in digital audio systems, that converts the digital code to analog for playback purposes.

Digital Tracking. A VCR feature, most notably found on some S-VHS models, that automatically compensates for disparities in videotapes recorded on different machines. Occasionally, a tape recorded on one machine may not play correctly on another VCR because the video heads or video head drums of the two machines may not be positioned at the same exact height. This difference causes video noise or bars to appear on the TV screen. The viewer may eliminate this interference by manually adjusting the tracking or skewing control dial. Digital tracking (sometimes called twin digital tracking), when activated, will correct the video signal automatically and optimize playback in any of the speed modes. Another advantage of digital tracking is that the user does not have to return the tracking control to its default position. This adjustment is necessary on VCRs without digital tracking to prevent future tapes from being recorded incorrectly.

Digital Transcoder. An accessory designed to convert digital signal formats from D-1 to D-2 videotape recorders and vice versa. The transcoder is used chiefly with industrial tape machines.

Digital Transfer. A technique that uses digital circuitry to transfer widescreen theatrical films to video. Digital transfer, which is superior to other, older methods, produces better image quality while providing special sophisticated effects such as enlarging images, zooming in and out and simulating the original movements of the movie camera. See SCANNING.

Digital Tuning. The use of quartz crystals that are tuned to each FM and AM frequency. Digital tuning, which replaces the conventional dial with a digital readout, does away with the need for a manual tuner. See TUNER.

Digital TV Receiver. A television set designed to produce superior audio and video through the use of digital encoding of the broadcast signal. Instead of the conventional method of analog signal processing, the digital system uses large-scale integrated circuitry. Some of these newly constructed chips digitally decode the

Digital VCR

video signal and process it while another handles the audio in a similar manner. Meanwhile, another special circuit in the TV receiver controls the tuning, power supply and other components. Digital TV reconstructs the signal more accurately than its counterpart, the analog process, which tends to lose much of the original quality. However, the relatively higher cost of this technique has affected its general acceptance in the marketplace. See ANALOG SIGNAL PROCESSING, DIGITAL SIGNAL PROCESSING.

Digital VCR. A videocassette recorder that uses special electronic circuitry to store and process information using electrical pulses to represent numbers. Video signals, which are waveforms, are normally recorded in "analog" form along with inherent video noise, or electronic interference. Digital machines, on the other hand, convert video signals into a predetermined code, or set of binary numbers (using multiple combinations of "0" and "1"). This method virtually eliminates video noise to produce an improved screen image and clearer special effects such as FREEZE FRAME, SLOW MOTION, PICTURE-IN-PICTURE, STROBE DISPLAY, MOSAICS and POSTERIZATION. Digital VCRs, first introduced to the general public in 1986, offer other benefits. For instance, viewers can call up two different programs simultaneously on their TV screens. In addition, signals converted into numbers not only are safe from distortion and noise, but can produce unlimited copies without degradation. Since the digital signal remains permanent, there is no loss in detail with successive generations of recordings.

Digital Video. A process of videotape recording with a potential for more accurate color renditions and pictures of higher resolution. Conventional recording uses the analog system of placing information on tape. On a professional/industrial level, there are two major digital formats—D-1, the component digital standard, and D-2, the composite digital standard. The former records the luminance and two color-difference channels digitally, while the latter, D-2, records such basic standard signals as NTSC and PAL. MII is another format competing for recognition. Digital compression, another technique, is still in the experimental stage. See DIGITAL VCR, THERMOMAGNETIC RECORDING.

Digital Video Camera. A professional video camera that has fewer moving parts and is generally lighter and smaller than its counterpart, the analog camera. First introduced by Panasonic in 1989, digital video cameras can filter out video cross-color and can make any necessary adjustments automatically. The company also introduced the first digital camcorder, which has a 1/2-inch format. Both models sell for more than $10,000 each.

Digital Video Effects System. A professional/industrial workstation designed to produce numerous special video effects during the editing process. These units can be used for transitional or non-transitional functions. Some transitional uses (dissolving from one scene to another) include warp, prism, curvilinear, montage, mirror, mosaic, sparkle, trailing, decay, drop shadow, multi-freeze and rotation effects. In addition, a DVE system can be utilized to perform non-transitional tasks, such as correcting errors in a source tape. An intrusive shadow, window glare or overhead microphone can be removed by simply enlarging the picture and trimming it until the unwanted image does not appear. DVE units generally cost thousands of dollars.

Digital Video Frame Storage. A laser videodisc player feature designed to retain one frame in memory, or in a "buffer." This single frame appears on the TV screen until random access search for a new segment is completed. Before the introduction of this feature, videodisc players would go to a black screen whenever random access was activated. Digital video frame storage operates automatically whenever the viewer presses the random access function. The last frame seen on screen is then stored in the buffer and appears until the new material is located.

Digital Video Interactive (DVI). A method of presenting on a CD-ROM disc more than an hour's worth of audio, video action, still pictures and graphics. These five-inch discs, developed by the Sarnoff Laboratories in Princeton, New Jersey, are able to hold an unusual amount of information because the material was compressed during recording. During playback, the information is automatically decompressed.

Digital Video Noise Reduction. Special electronic technology that improves signal-to-noise ratios by combining two video fields, thereby doubling the amount of video information displayed on the TV screen. First introduced into S-VHS videocassette recorders, this electronic video noise reduction system differs greatly from other attempts to reduce image interference from video signals. Using special comb and notch filters, this new system, also known as field correlation, adds a complete field of information to each succeeding field. Since both fields are almost identical, the signal power is therefore doubled. Although distortion and video noise is inherent in all video signals, the interference does not double along with the enhanced dual-field video image. This results in a clearer screen image without the usual additional video noise and wavy lines. Several sophisticated videocassette recorders with this digital feature, sometimes listed as DVNR, offer several settings. They include one for use with prerecorded tapes, another to compensate for poor reception caused by a weak signal, and the usual on/off switch. The digital system, introduced by NEC in 1986, works with both broadcast signals and prerecorded tapes. See DIGITAL SIGNAL PROCESSING, FIELD, FIELD CORRELATION, FRAME, VIDEO NOISE, VIDEO SIGNAL.

Digital Video Stabilizer. See COPY-PROTECTION REMOVER, IMAGE STABILIZER.

Digital Video Transmission. An experimental method that permits specially equipped VCRs to receive

Digital Zoom

rented or purchased movies by way of satellite. By calling on the telephone, the consumer can order a movie to be transmitted to his or her VCR. Because of the digital process, the rented film can usually be recorded in about 10 minutes and can be played back two times. If the film is purchased by this method, it can be played countless times. The only limitation, albeit a major one, is that at the present time a satellite dish is required. Developers hope that in the future they can build receiving capabilities directly into the VCR.

Digital Zoom. A VCR feature that employs special electronic circuitry to produce various effects such as enlarging one portion of a screen image to full screen size. Some viewers may find this a useful method of examining an image more closely. The digital zoom feature, which appears only on digital-type videocassette recorders, can be used on any segment of the screen. For example, some digital VCRs can divide a screen picture into four parts with a fifth image appearing in the center. This last image can enlarged four times its original size.

DIN (Deutsche Industrie-Norm). A German electronic standard for connections. A DIN connector was used in earlier Sony professional equipment. Today DIN plugs and connectors are still available in 3-pin and 6-pin configurations. See MULTI-PIN CONNECTORS.

Diplexing. A process in which video circuits carry two full frequency audio channels, thereby making STEREO TV possible. The two audio channels are transmitted above the video signal, the main one at 5.8 MHz and the other at 6.4 MHz. Diplexing is capable of operating on telephone and microwave transmissions. See COMPOSITE VIDEO SIGNAL, MICROWAVE, VIDEO SIGNAL.

Direct Audio and Video Inputs. A feature on some video components which produce improved sound and picture by avoiding superfluous circuitry. For example, if a VCR signal were fed to a conventional TV set, it would have to be modulated to an RF signal. But if the same signal were directly connected to the audio/video inputs of a monitor or a TV set with this feature, it would produce clearer sound and a better picture. A PROJECTION TV system with direct inputs saves the VCR signal from passing through three circuit systems: (1) the modulator that directs the signal to a TV channel, (2) the circuitry of the TV section which receives and isolates the VCR signal and (3) the circuits which convert it back to an audio and video signal. Direct inputs are also recommended for COPYING tapes. More TV sets are becoming available with direct inputs as well as the standard RF input. See MONITOR, VIDEO SIGNAL.

Direct Broadcast Satellite Service (DBS). A video service designed to deliver up to three channels to roof dish-type antennas 12–36 inches in diameter. The service may be by subscription or advertiser-supported. The DBS system consists of a parabolic antenna and feed horn, an amplifier/down converter which modifies the incoming signal and a tuner/converter

Display

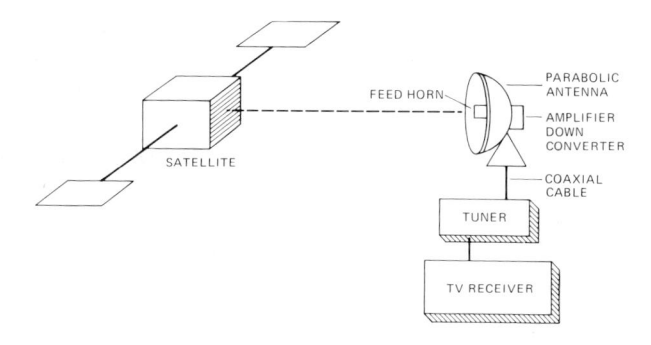

Figure 12. Direct broadcast satellite.

which again alters the signal for a standard TV set. The smaller-diameter dishes are possible because the satellite transponder's power is increased from 4 or 5 watts to 70 or more. By 1990, experiments with high-power satellites using digital transmission technology were in progress to create 108 channels by compressing 4 channels into each of the 27 transponders ordinarily used to transmit 1 channel.

Direct Channel Access. See ELECTRONIC CHANNEL SELECTION.

Direct Chapter Search. A videodisc feature, usually found on the wireless remote control component, that allows the user to access different programs on the disc. See AUTOMATIC CHAPTER SEARCH, FRAME/CHAPTER SEARCH.

Direct Drive Capstan Servo Motor. See VTR.

Direct Video/Stereo Audio Inputs. A TV monitor/receiver feature that permits the connection of a stereo videocassette recorder directly into the TV's A/V inputs. This technique improves picture quality. In addition, the hookup provides stereo sound by way of the built-in speakers of the monitor/receiver. Conventional TV sets do not ordinarily provide direct video/stereo audio inputs.

Direct View TV. Refers to conventional television as opposed to rear or front projection TV. Some large-screen direct view monitor/receivers feature screens ranging from 27 to 40 inches, measured diagonally.

Directional Microphone. A microphone which is more sensitive to the sound in front of it, blocking any sound coming from its rear. It is preferred for recording from a distance. See MICROPHONE.

Dish. See PARABOLIC ANTENNA.

Display. In video, a method of presenting a picture from a received signal. The display may take the form of a cathode ray tube (CRT), liquid crystal display (LCD) or beam indexing. Hand-held or pocket TVs may use any one of these three types of display. See BEAM INDEXING.

Display Screen

Display Screen. Refers to the presentation of various kinds of information on the TV screen other than prerecorded or over-the-air broadcast programs. VCR displays may take the form of menus that permit the viewer to program the unit to record events for later viewing. Displays of some TV receivers provide on-screen menus for adjusting the color or audio portions of a picture, switching from one antenna source to another or confirming a channel selection. Many videodisc players display, often by means of superimposed images, such functions as chapter, track, frame, speed, playback or scan direction. See NON-INTERLACED DISPLAY, ON-SCREEN DISPLAY, ON-SCREEN PROGRAMMING, STROBE DISPLAY.

Display Window. That part of a VCR, located on the face of the unit, that provides the user with a variety of information, such as tape counter, current time, etc. If the VCR is programmed, the window will display the time, day, channel and number of shows to be recorded as well as the recording speed. Also, the display window shows Play, Fast Forward, Reverse and Rewind modes when these have been activated. Different models, depending on sophistication, offer other features, such as auto-rewind when tape has reached its end and an indicator light when a cassette is inserted.

Dissolve. In video, a special editing effect in which a new picture slowly fades in while the old one fades out. Many camcorders and several recent VCRs offer an array of special editing effects, including a fade feature which can be used to create dissolves. A SPECIAL EFFECTS GENERATOR may be used to produce this effect on cameras not equipped with this feature, while EDITING CONSOLES can be purchased to create similar effects when editing with VCRs. See FADE, WIPE.

Distortion. An electronic interference in the video image, usually lasting for only a few seconds. Distortion can be caused when editing from one machine to another, when placing a VCR in the Pause mode during recording to edit out commercials or other unwanted material or when a video camera is stopped between shots or scenes. Newer VCRs and cameras are better equipped to minimize this type of distortion. See BREAKUP, GLITCH.

Distribution Amplifier. An accessory that increases the strength of a video signal and sends one audio or video signal to several VCRs. The video distribution amplifier (VDA), as it is sometimes called, usually consists of a box with one high impedance video input and more than one isolated 75-ohm output. At times the distribution amplifier is part of a switcher, but in either case, its functions are the same: to restore the video signal strength back to normal and to isolate the outputs from each other. It is especially useful when playing a VCR through more than one TV RECEIVER, since splitting a VCR signal tends to weaken it. The amplifier increases the signal level only slightly more than one decibel from the original, causing no significant change. The VDA is also known as an in-line amplifier, signal amplifier, distribution amplifier and video amplifier. This ampifier differs from an RF AMPLIFIER.

D-MPX. See DIGITAL S-VHS.

DNR. See NOISE REDUCTION SYSTEM.

DOC. A professional term or abbreviation that refers to dropout compensation. Several industrial products, such as the time base corrector, provide this feature in addition to their major functions. See DROPOUT COMPENSATOR CIRCUITRY.

Dockable Video Camera. A professional/industrial video camera designed to connect to a specially designed unit so that the camera can be operated in several formats. Cameras so equipped usually have a MULTI-STANDARD SWITCHABLE ENCODER that offers special outlets for Beta, S-VHS and other formats.

Dolby Noise Reduction System. A technique invented by RAY DOLBY designed to increase high fidelity signals during recording and condense them in reproduction. This results in weak segments of the signal being boosted during recording and decreased to normal when they are played back, thereby minimizing background noise. DNR provides an inexpensive method of adding quality stereo to videotape in both professional machines and home VCRs. Variations of the system include Dolby A, B and C. Dolby B provides an increase in noise reduction to 10 dB (decibels) above 7,000 Hz, while Dolby C is said to offer an increase of 20 dB over a wider range than B. See NOISE REDUCTION SYSTEM.

Dolby, Ray. Engineer; co-developer (with CHARLES GINSBURG) of the first video recorder, demonstrated in 1956; inventor of the famous DOLBY NOISE REDUCTION SYSTEM used in almost all quality audio recorders and many VTR and VCR machines.

Domestic Satellite. See SATELLITE FOCUS.

Donald Duck Effect. An unintelligible sound track caused by speeding up videotape playback without modifying the audio pitch. Some VCRs have double- or triple-speed play, in which the unmuted audio track emerges sounding like the Walt Disney character. Some machines, on the other hand, disengage the sound track in this mode, while others use special digital processing which makes the sound track intelligible. See DOUBLE SPEED PLAY.

D-1 VTR. A professional/industrial component digital videotape recording system that is capable of producing sophisticated visual effects. D-1 VTRs offer a separate black & white image from the color image, a feature that allows the professional to experiment with color difference matting, color correction and perspective moves. In addition, D-1 can handle simple and three-dimensional graphics and accurate scene matches. This flexibility in picture manipulation has made the D-1 format a popular tool in turning out visual effects for broadcast television.

Dot Pattern

Dot Pattern. Tiny dots of light made by the signal of a dot generator and appearing on the screen of a color picture tube. The three color-dot patterns (red, green, blue) merge into one white-dot pattern once the beam convergence is attained. See DYNAMIC CONVERGENCE.

Dot Pitch. The size of the phosphor pixel (picture element) or dot. The smaller the dot pitch (or the increased number of pixels), the better the detail in the screen image. The average size of a current dot is .81mm. Top-of-the-line computer monitors often have a .31mm or smaller dot pitch. Fine dot pitch screens make letters and graphics more distinct.

Double-Azimuth Four-Head S-VHS Camcorder. See AZIMUTH.

Double-Side Videodisc Player. A relatively high-priced unit that permits both sides of a CLV (Constant Linear Velocity) or CAV (Constant Angular Velocity) 12-inch videodisc to play continuously without the viewer's having to change or turn over the disc. These players usually provide a quick transition from one side to the other with little interruption.

Double Speed Play. A technique that permits viewing a program at two times the normal speed and listening to an intelligible sound track without the DONALD DUCK EFFECT which usually accompanies audio tracks that are speeded up. This feature was first used on a VCR manufactured by JVC.

Down Converter. A device which decreases the frequency of a signal. It is used in conjunction with single PAY TV systems that send their signals via MULTIPOINT DISTRIBUTION SERVICE (MDS) MICROWAVE. The down converter lowers the frequency of the signal so that it can be received by the TV set.

Downlink. In SATELLITE TV, the spherical dish which receives the return signal from a satellite. The original signal is sent from an earth station called the UPLINK to the satellite 22,300 miles above the equator, where the signal is then transmitted to various (downlink) receiving stations. The downlink signal usually ranges from 3.7 to 4.2 GHz. See EARTH STATION, SATELLITE SIGNAL, SATELLITE TV, UPLINK.

Downloading. Recording an off-the-air program for viewing at a more convenient time. Downloading, or using the time-shift capacity of the VCR, is one of the machine's strongest selling points. Home View Network, the now-defunct brainchild of ABC, introduced broadcast downloading—for a monthly fee—as an alternate pay TV service. Today, the term "downloading" is more often used with computers, particularly with users who have modems and can download data from other sources. See TIME SHIFT.

Dropout. A white fleck or streak on the screen during playback of a videotape caused by dirt on the tape or video head. Dropouts may also occur as the result of a particle of coating either worn away or falling from the tape. Dropouts can originate from inferior tape, old tape that has become brittle and dried up or clogged heads affecting the sensitive coating on the tape. The problem is permanent and irreversible, except when caused by dirty heads—which should then be cleaned. Some dropout is inevitable, but if excessive, the tape brand should be changed. To minimize the problem, VCR manufacturers have built DROPOUT COMPENSATORS into their units.

Dropout Compensator Circuitry. A feature built into virtually all VCRs and videodisc players and designed to forestall signal loss which causes horizontal lines to appear across the TV screen. A dropout compensator circuit fills in dropouts with a color or a shade of gray which will blend in the the TV image. Similar circuitry, but on a more sophisticated level, has been incorporated into several professional/industrial instruments, such as TIME BASE CORRECTORS. They provide DOC (dropout compensation) in several modes, including, among others, VTRSC/direct mode and S-VHS. The TBC circuitry uses RF reference feed from a VCR, whose signal flags the TBC when a dropout occurs. The unit then corrects the dropout error by replacing it with a video image from the previously stored TBC memory. See DROPOUT.

Dropout Measurement. Refers to the length of time the signal does not appear on the TV screen and the loss of signal strength caused by the dropout. The time-length of dropouts, measured in microseconds, varies. The dropout may last for as short a time as a fraction of one line scan or continue for the length of a few lines. If this dropout time is relatively short, built-in dropout compensator circuitry in later-model VCRs corrects the problem automatically; otherwise, streaks appear across the screen. Signal strength loss resulting from dropouts is measured in decibels (dB). See LINE SCAN.

D-2 Format. A professional/industrial recording format that offers composite digital recording. On a professional/industrial level, there are two major digital formats—D-1, the component digital standard, and D-2, the composite digital standard. The former records the luminance and two color-difference channels digitally, while the latter, D-2, records such basic standard signals as NTSC and PAL. D-2 VTRs (videotape recorders), which serve as playback and post-production machines, offer several state-of-the-art advances as well as compactness and higher quality than their rival units. See SERIAL DIGITAL INTERFACE.

Dual Camera Recording System. A video camera feature that permits the integration of two different pictures from two cameras. First introduced by Panasonic in 1989, the dual camera recording system allows the user to record one video image with one camera and superimpose another picture with a second camera mounted on top and connected with a multipin attachment. Video from the mounted camera can be blended into the main image as an inset, as a dissolve or as a superimposed picture.

Dual Deck VCR

Figure 13. A dual deck VCR for simplified videocassette recording, tape duplication and editing. (Courtesy GO-Video, Inc.)

Dual Deck VCR. A videocassette recorder with two slots designed for editing or duplicating tapes. Companies experimenting with these units boast that tapes copied on these machines cannot be distinguished from the original. Dual deck VCRs offer several advantages. They can record one program while playing a second cassette, record two different programs simultaneously and edit from one tape to another. As expected, manufacturers who have announced production of dual deck VCRs have come under heavy fire from several sources, including the Motion Picture Association of America. The unit, critics charge, will promote piracy of copyright material and cost the software industry a loss of millions of dollars. Although several Japanese firms have dismissed the dual deck VCR as an unprofitable product, several American companies have demonstrated such units.

Dual Digital/Analog Converter. A videodisc player feature designed to improve the stereo output. These converters usually come in pairs for left and right channels and permit concurrent decoding, resulting in enhanced stereo that simulates a live concert hall.

Dual Electromagnetic Focus. A feature used with some projection-TV systems to produce a very small beam for better horizontal resolution. In conventional television, electromagnetic focusing is accomplished by a single coil attached to the neck of the cathode ray tube. As direct current passes through the coil, magnetic field lines are produced parallel to the axis of the tube.

Dual Field Auto Exposure System. The ability of the CCD image sensor of a camcorder to simultaneously measure the light levels of an entire image and the central zone, calculate these and adjust the exposure with emphasis upon the central zone. Some cameras allot about 35 percent of the image to the central area.

Dual Image Effects. A video camera feature that permits the user to mix a still image with live images recorded by the camera. Several types of effects are possible with this feature, such as creating a split-screen with a still frame on one side of the screen; inserting a still image in the center of an action scene; or putting an active image in the middle of a full-screen still picture. These effects can be accomplished only with a digital-type camera.

Dual-Loading System. A method of loading tape, employed by some VCRs, that combines the standard full-loading procedure with half-loading. Dual-loading is designed to speed up the Fast Forward and Rewind modes. The high-speed tape transport also results in accelerated searches in all three playback modes.

Dual-Side Play. See DOUBLE-SIDE VIDEODISC PLAYER.

Dub In/Dub Out Connectors. Used by Sony on its professional VCRs to provide high-quality duplicated tapes. Other manufacturers also utilize special connectors for this purpose. The technique should not be confused with the audio/video inputs/outputs used with home VCRs for copying.

Dubbing. Refers to re-recording a new section of video over existing footage without affecting the audio track. Some camcorders provide a FLYING ERASE HEAD, a few of which are designed to produce this video dubbing feature. Dubbing also refers to copying a tape. See COPYING, VIDEO DUBBING.

DuMont, Allen B. Inventor who in 1939 marketed the first all-electronic TV receivers.

Duplicating of Tapes. See COPYING.

DVE. See DIGITAL VIDEO EFFECTS SYSTEM.

DVI. See DIGITAL VIDEO INTERACTIVE.

DVNR. See DIGITAL VIDEO NOISE REDUCTION.

Dynamic Convergence. Refers to the joining of the three electron beams (red, green, blue) of a color picture tube at the aperture mask where they are reflected both horizontally and vertically. Horizontal dynamic convergence and vertical dynamic convergence each refers to the scanning direction.

Dynamic Picture Control. See HORIZONTAL IMAGE DELINEATION.

Dynamic Track Following. A feature first used on Grundig V-2000 VCRs which eliminated white noise bars during slow motion, freeze frame and visual scan. On conventional Beta and VHS machines the video heads in Fast Scan cannot accurately retrace the diagonal tracks laid down during recording; instead, the heads cross over to adjacent tracks or signals, causing interference or white bars. Grundig solved this problem by having the two video heads move up or down to avoid inaccurate retracing and by

Dynamic Track Following

keeping them on the full width of the recorded track. Dynamic track following also assures noise-free pictures in slow motion and freeze frame modes without the use of additional or over- sized heads as found in some Beta and VHS machines. Digital video equipment has overcome the problem in its own unique way. See DIGITAL VIDEO, TRACKING.

E

Earth Station. The receiving base on earth where the PARABOLIC or SPHERICAL ANTENNA is located. Originally, the term "earth station" applied to both a receiving and sending station but has since been accepted to refer to the DOWNLINK or receiving only. There are more than 5,000 earth stations in the United States, most owned by private citizens. See SATELLITE TV.

ED-Beta. A professional-quality format for VCRs or camcorders that can deliver over 500 lines of resolution through the use of an extended bandwidth. Introduced by Sony in 1987, ED-Beta (Extended Definition) generates superior resolution and offers digital special effects, specially designed video heads for high-density metal particle tape, flying erase head, eight-segment assemble editing, automatic audio and video insert editing, linear time and frame counter, jog/shuttle variable speed search and various on-screen displays. The format, relatively more costly than the average consumer unit, records at a very high bandwidth (up to 9.3 MHz), compared to 5.6 MHz for many of its conventional counterparts, and requires special metal videotape. ED-Beta can record and play back regular Beta tapes.

Edge Noise. Refers to the grain or fuzz that appears on the edge of objects. Edge noise reduces the sharply defined look of objects in a screen image. Some VCRs offer a built-in digital time-base corrector to compensate for edge noise. The color version of edge noise is referred to as EDGING.

Edge Track Recording. The placement of the audio track in a linear position on videotape. Edge track, or linear, recording differs from diagonal recording. See DIAGONAL AUDIO RECORDING, LINEAR AUDIO, LINEAR STEREO.

Edging. Refers to unwanted edge noise in the form of color that appears around objects of different colors. Similar to edge noise in monochrome television, edging affects the otherwise sharply defined color of objects on screen.

Edit Control Jack

Edit Control Jack. A special VCR receptacle that permits two units to be connected for synchronized operation during editing. See EDIT CONTROLLER.

Edit Controller. A VCR feature, either built into or added externally to the machine, designed to operate the controls of two videocassette recorders during the editing process. Chiefly made up of miniature computers capable of storing memory, the edit controller allows the user to assign to it several edits at one time. Once the controller has stored the information in its memory, it automatically performs assemble edits. Beta and 8mm units have incorporated the concept of the edit controller into their systems for several years. Sony, which offers the controller as an accessory and calls the unit a remote editing control, allows the user to assemble eight video sections before it automatically executes the editing sequence. To be effective, an edit controller or similar device should be able to find and return to edit points. Some units use the VCR counter, others rely on a vertical interval time code recorded between video fields, and still others that depend on coded track different from the video. A more sophisticated—and more costly—version of the controller is the editing console. See ASSEMBLE EDITING, EDITING CONSOLE, EDITOR.

Edit Fader. A device used in editing to fade a video signal to black. It is utilized chiefly by professionals and in conjunction with industrial video recorders.

Edit Master. The finished version of an edited videotape. Usually the first duplicated tape, or second generation, from the original footage, the edit master contains all pictures, music, dialogue and special video and sound effects.

Edit/Start Control. See AUTOMATIC TRANSITION EDITING.

Edit Switch. In VCRs, a feature designed to switch a low-pass filter out of the video output circuitry as a means of cutting back on video noise. However, the function presents one major drawback. Since the process simultaneously lessens image resolution, it also accelerates generation loss during editing. An edit switch on a video camera is designed to change the video signal that the camera transmits to compensate for any degradation that may occur during the editing or copying process. Relatively few cameras offer this feature.

Editing. In video, an electronic process of transferring or duplicating selected recorded segments of tape onto another tape. A second VCR is required, one functioning as a "player" and the other as the "recorder." For best results they are connected through their audio and video inputs and outputs. The recording machine is also hooked up to a TV receiver which acts as a monitor. Industrial machines in either 1/2-inch or 3/4-inch format are best suited for this purpose. They provide more accuracy in eliminating unwanted video frames and guarantee glitch-free edits. However, some top-of-the-line Beta and VHS machines can produce very sat-

Effective Isotropic Radiation Power (EIRP)

isfactory editing results. Some VCRs come equipped with built-in EDIT CONTROLLERS, a computerized system complete with memory, that can perform accurate and professional-looking assemble edits. In addition, edit controllers, and their more costly and sophisticated big brothers, EDITING CONSOLES, complete with readout panels, can be added as an external accessory for even more fancy editing. The basic methods of editing, or electronic editing as it is sometimes called, are ASSEMBLE EDITING, CRASH EDITING and INSERT EDITING. See AUTOMATIC TRANSITION EDITING, CLEAN EDITING.

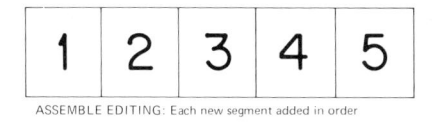

| 1 | 2 | 3 | 4 | 5 |

ASSEMBLE EDITING: Each new segment added in order

| 1 | 6 | 3 | 7 | 5 |

INSERT EDITING: New material inserted to replace old segments (2 and 4)

Figure 14. Editing techniques.

Editing Console. A sophisticated accessory, used in conjunction with a video camera and a VCR, that automatically synchronizes and controls both units during the editing and copying processes. These more costly components offer several advantages. They provide a variety of special-effects transitions, including wipes, fades and dissolves, which can be inserted between edited scenes. In addition, these consoles, ranging in price from several hundred to several thousand dollars, allow different titles and fonts to be added—in an array of colors. Editing is usually accomplished by first previewing tapes, then entering into a keyboard the selected editing points. The computerized console, which contains its own memory bank, stores the information and predetermined sequence in its multi-scene memory and then automatically carries out the transfers with precise edits. Some consoles, also known as editing controllers, operate between camera and VCR by infrared beam while others must be connected using the appropriate cables and jacks. Some models include a digital effects generator, an audio processor, a video distribution amplifier, a color processor and an enhancer. See EDIT CONTROLLER.

Editing Controller. See EDITING CONSOLE.

Editor. An electronic device used by professionals to control synchronization of at least two devices for the purpose of switching video and audio material to a specific point in a program. The editor permits merging pictures and sound from two videocassette recorders or one video camera and one VCR into one master tape. Most of these devices have various controls for different functions such as Play, Forward, Reverse, Preview, etc. Some less costly models are available for the nonprofessional. See EDIT CONTROLLER, EDITING CONSOLE.

Effective Isotropic Radiation Power (EIRP). Refers to the satellite signal level strength that reaches earth. EIRP is described in decibels per watt (dBW). Satellites transmit relatively

EFP

Figure 15. An 8mm camcorder that takes 8mm videocassettes of special metal particle tapes, can record up to two hours, has an 8:1 zoom lens and weighs 3.5 pounds complete with battery and cassette. (Courtesy Canon U.S.A.)

narrow beams which widen as they approach Earth. The pattern or surface area they cover is called the FOOTPRINT. It is this footprint shape that determines the EIRP. See FOOTPRINT, SATELLITE SIGNAL.

EFP. See ELECTRONIC FIELD PRODUCTION.

EIA Linearity Chart. A test chart available from the Electronic Industries Association with configurations to equal the bar and dot patterns from a sync generator in the average broadcast studio. The function of the chart is to test and measure the scan linearity of professional, broadcast-type equipment. It is designed to duplicate the normal broadcast configurations of 14 horizontal bars and 17 vertical bars. It is also known as a ball chart. See VIDEO TEST CHART.

8mm Hi-Fi. High sound quality built into the 8mm video recording format. This format was originally designed to automatically incorporate AFM high-fidelity recording in all 8mm camcorders and VCRs. Unlike standard VCRs that place the separate audio track longitudinally on the tape, 8mm AFM units "write" the audio track on the tape diagonally along with the video information. The high quality sound, however, is restricted to one monophonic track, thereby not necessarily producing stereo. Some 8mm units are equipped with Pulse Code Modulation, a digital audio recording process that can produce stereo audio. These units are usually more costly than models without PCM.

8mm Video. A relatively new mini-video camcorder format that uses a compact cassette and is capable of producing high-fidelity audio. Having captured almost 70 percent of the compact camcorder market in 1988, this popular format offers several advantages over its rivals. Its small size and light weight make it convenient to take almost anywhere. Its 8mm cassettes offer 60-, 90- or 120-minute lengths. Its flying erase heads provide smooth edits and clean scene transitions. The video quality of the 8mm format equals that of VHS in many respects and surpasses it, although only slightly, in color reproduction. In addition, its built-in hi-fi audio capability offers superior sound to competing formats. However, 8mm video is not compatible with most home VCRs. Introduced by Sony in 1985, the 8mm format has gained adherents among other companies, including, among others, Canon, Olympus, Kodak and NEC. Some models have added advanced features, such as automatic focus, glitch-free editing and the capability of superimposing time and date upon an image. Other competitive formats include Hi8, S-VHS, S-VHS-C, VHS-C. See CAMCORDER, FLYING ERASE HEAD.

EITV (Educational/Industrial Television). A monthly magazine targeted at those who use professional video equipment. The articles include how-to information such as testing for scan linearity and using a COLOR VECTORSCOPE. Other feature stories relate how a particular company intends to adopt the latest development in the video field to its own needs. The magazine, which is abundantly illustrated, has several columns.

Eject. A button or control on a VCR to lift the cassette housing on top-loading machines to permit either the removal or loading of a videocassette.

Electret-Condenser Microphone. A moderately priced microphone known for its sensitivity and accuracy. It requires a power supply and an amplifier built into its housing. See MICROPHONE.

Electroluminescent TV Screen. A system of FLAT SCREEN TV utilizing numerous microscopic elements. Electroluminescent TV, or EL, is one of a few systems considered as a replacement for the CATHODE RAY TUBE in large-screen TV. EL can be compared in effect to a large array of light bulbs.

Electromagnetic Frequency Spectrum. The broadcasting frequency which holds the various television signals transmitted from broadcast stations. A certain part or band of this spectrum is allotted to each signal. The Federal Communications Commission apportions each slot to a channel, thereby permitting many signals while avoiding interference. For instance, a TV channel is given a band 6 MHz (megahertz) wide and must carry its video, audio and color information within these parameters. New forms of television transmission may require a wider slot than the standard 6 MHz. HIGH DEFINITION BROADCASTING, for example, whose signal transmits much more information and therefore requires an area five times greater than the above standard, will have to utilize either cable systems or the 12 GHz (gigahertz) portion of the spectrum via satellite broadcast.

Electron Beam Recording. A technique employed in video-to-film processes in which movie film images are created by electronic impulses. The electronic beam replaces the light which normally exposes motion picture film.

Electron Gun. An electronic component which gives off a continuous beam of electrons that are focused on the screen of a CATHODE RAY TUBE.

Electronic Camera. The use of video cameras recording simultaneously with film cameras. The electronic images are delivered to a central control board for a director's perusal and editing. Electronic cinema was first used successfully by FRANCIS FORD COPPOLA during the production of his 1982 film *One From the Heart*.

Electronic Channel Selection. A feature on a VCR permitting remote control, faster electronic bypassing of channels and multiple channel recording. The mechanical rotary-type tuner allowed only one-channel recording with the VCR timer. Electronic channel selection combines the benefits of the versatile microprocessor chip with those of the Varactor tuner. See ELECTRONIC TUNER.

Electronic Classroom. A supplemental system of education using interactive video so that teachers and students can listen to and interact with one another. Although the video portion is restricted to one-way (the students can see the instructor), the electronic classroom offers several benefits. Federal and state governments, universities and public and commercial television networks can provide a vast array of courses. Small, budget-tight schools and systems can avail themselves of sophisticated subject matter and lessons while retaining the all-important communication between instructor and students. The system usually involves a satellite dish, electronic keypads and cordless telephones. The initial outlay for the basic equipment comes to much less than the annual salary for one teacher. Some universities offer as many as 20 different courses to dozens of schools while some private networks are feeding instructional material to more than 700 schools in more than two dozen states. See CHANNEL ONE.

Electronic Darkroom. In still video photography, a facility capable of using a laser beam to scan photographs or art work and converting the images into digital signals that are then transmitted to a computer. The electronic darkroom, usually part of a newspaper, periodical or other publishing enterprise, permits the image to be edited for proper color, size and detail. The picture is then sent to another unit that converts it into a film image ready for the printed page. Since images can be manipulated quite easily, professionals in various fields have found the technique effective in removing distracting materials such as blurred objects, poles or overhead wires. But some capabilities, such as inserting one image into another existing one to form a completely different picture, has caused critics to voice concern about the ethics of reconstructing electronic images. Another related consideration is that no original negative will exist to confirm the authenticity of a picture under question.

Electronic Editing. See EDITING.

Electronic Field Production (EFP). Refers to professional video equipment which usually features more sophisticated capabilities and a more rugged housing than equivalent home video units. For example, EFP cameras operate more efficiently in low light levels, hold registration better, contain special prism optics, provide superior electronic stability and offer more durability. Some companies which sell home video equipment have a professional products division which specializes in EFP components. Other companies produce only professional equipment. See REGISTRATION.

Electronic Frequency Synthesizing Tuner. See FREQUENCY-SYNTHESIS TUNER.

Electronic Game. See VIDEO GAME SYSTEM.

Electronic Indexing Signal. See ELECTRONIC PROGRAM INDEXING.

Electronic Industries Association (EIA). The committee that sets electronic standards in the United States.

Electronic Industries Association of Japan (EIAJ). The Japanese committee which sets electronic standards for 1/2-inch helical scan video recorders, etc.

Electronic Program Indexing. In a VCR, a feature designed to place an electronic signal at the start of each recording. When the machine is switched to Fast Forward or Rewind and the index control is turned on, the tape stops automatically at each point that a program was recorded. With the switch off, the tape continues to rewind or move forward. On some VCRs the signal is not automatically encoded on the tape unless so desired. There are other techniques for locating segments of tape. One system designates numbers to the portions of tape. In the LV videodisc system, electronic encoding permits the viewer rapid access to marked sections. Each CHAPTER or FRAME on the disc can be recalled. More prerecorded programs, including sports discs and music-oriented shows, are utilizing this feature. The indexing feature is also known as picture indexing, electronic indexing, etc.

Electronic Still Camera Standardization Committee (ESCC). An organization composed of more than 40 manufacturers of still video cameras dedicated to promoting the relatively new cameras and accessories and to setsing universal standards for the products. The ESCC, founded in 1983, focused on the still video floppy disk

Electronic Still Video

(VF) as the standard medium for the emerging system of electronic still photography (ESP). See STILL VIDEO.

Electronic Still Video. See STILL VIDEO.

Electronic Switching. In video, a process used with certain accessories such as switchers to avoid problems of noise usually associated with electrical signals and mechanical switching. Electronic switching is performed by diodes and integrated circuits built into these accessories. The use of electronics in signal switching provides another major advantage—no deterioration of parts. See SWITCHER CONTACT.

Electronic Text Generation. Refers to the electronic production of characters for television and broadcast purposes. The first successful TV character generator appeared during the early years of television. The characters were relatively crude, and the technique was unable to produce proportionally spaced letters and numbers. During the 1960s CBS Laboratories experimented with an electronic character generator that turned out Helvetica Medium typeface in 18- and 28-line sizes and proportional characters of graphic-arts quality. In 1964 RCA had the capability to produce by way of a twin-channel generator upper-case characters in two sizes. With the advent of personal computers, character generators combined with software to offer scores of print styles by the early 1980s. Today, a host of manufacturers offer compact, sophisticated, less expensive character generators with such features as italics, drop shadows, anti-aliasing, character sizing, graphic symbols, animated characters, etc. See CHARACTER GENERATOR, SPECIAL EFFECTS GENERATOR.

Electronic Tuner. A feature which allows the tuning of TV channels electronically via a keyboard-type panel instead of the conventional rotary knobs. By employing a microprocessor, the electronic tuner can adjust itself to receive local channels in numerical order, can be programmed with various instructions and can offer cable-ready TV sets and VCRs. Another advantage of electronic tuning is the remote control accessory for changing channels, a feature not possible with the mechanical knobs. In addition, since hundreds of mechanical parts are eliminated, tuner reliability has vastly improved. See CABLE-READY, MEAN TIME BETWEEN FAILURES, TUNER.

Electronic Viewfinder. A video camera viewing system that displays exactly what the camera lens sees. The electronic finder, virtually a one-and-a-half-inch miniature black-and-white TV screen, is used for focusing and composing the picture. It can also be used to view what has been recorded by simply playing back the tape on the VCR. Some finders contain indicators for battery level, VCR recording/pause mode and under- and overexposure. Most EVFs on cameras intended for home videos come with a monochrome monitor because the cost of a color system is more costly. Some manufacturers provide a viewfinder with a built-in playback speaker so that both sound and picture can be monitored.

Emshwiller, Ed. VIDEO ARTIST who uses digital cameras, minicomputers, video synthesizers, chromokeyers and character generators to create relatively new concepts in art. See VIDEO ART, VIDEO ARTIST.

Encoder. See MULTI-STANDARD SWITCHABLE ENCODER.

Encoder Adjustment. A necessary function if multiple camera setups are to operate accurately. Precise phase matching, or GENLOCK, adjustments or realignments can be made in the field by experienced technicians with a special portable instrument known as a VECTORSCOPE.

Encryption. A term applied to the process which places codes on signals which are transmitted by satellite-to-cable, microwave, etc. Copyright owners who want to guard their material view this process as a means of protection, especially against those with their own backyard satellite earth stations. Encryption would have to be employed by the cable operators who may choose from several techniques, including analog or digital encryption. These encoding techniques are presently used by government and military communications. See ANALOG ENCRYPTION, DIGITAL ENCRYPTION, VIDEO ENCRYPTION SYSTEM.

End-of-Tape Marker. An indication placed on a videotape to specify that the tape is reaching its physical end. When the marker is reached, most VCR machines automatically begin to rewind if in the Record or Play mode. The marker may take several forms, including among others a photo-reflective strip or a transparent segment of tape.

End User. A commercial term that refers to the consumer who ultimately puts a specific piece of equipment to use.

ENG. Refers to Electronic New Gathering field cameras used by various television news stations. These professional video cameras are relatively light and portable.

Enhance Control. A function on an IMAGE ENHANCER (or MINI-ENHANCER) or IMAGE PROCESSOR designed to increase the sharpness of high frequency information. The control adjusts the amount of "boost" applied to the high frequency signal. Enhancers are signal processors and are used to increase the detail in a video picture. Therefore, they exaggerate the high frequency information in the video signal. Rotating the enhance control helps to improve sharpness. The control often has a range capable of increasing 2 MHz signals from zero to about three times. For maximum effectiveness, the enhance (sharpness) control usually operates in conjunction with a response (noise reduction) control. See RESPONSE CONTROL, SIGNAL PROCESSOR.

Enhanced TV. A general term for several experimental systems intended as an alternative to high definition television (HDTV). The more formal term for enhanced television is IDTV, or

Enhancement Hardware

improved definition television. The search for an enhanced system grew out of the need for a compatible process that would not make the millions of current TV receivers obsolete. The topic, however, has become academic since the Federal Communications Committee early in 1990 decided to opt for a true HDTV system as a national standard before considering any enhanced TV system. Manufacturers of HDTV have developed simulcast systems in which TV stations would use one channel for standard NTSC broadcasts while part of an empty adjacent channel would transmit information required for HDTV's 1,050-line wide-screen image. See HIGH DEFINITION TV, IDTV.

Enhancement Hardware. In video, accessory equipment designed to electronically improve the video or audio signal of a recording. Image stabilizers, image detailers, line doublers and noise reduction systems are examples of enhancement hardware.

Enhancement Light. See COLOR ENHANCEMENT LIGHT.

Enhancer. See IMAGE ENHANCER.

Entertainment and Sports Programming Network (ESPN). An advertiser-supported TV service which offers a 24-hour schedule of sports. ESPN has been broadcasting since September 1979, offering a variety of sports including, among others, golf, tennis, lacrosse, rugby, karate, bowling, as well as the more conventional games. ESPN, which is owned by Getty Oil, led the list of cable TV networks by the end of 1989, boasting more than 54 subscribers.

Envelope Delay. A type of distortion that takes place during transmission when a phase shift fails to maintain its constancy over the frequency range. Envelope delay is one of several types of degradation that affects NTSC picture quality. See DEGRADATION.

EP Speed. See EXTENDED PLAY.

Equalizer. In audio, a unit designed to selectively increase or decrease portions of the frequency range, thereby customizing the sound to personal taste or to accommodate the ambience of a specific room. The equalizer, a feature on many audio/video receivers, may be positioned between the output of a microphone MIXER and the input of a VCR. It is also known as a graphic equalizer. In video, it acts as a multi-function video processor which provides (1) color and definition control during dubbing, (2) its own modulator to direct viewing, (3) a distribution amplifier to permit making more than one tape copy simultaneously and (4) a stabilized image for copying. In general, a video equalizer is used to control video signal loss.

Erase. In video, to clear program material or information electronically each time new material is recorded. Tape may also be erased by using a BULK VIDEOTAPE ERASER. See ERASE HEAD.

Erase Head. A special magnetic head which automatically erases previously recorded material on the tape.

The erase head is the first process that affects the videotape as it makes its way through the VCR. After the tape is cleared of information by the head, it passes the video heads which place new information on the magnetic coating. Finally, the tape passes the audio/control head for further information. The erase head, which is activated only during recording, remains stationary in contrast to the video heads which rotate on the drum assembly. Flying erase heads, found on current 8mm camcorders and some VCRs, provide seamless edits between scenes. In addition, they save space by eliminating the larger erase heads. See AUDIO HEAD, FLYING ERASE HEADS, VIDEO HEAD, VIDEO HEAD DRUM.

ESCC. See ELECTRONIC STILL CAMERA STANDARDIZATION COMMITTEE.

ESPN. See ENTERTAINMENT AND SPORTS PROGRAMMING NETWORK.

Event. In video, refers to a timed, recorded, single-continuous program. The first VCRs permitted only a single event to be programmed for recording per 24-hour period. Today's VCRs with their PROGRAMMABLE TIMERS and ELECTRONIC TUNERS allow multiple programming of several events over a period of as many weeks. The greater the programmability, the more expensive the machine.

Excessive Amtec Error. Refers to the improper duplication of master tapes. Faulty equipment at the copying plant can cause a pattern of dark horizontal streaks on the prerecorded tapes. The term is often used in conjunction with the recording of professional two-inch tape.

Exposure Level Lock. A video camera feature that permits the operator to "lock in" the proper exposure setting for a predetermined scene before shooting. This helps to compensate for sudden lighting changes, such as the camera panning or the subject moving to where backlighting would otherwise affect the exposure.

Extended Play (EP). The six-hour tape speed with T-120 VHS cassettes or eight-hour speed with a T-160 cassette. EP is also known as Super Long Play (SLP). It is the slowest speed on VHS videocassette recorders. The other speeds are Standard Play (SP) and Long Play (LP).

Extended Resolution Format. Refers to such videocassette or camcorder recording systems as Hi8, S-VHS and ED Beta that employ special techniques to enhance the picture quality, thereby producing a video image with 400 lines of resolution. One popular method, for instance, uses an advanced form of comb filtering to separate signals into groups which are then placed into predetermined spaces to prevent crosstalk. In addition, some Hi8 camcorders use a CCD image sensor capable of generating more than 400,000 pixels (picture elements). These extended resolution formats usually require more costly, specially prepared tapes to capture the additional video information.

Extender Lens

Extender Lens. An accessory lens placed between the video camera lens mount and the standard lens to change the latter's viewing range. Extender lenses can be used in conjunction with telephotos and wide angles to increase their ranges. Other extenders permit 35mm lenses normally used with film cameras to operate with video cameras. Extender lenses can only work with cameras that accept interchangeable lenses. See ANGLE OF VIEW.

External Sync Input. A jack, chiefly found on industrial video recorder decks, designed to allow the unit to follow the commands of a TIME BASE CORRECTOR. The inputs help to synchronize signals so that they can be controlled and mixed with other signals from different units.

F

F Connector. A threaded barrel-type metal fitting used with coaxial cable. The F connector is designed basically for RF signals which transmit both audio and video signals. An F-barrel connector, with a female thread on both ends, is used to join together two coaxial cables. The F connector is also known as coaxial connector. See COAXIAL, CONNECTOR.

Fade. In video, a special editing effect which involves the slow eclipse of an image until the screen is black or another scene appears. In the latter case, the effect is known as a dissolve. The fade is the conventional method used to end a film. It may also be use as a transition. Fades and other editing effects have become more readily available to home video users with the introduction of these features on many recent VCRs. In addition, accessory editing modules can be purchased to perform similar tasks. See EDIT CONTROLLER, EDITING, EDITING CONSOLE.

Fade Duration Control. A feature designed to adjust the length of a specific fade. Found chiefly on image enhancers/processors, the fade duration control operates with both audio and video signals. See CROSS-CHANNEL FADE, FADER.

Fade In/Fade Out. A video camera editing feature that allows for smooth transitions between scenes. Fade In usually refers to going from dark to light while Fade Out implies going to dark or black. See FADE THROUGH BLACK.

Fade Through Black. A technique used in video in which a picture or image is made to fade to black before a second picture appears. Fade through black provides smoother transitions between scenes. Some video cameras offer this as a built-in feature; not only do they permit fade to black, but they can also go from black to full video. SPECIAL EFFECTS GENERATORS, SIGNAL PROCESSORS and other devices also provide fade through black. See CROSS-FADE, FADER.

Fader. A signal processing device which permits fading to black or from black to full video. It usually operates with color as well as black-and-white

Family Channel

video. In home video the fader is generally employed during editing from one VCR to another or from a video camera to a VCR for smoother transitional scenes. Many video cameras have a built-in automatic fade-in and fade-out feature, making an external fader accessory unnecessary with these models. See CROSS-CHANNEL FADE.

Family Channel. A CABLE TV advertiser-supported network featuring television rerun shows, TV and theatrical films, and several original programs around the clock. By 1990 it was listed as the seventh most popular cable network reaching 48 million subscribers.

Farnsworth, Philo Taylor. Received patent in 1930, at 24 years of age, for his electronic television system which was capable of transmitting visual images. He also invented an early camera tube in 1927, the IMAGE DISSECTOR TUBE, which is no longer in use. TV historian Albert Abramson in *The History of Television, 1880 to 1941* (1987) attributes Farnsworth with constructing and operating the first electronic television system. See CAMERA TUBE.

Faroudja, Yves. California-based French engineer and experimenter in improved, large-screen video; inventor who contributed to the development of Super-VHS video recording and the 8mm camcorder. He has developed a system of enhancing the present television picture that, unlike the proposed high definition TV, does not make current home TV sets obsolete or alter current broadcasting standards. His method involves doubling the number of lines per picture for more image detail, resulting in a theater-size video image that almost matches the quality of 35mm film projection. Although his system requires the use of additional image processors—at relatively little cost—present television sets may still derive some benefits from the system without upgrading. See ACTV, HDTV, HDSNA, LINE DOUBLING.

Fast Forward. A feature on virtually all VCRs to move the tape ahead without picture or sound to a predetermined location. Earlier industrial VTRs by Sony displayed a picture while the machine was in Fast Forward mode. In some advanced VCRs, if Fast Forward is pressed half-way while the machine is in Play mode, the VISUAL SCAN special effects feature is activated. The VCR returns to Play when FF is released. Some machines identify the Fast Forward mode simply with an arrow (—>).

Fast Motion. See FAST FORWARD.

Fast Scan. See VISUAL SCAN.

Fast Search. See VISUAL SCAN.

F.B.I. Warning. See TITLE 17.

FCC. See FEDERAL COMMUNICATIONS COMMISSION.

FD Tube. See FLAT DISPLAY TUBE.

Federal Communications Act: Section 605. The section quoted in reference to receiving Pay TV signals without permission or without paying for the service:

No person not being authorized by the sender shall intercept any radio communication and divulge or publish the existence, contents, substance, purport, effect, or meaning of such intercepted communication to any person. No person not being entitled thereto shall receive or assist in receiving any interstate or foreign communication by radio and use such communication (or any information therein contained) for his own benefit or for the benefit of another not entitled thereto. No person having received any intercepted radio communication or having become acquainted with the contents, substance, purport, effect, or meaning of such communication (or any part thereof) knowing that such communication was intercepted, shall divulge or publish the existence, contents, substance, purpose, effect, or meaning of such communication (or any part thereof) or use which communication (or any information therein contained) for his own benefit or for the benefit of another not entitled thereto. This section shall not apply to the receiving, divulging, publishing, or utilizing the contents of any radio communication which is broadcast or transmitted by amateurs or others for the use of the general public, or which relates to ships in distress.

Federal Communications Commission (FCC). The agency assigned by the U.S. federal government to regulate radio and television broadcasting under the assumption that radio waves (which include television) always travel across state lines. The FCC emerged as a result of the findings of the Federal Radio Commission of the early 1930s and the Communications Act of 1934. In reality, it was created to settle differences between private radio station owners whose broadcast signals were interfering with each other's. Today, the FCC is committed to the concept that any new standard of TV broadcasting or television system (such as high definition TV) must be compatible with present TV sets. This has made the introduction of TV systems with better or higher definition (more horizontal lines per picture) difficult.

Feed Horn. The component of a SATELLITE TV system that is attached to the LOW NOISE AMPLIFIER and faces the center of the antenna. The feed horn, usually rectangular or circular in shape, performs various functions: (1) it captures the signals reflected from the PARABOLIC or SPHERICAL ANTENNA; (2) it shuts out other, interfering signals; (3) it eliminates noise from land sources; and (4) it directs the received signal into the LNA. It is also known as a feeder horn.

Feedback. See VIDEO FEEDBACK.

Ferguson Decision, The. The ruling on October 1, 1979, of Warren J. Ferguson, judge for Los Angeles' Federal District Court, concerning the MCA/DISNEY VS. SONY LAWSUIT. The studios charged that Sony, along with others, was breaking copyright laws as well as adversely affecting their financial status. Judge Ferguson, however, ruled in favor of Sony, stating that "such recording is permissible under the copyright acts of 1909 and 1975." He based his ruling on (1) the existing exclusion of home audiotap-

Ferric Oxide

ing, (2) the Fair Use doctrine and (3) the use of VCRs for other than off-the-air purposes. Furthermore, Judge Ferguson reasoned, broadcast material was already paid for by sponsors and was being transmitted over public-owned airwaves. The judge's decision was eventually overturned in 1981 by a higher appellate court. Meanwhile, Sony appealed this reversal, leaving VCR owners in a state of confusion regarding the legalities of recording. The Ferguson decision made a point of excluding the duplication of Pay TV material. See MCA/DISNEY VS. SONY LAWSUIT.

Ferric Oxide. A coating used mainly on VHS videotapes designed for better frequency response. The iron oxide composition, however, may cause excessive snow or distortion. See FERRITE, COATING.

Ferrite. A ceramic-structured, magnetic substance formed by combining iron oxide and other metallic oxides under high temperatures. The quality of a videotape depends upon the density of its ferrite particles. Ferrite was developed in the 1930s by the Japanese.

Fiber Optics. A process that permits special glass fibers to transmit program material in the form of light. Instead of individual wires, each carrying specific information such as that used by the telephone and CABLE TV, fiber optics cables, or light transmission, can carry a multiplicity of broadcasting channels as well as other types of information. With a diameter of less than 1/5 of an inch, these optically pure glass fibers are resistant to electromagnetic interference and other forces which now plague ordinary wire. They are also capable of transmitting laser-produced light beams over vast distances. The use of fiber optics is becoming increasingly important in such areas as telecomputing, which requires digital, rather than analog, signals. Since these signals hold much more information, they need a fiber optics cable, which can transmit thousands of times more information than conventional carriers such as copper wire or radio waves. The United States holds a strong lead in the production and installation of fiber optics.

Field. Half the FRAME of a television picture. American TV, which follows the NATIONAL TELEVISION STANDARDS COMMITTEE (NTSC), broadcasts two 262 1/2-line picture fields which compose one frame. Since TV works at 30 frames per second, 60 fields of video must be generated to create these 30 frames. Each video head (of one pair) records one of these fields. Some VCRs feature freeze frame by locking into two fields or one frame while others show one field twice for the same effect. See FIELD TRACK, FRAME.

Field Blanking. The period of time in which the scanning dot or spot is turned off as it returns from the bottom to the top of a VIDEO CAMERA TARGET AREA. The scanning dot moves horizontally from left to right, scanning the 525 lines which form the NTSC standard. During its return from the bottom right of the last line scan to the top, ready to repeat its scan pattern, the video camera emits

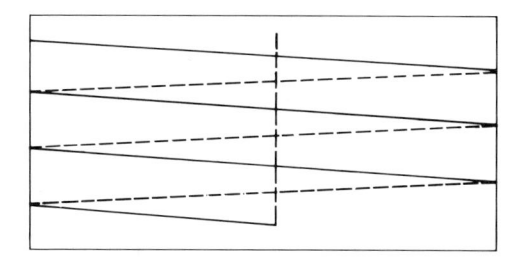

Figure 16. One field (262 1/2 lines) with field retrace.

no signal. This field blanking period, also called field retrace period, permits control pulses to be added to the video signal. Field blanking is often referred to as VERTICAL BLANKING INTERVAL. See HORIZONTAL SYNC, LINE BLANKING.

Field Correlation. A video noise reduction system that produces an improved screen image by storing signal information from a line or FIELD and adding it to a previous line or field, thereby increasing the signal-to-noise ratio by 3 dB. A video image contains the necessary picture information along with unwanted video noise in the form of specks and blotches. Since very little information changes from one line or field to the next, two adjacent fields can be added together to double the information. On the other hand, video noise is more random. So when this interference is necessarily added along with the field, it does not double in quantity. This results in an enhanced video image without appreciable gain in video noise. See DIGITAL VIDEO NOISE REDUCTION SYSTEM.

Field Scanning. A process designed to obtain noiseless special effects with a videocassette recorder. When certain effects such as SLOW MOTION and FREEZE FRAME appear on the screen, they are usually accompanied by VIDEO NOISE, such as horizontal NOISE BARS. To minimize or eliminate this type of interference, some VCR manufacturers have found that by reproducing two FIELDS instead of one complete picture or FRAME of a video signal, the final TV picture is cleaner and steadier. A picture is composed of one frame of 525 lines or two fields of 262 1/2 lines each, one containing the odd-number lines, the other the even. Although the field contains only one-half the information, it is easier to produce an interference-free picture by using this procedure of field scanning. See FRAME SCANNING, FREEZE FIELD, SCANNING.

Field Strength Meter. An instrument used in the field by CABLE TV service technicians to test the signal strength in various locations where subscribers complain of weak reception.

Field Track. A diagonal line of information or programming placed down on videotape as it passes over one video head during recording. Since two fields make up one FRAME, sixty

Fill Light

field tracks comprise the 30 frames per second required for the NATIONAL TELEVISION STANDARDS COMMITTEE (NTSC) standard. See FIELD, FRAME.

Fill Light. The addition of lighting in shadowy areas within a scene to help balance the proper brightness and contrast. See LIGHTING.

Film Chain. See FILM-TO-TAPE TRANSFER, TELECINE ADAPTER.

Film-to-Tape Transfer. A technique of placing films or slides onto videotape. The image from a standard movie projector (or slide projector) is projected onto a TELECINE adapter and recorded with a video camera. Sound can be added later with the Audio Dub mode or the VCR. Another method employs a special sheeting called POLACOAT, which is mounted on a home-made wooden frame and used as a projection screen. As the movie or slide projector image is projected on the screen, a video camera placed behind the screen records the projected image. One disadvantage of this procedure is that titles and printing come out reversed. Professional film-to-tape transfer—converting theatrical releases to the home video rental market—presents certain problems that render only a fairly reasonable facsimile of the original film rather than the more desirable true replica. First, the ASPECT RATIO of the TV screen differs from that of the theatrical wide-screen, resulting either in part of the image not appearing on the TV screen or a further-reduced image if LETTERBOXING is used. Second, producers of videocassettes may not always be fortunate enough to obtain the master positive of the film. Second- or third-generation film prints often suffer in contrast, detail and sharpness. Furthermore, the nature of present-day videocassette systems restricts the contrast range of the original material during the copying process. This affects color, shadows and the darker shades of the film image which a director and his or her creative assistants may have endowed with delicate subtleties. Film enthusiasts and those who work with film-to-tape transfer hope that HIGH DEFINITION TV, when it is introduced, will solve these problems.

Filmless Photography. See STILL VIDEO.

Filter. A device that isolates data, signals or other materials according to specific criteria. Used with a video camera, a filter refers to the piece of transparent material designed to change the light which enters the lens. This type of filter is usually made from special glass. The term "filter" is used in video and/or audio to describe a circuit that damps signals greater or smaller than a specific frequency without affecting signals in its pass-band. A high-pass filter is used to reduce interference on all channels and a low-pass filter is attached at the CB source which usually affects channels 2, 5 and 6. Also known as a TV interference filter, this type prevents interference from other outside sources such as motors, computers and amateur radio transmissions.

Filter Kit. A package containing various filters to be used with a video camera. Video filter kits can range from five to 75 filters. Some special effects filters include a fog filter to give a fog effect to a scene; a multiple image filter; a prism filter to distort light for surrealistic effects; a star filter to turn conventional lights into "starry" images; a sepia filter to add an old-fashioned look. There are also color correction filters. Besides balancing light sources, these can be used to create particular moods; an orange filter can add a feeling of warmth to a scene while a blue filter can create a cool effect.

Fine Editing. See AUTOMATIC TRANSITION EDITING.

Fine-Pitch Picture Tube. A technique to improve image resolution by permitting smaller details to appear on the TV screen. This is accomplished by using a series of continuous vertical stripes that produce a larger number of smaller pixels, or picture elements. These tinier red/green/blue color phosphors that form the final image appear on the inner face of the picture tube to produce higher resolution. Also known as Fine-Pitch Aperture Grille, this process is similar to that which utilizes a shadow mask. See SHADOW MASK.

Fine Slow. A digitally produced slow motion effect that appears more natural than the conventional slow motion which, on many VCRs, is usually accompanied by horizontal noise bars and other interferences. See DIGITAL EFFECTS, SLOW MOTION.

First National Kidisc, The. An LV INTERACTIVE VIDEODISC released in 1981 offering multiple games, activities, puzzles and educational programs in which children from ages 6 to 12 can participate. It was one of the first to venture into NON-LINEAR PROGRAMMING (NFL's "How to Watch Pro Football" was the first interactive videodisc). Produced by Optical Programming Associates, the Kidisc contained 24 chapters, each a complete unit that could be located quickly on a videodisc player. The disc was designed to make use of the many special effects of the VDP.

First Sale Rights. Regarding prerecorded videotapes, the principle that every copy of a work sold outright by a copyright owner can be disposed of by the purchaser in any way he or she desires. This concept suggests that a first sale copy can be lent out, leased or sold, but it cannot be duplicated or displayed at public performances, even if the showing is free. Not all copies of films or other copyrighted material are authorized first items. The rule of first sale rights is important to the thousands of video stores that rent and sell tapes. It permits the retailers the right to buy the tape and do whatever they want with it—within the above strictures of the law.

Fisheye Lens. An extreme wide-angle lens whose view ultimately appears curved. This type of lens is chiefly used for special effects.

Fixed Focus Lens. In video, a lens of one size, called a fixed focal length, usually listed in millimeters, such as 16mm. The fixed focus lens lens is of-

Flagging

ten called the "normal" lens if it is in the 16mm range. Other basic types of fixed lenses are the wide angle and telephoto. The shorter the focal length, the wider the area the lens encompasses. Most video cameras come equipped with a ZOOM LENS for more versatility rather than with a fixed focus type that has a limited ANGLE OF VIEW. See LENS.

Flagging. A defective image characterized by a bent, pulled or slanted picture at the top of the TV screen. It is usually a result of too little or too much TAPE TENSION, a problem known as SKEW ERROR; an older model TV set; or a tracking control problem resulting from tape that has been stretched.

Flashback. A feature, usually found on TV monitor/receivers, that allows the viewer to move back and forth between a current channel and the previously tuned one.

Flat Display Tube. A CATHODE RAY TUBE that is paddle-shaped rather than cone-shaped as in the traditional CRT. In the FD tube the RASTER is even with the neck of the CRT instead of facing it. Because of this configuration, the path of the electron beam must make a right-angle turn to scan the phosphor surface of the screen. The major advantage of the compact flat display tube is its use in small, hand-held portable TV sets such as the Sony Watchman. The flat display tube differs from the flat square tube (FST) which is used in large television sets and monitors/receivers.

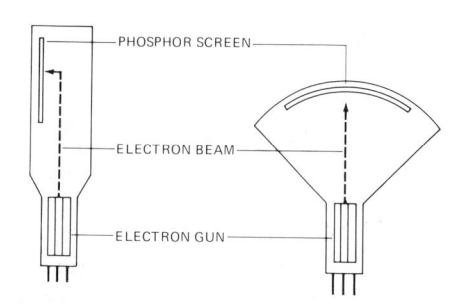

Figure 17. Flat display tube and cathode ray tube.

Flat-Field Signal. A special signal that can be adjusted from 0 percent to 100 percent white in steps of 10 percent. The flat-field signal, which produces a TV image of uniform brightness, is used for checking color temperature. This is performed by means of a color analyzer which measures the radiated light from the TV screen. The signal also helps in setting color balances at black level and at near white level. The flat-field signal consists of the composite video with sync and color burst. It is usually one of the functions on a TV SIGNAL GENERATOR. See COMPOSITE VIDEO SIGNAL, COLOR BURST, SYNC PULSE.

Flat Screen TV. A thin, almost flat screen television system utilizing one or more of the various advanced technologies to produce very high image resolution. Some possible processes include electroluminescent (EL), gas discharge and liquid crystal display (LCD). The first two make possible wall-sized images while the LCD type can be used for small, compact TVs. EL display consists of a microprocessor-chip-controlled inner layer of light-emitting material covered by a fluorescent coating to augment the light output. Another method of attaining flat screen TV is by using gas discharge panels composed of red, green and blue cells. They can produce large images at low-power consumption. Another, more recent, development has come from Matsushita. Operating by way of a beam matrix, the new system has many parallel beams striking the display screen. This replaces the conventional single scanning beam. Using a screen only two inches thick, the experimental flat screen TV system offers several benefits, including increased brightness, continuous sharpness over the entire screen area and considerably less distortion when the image is seen from the side. The new system has one major drawback, however—it is very costly to produce.

Flat Square Tube (FST). See FLAT DISPLAY TUBE.

Flat Tension Mask. A flat-screen picture tube, developed in 1986 by Zenith, that significantly improves the screen image. The new tube significantly increases brightness and contrast and eliminates glare from ambient light sources. In addition, the more-closely arranged red, green and blue dots that make up the screen image provide improved sharpness. The Flat Tension Mask is chiefly the brainchild of Kazimir Palac, a Zenith engineer.

Flesh Tone Reference Chart. See VIDEO TEST CHART.

Flickering. An effect resulting from copying a projected movie image with a video camera. Flickering is caused by the difference in frames-per-second. Silent films travel through the projector at approximately 18 fps, sound movies at 24 fps—both different from video which is synchronized at 30 fps. The discrepancy between the two systems causes flickering. Professional equipment has variable speed controls which can remedy this problem. See FILM-TO-TAPE TRANSFER.

Fluid Head Tripod. An expensive, professional video tripod with a camera mount or head which operates hydraulically. Fluid head tripods are superior to the more conventional and less costly types that employ counter-springs for the smooth performance of the pan and tilt features. See TRIPOD.

Fluorescent Screen. The coating placed on the face of a picture tube or cathode ray tube that glows when electrons hit it.

Flutter. In video, an effect with videocassette recorders that is caused by very brief and rapid tape speed variations.

Flying Erase Head. An additional erase head mounted on the rotating head drum of a video recorder or camera that is used during editing to accurately erase unwanted individual video frames and insert a new scene over a previously recorded one without glitches or picture breakup. The flying erase head scans the video tracks in front of the video head to ensure that only the precise parts of the video to be replaced are erased and written over. A new segment can be inserted into a prerecorded video-

Flying Spot Scanner

tape with virtually no distortion or video noise both at the beginning and end points of the edit. These erase heads originally were found only on industrial machines used by professionals. Some better-quality and more costly home video VCRs have incorporated flying erase heads into their array of features. Current 8mm camcorders and some Super-VHS-C models use only flying erase heads not only for seamless editing but to replace the more bulky full erase heads. Other camcorders and VCRs often use flying erase heads in conjunction with full erase heads. See BREAKUP, EDITING, ERASE HEAD, FRAME.

Flying Spot Scanner. A professional, sophisticated and computerized unit used in transferring commercial films to tape. Since the ASPECT RATIO (height × width) of a wide-screen movie is different from that of the TV screen, the film must be "scanned" to decide what should be eliminated from a scene. The scanner provides a range of features to help the technician plan and execute the transfer smoothly and artistically. See SCANNING.

FM Deviation. The spectral space or area occupied on a bandwidth by an FM (Frequency Modulation) signal. The conventional VHS luminance (brightness) signal recorded on videotape takes up 1 MHz of space (between 3.4 and 4.4 MHz). But when FM deviation is expanded on a higher frequency (between 5.4 and 7 MHz) to 1.6 MHz, the Super-VHS format, which can record at higher frequencies, takes advantage of this by providing increased picture detail. Since the recording density is about 60 percent greater with S-VHS machines than for standard VCRs, special high-resolution recording tapes must be used.

FM Microphone. A wireless microphone designed to transmit its signal through the air and extend the range between the subject and the video camera. The signal of an FM (Frequency Modulation) mic is picked up by an FM radio or receiver which is connected to the audio input of a VCR. Because it is wireless, the mic allows the subject to wander within telephoto range of the camera and still be recorded. See MICROPHONE.

FM Simulcast. An external audio mode that permits the picture of a TV the screen to be combined with a stereo audio input from an FM tuner. The TV/FM simulcast sound is then transmitted through the speaker system of a TV monitor/receiver. This feature is found more often on TV receiver/monitors than on conventional TV sets.

FNN (Financial News Network). An advertiser-supported CABLE TV network specializing in business news during the daytime hours and business and money-related programs in the evening. With an annual advertising and promotional budget of about $5 million, FNN is a relatively smaller operation than many other cable networks, including CNBC, its most recent competitor. The business network produces daily financial news spots for various radio stations and a syndicated 30-minute business program for morning television.

Focal Range. A measurement in millimeters (mm) that describes the parameters of a zoom lens. For example, the average video camera has a zoom lens with a focal range of from 12.5 to 75mm. This lens is also referred to as having a 6:1 ZOOM RATIO. See ANGLE OF VIEW, LENS.

Focal Zone Indicator. A camcorder feature that is designed to flash when the subject to be recorded within a particular frame is not in focus.

Focus. The convergence of an electron beam or light rays at a predetermined point. Many TV receivers feature a focus control that can sharpen the focus of the electron beam.

Focus Lock. A video camera feature that, when engaged, retains a predetermined focus position. This function is useful in preventing the focus from automatically changing in the event someone or something moves between the camera lens and the main subject. This feature is similar to the FOCUS PRIORITY ZOOMING.

Focus Priority Zooming. A video camera feature that keeps the subject in proper focus despite the number of ensuing zoom positions. With focus priority zooming activated, the camera operator can follow his or her subject from the maximum focal distance to the front of the lens without shifting to the macro feature.

Footcandle. A unit that measures the amount of light on an object emitted from one candle at a distance of one foot. One hundred footcandles, therefore, would equal the amount of light 100 candles would cast on an object from a distance of one foot. In video, usually 10 footcandles are required for a clear black-and-white picture and 20 for color. However, black-and-white cameras can operate with as little as 1/2 footcandle.

Footlambert. A unit of measure in reference to screen brightness. Footlamberts are more critical with components such as projection TV screens. A reading of 70 and above is considered very good for these screens. A movie theater screen has a rating of approximately 15 footlamberts.

Footprint. A signal pattern by a satellite or the area of the earth that can receive a satellite signal. Each satellite provides or casts its own footprint. If a SATELLITE TV system owner expects good reception from a particular satellite, he/she has to be in its footprint path. A satellite dish in one state may receive an excellent signal from a particular satellite whereas a dish in another state may not get that signal at all.

Format. In video, the mechanical configuration and width of videotape. The different tape formats include 1/4-inch as used in several camcorders, 1/2-inch, for both home and industrial use, 3/4-inch industrial, one-inch and two-inch professional/industrial tapes. Format may also apply to machines, such as Beta, VHS, 8mm, etc. Formats are not compatible with each other.

Forward Compatibility. The ability of a piece of equipment to handle later advances in software (tapes or discs).

Forward Scan

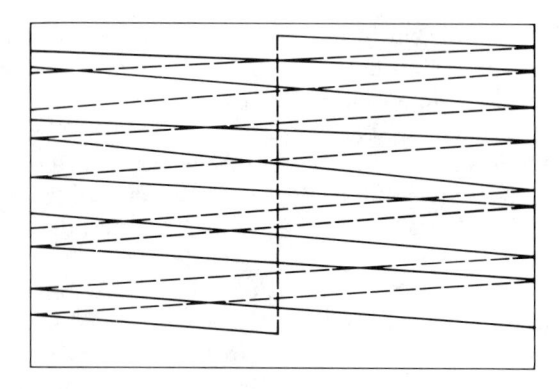

Figure 18. One frame (or two fields) of 525 lines.

For example, a VHS videocassette recorder is considered forward compatible if it is equipped to play back later innovations in videocassettes, such as improved Super-VHS tapes. Conversely, backward compatible equipment can play the latest tapes or discs as well as standard tape and disc formats. See BACKWARD COMPATIBILITY.

Forward Scan. See VISUAL SCAN.

Frame. A television picture consisting of 525 horizontal lines (NTSC standard) or one frame of information made by a pair of video heads each of which lays down one diagonal field on magnetic video tape. Each head records one field containing 262 1/2 lines. Two fields make up one frame of 525 lines. The TV picture is transmitted at 30 frames (60 fields) per second. With an LV videodisc player the disc spins at 1800 rpm or 30 frames per second. At standard speed one frame represents one revolution of the disc. With a CED player, whose disc spins at 450 rpm, each revolution represents four television frames. See CED and LV VIDEODISC SYSTEMS, FIELD, FRAME FREQUENCY, NATIONAL TELEVISION STANDARDS COMMITTEE.

Frame Advance. Part of a freeze-frame or still-frame feature found on many VCRs. Once in this mode, a second button or control is pressed to move the picture forward one frame at a time. If the button is held down, a succession of frames or individual pictures appears on the TV screen. Some VCRs are capable of producing noiseless frame advance; that is, still frames without NOISE BARS. Frame advance also refers to the particular method used to move a picture forward, such as a jog dial, buttons or skip button. See FREEZE FRAME.

Frame Buffer. See DIGITAL VIDEO FRAME STORAGE.

Frame-by-Frame Advance. See FRAME ADVANCE.

Frame-by-Frame Recording. An editing feature, found chiefly on more advanced VCRs, that permits the user to record specific frames in a range of increments. These may vary from as few as three to as many as 33 frames at any one time. This feature introduces animation, frame-by-frame editing and other sophisticated editing techniques to the home enthusiast.

Frame/Chapter Search. A videodisc player feature, found on many machines, that allows the viewer to scan frames and chapters of a disc. This is usually accomplished with a multiple-key direct access search function.

Frame Frequency. Refers to the number of times each second that the picture area of a television set is completely scanned. According to the NTSC (National Television Standards Committee), the frame frequency in the United States is 30 per second. See FRAME.

Frame Scanning. A process utilized in a videocassette recorder to produce special effects such as SLOW MOTION and FREEZE FRAME. For example, a freeze frame is created by stopping the tape while the video heads continue to reproduce the same frame of a video signal. (A frame consists of two fields, each consisting of 262 1/2 lines, one made up of odd-number, the other of even-number lines.) This technique, however, often produces VIDEO NOISE in the form of horizontal NOISE BARS along with the TV picture. To eliminate this type of interference, VCR manufacturers have come up with various innovations. Some Beta-format machines, for instance, utilize four video heads, one pair for producing special effects, the other set for the tape speeds. Some VHS machines, on the other hand, use field scanning to produce noiseless special effects. See FIELD SCANNING, FRAME, FREEZE FRAME, SCANNING, SEQUENTIAL SCANNING.

Frame Servo. A feature, found chiefly on industrial video recorder decks, that controls the odd or even scanning field of an edit. This results in a smooth, continuous on-screen image at the point where the edit occurs.

Frame Synchronizer. A control employed by networks to scan each video signal line and check the entire image for interference. When the transmitted picture fades, the synchronizer "freezes" the last frame and presents this on tube instead of a poor picture or blank screen (called "dead air"). Frame synchronizers are sold separately as professional/industrial units or as part of a TIME BASE CORRECTOR. In 1980 a frame synchronizer/time base corrector retailed for about $20,000; currently, they sell for a fraction of that price. As a separate accessory, each synchronizer differs in the number and type of features, which may include compatibility with S-VHS and Hi8, infinite window correction, shuttle handling and chroma line shift.

Frame/Time Search. A videodisc player feature that permits the viewer to shift to any point on a laser videodisc. The feature is often accompanied by an on-screen display that reveals the exact location on the disc and length of time it took to reach that point. Some frame/time search modes offer frame counts with the CAV for-

Framestore

mat while others provide real-time display with the CLV format.

Framestore. A digital process designed to hold a video image in memory. For instance, professionally designed VIDEOWALLS use semiconductor memory to store digital video signals. Each of these "memories" is considered a framestore. Each TV monitor in a videowall contains its own framestore, which can be a separate image or part of a larger image.

Freeze Field. A special effect, similar to that of FREEZE FRAME, found on some VCRs. A TV picture consists of 525 line scans, composing one FRAME. Each frame is made up of two fields. Many machines in the "freeze" position provide a "still" of two fields or one frame, each video head of a pair recording one frame. This is known as a freeze frame. Some VCRs, however, freeze only one field; this tends to present a clearer, more steady picture with fewer NOISE BARS. The freeze field is also known as still field. See FREEZE FRAME, LINE SCAN.

Freeze Frame. A special effect, found on VCRs, that locks one frame onto the TV screen. If the Pause mode (sometimes called Still) is pressed while the VCR is in Play, the tape "freezes" or becomes motionless and a frame or field (half a frame in Beta machines) appears on the screen. Because some VCRs do not scan an entire frame, NOISE BARS and JITTER often accompany the frozen image. Digital technology, introduced into VCRs in 1986, offers noise-free freeze frames. The technique operates by storing a single frame from the otherwise moving tape sequence in a computerized memory bank in the form of digital numbers, which are not affected by video noise. Digital freeze frame can also work with images from a live broadcast or cable TV. Keeping the VCR in the Freeze Frame mode for long periods of time may damage the video heads or the tape since the tape is stationary and pressing against the rotating heads. See DIGITAL STILL-FRAME MEMORY, FIELD, FRAME, PAUSE, VIDEO HEAD.

Frequency. The number of times a signal pulsates or vibrates per second; also, the speed of vibration of a sound wave. Noise causes sound waves whose energy can be measured in terms of its amplitude (strength) and frequency (vibration speed).

Frequency Band. A specified and continuous range of frequencies. Different types of transmission require different frequency bands.

Frequency Response. In video, a term which describes the capacity of a system to produce detail or resolution in its picture. Technically, video frequency response refers to the number of times electron beams can turn on and off during one full scan from left to right on a TV screen. These beams scan the raster, or face, of the picture tube in an exact amount of time, shut off and return to the left side of the screen. This procedure recurs until a complete image appears on screen from top to bottom. The greater the frequency response, the better the detail in the picture. Frequency response is expressed as a range; e.g.,

Front Projection TV

50–9,000 Hz. VCR manufacturers claim a frequency response of up to 12 KHz (12,000 Hz). See RESOLUTION.

Frequency-Synthesis Tuner. A special feature built into many CABLE READY TV sets designed to receive as many channels as CABLE TV carries. Unlike the ANALOG TUNING system, the TV receiver with frequency-synthesis is pretuned to all the anticipated channels, disregarding the frequencies without channels. With the FS tuner, fine tuning is unnecessary but, like many other automatic features which allow no leeway, the tuner presents difficulties with channels of minimally different frequencies. VCRs at the present time do not have this full-range cable-ready capability. The FS tuner is also known as electronic frequency synthesizing tuner. See ANALOG TUNING, TUNER.

Fringe Area. The region just beyond the limits reached by a television transmitter. Reception in fringe areas usually results in weak and unreliable signals which need additional boosting from such devices as a high-gain directional antenna or more sensitive receivers.

Fringing. An effect which is caused by one color "spilling over" into another. In a TV set, the convergence alignment would need adjusting; in projection TV, it would be the convergence controls that need resetting. In video, excessive chrominance/luminance delay inequality would cause color fringing. See CHROMINANCE/LUMINANCE DELAY INEQUALITY, CONVERGENCE ALIGNMENT, CONVERGENCE CONTROL.

Front Projection LCD TV. A relatively recent technology applied to front projection TV which uses liquid crystal display panels instead of conventional image tubes for projecting a large screen image. This system offers several advantages over conventional front projection TV. First, it is more compact and portable. Second, a screen or TV monitor is not essential since the picture can be projected onto any wall. Finally, the LCD system can accommodate special anamorphic lenses that can convert images into wide-screen pictures similar to those seen on theatrical screens. Some models provide a zoom lens which allows the user to adjust the size of the image to fit a given wall space or special screen. This system generally is more expensive than either rear projection or conventional TV.

Front Projection TV. A projection TV system in which light is projected onto a high gain, silver-like screen and reflected back to the audience. The system contains a TV set/projector, a one- or three-lens format and a special screen which reflects light much like a mirror. The process is similar to that which is used in a movie theater. Early front projection TV systems were often cumbersome and difficult to operate. Two-piece systems, for example, require that the projector be stationed on the floor in the middle of the room while the projection screen reclines near an opposite wall. Some manufacturers, however, disguise the projector within a

FST (Flat Square Tube)

coffee table and the obtrusive screen is concealed behind curtains. FRONT PROJECTION LCD TV systems, which use liquid crystal display panels instead of conventional tubes, are more compact, easier to handle and lighter in weight than those models using image tubes. The REAR PROJECTION TV system has virtually captured the entire market for projection TV systems. Both front and rear systems have certain drawbacks. Aside from generally costing more than conventional TV systems, they project an image that lacks the subtle details, contrast and sharpness, qualities that have become standard in direct television.

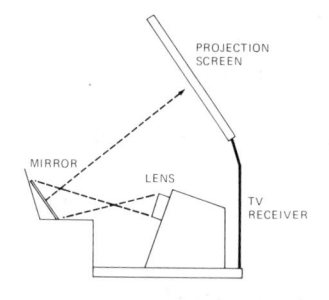

Figure 19. One-piece front projection TV system.

FST (Flat Square Tube). See FLAT DISPLAY TUBE.

F-Stop. A calibrated measurement referring to the amount of light a lens admits. All lenses used on video cameras have such f-stops as f/1.4, f/1.8, f/2, etc. The f-stop can be changed, depending on the lighting conditions. The smaller the f number, the greater the amount of light entering the camera. Most low-light video cameras feature an f/1.4 lens. See LENS.

F-Type Connector. See CONNECTOR.

Full-Color Transform. Refers to a special visual effect that produces a combination of two-dimensional animation and three-dimensional space. Full-color transform is usually one of many sophisticated features of a professional/industrial CHARACTER GENERATOR unit.

Full Function Wireless Remote Control. See REMOTE CONTROL.

Full Load. A VCR videotape transport system that wraps the tape against the video head drum and places the medium tautly around the appropriate rollers and capstans. The full-load technique allows for quicker playing and recording since the tape is in proper position for either function. In addition, full load permits the use of Real-Time Counter, a feature that accurately measures tape usage in hours, minutes and seconds. Full load differs from the half-load method which slows down the initial playing and recording procedures. The original Beta machines utilized full load whereas VHS units used M-Load, a system that drew tape from the cassette only in play and record modes. See HALF LOAD, M-LOAD, U-LOAD.

Full Tape Interchange. The capability of a tape recorded on one machine to play back properly on another unit. Videotapes, especially those recorded at the slowest speed, occasionally encounter problems when they are played back on other VCRs. The picture breaks up, rolls, or contains line or noise bars. Adjusting the tracking control frequently corrects the anomaly. Tapes recorded as slow speeds tend to be more sensitive to the slight differences in the position of the video heads in other machines. Some video technicians attribute the inability of a tape to play accurately on other units to such additional causes as variations in the tape transport and width of the video heads of both units.

Functional Integrated Circuit. Electronic circuitry that compensates for discrepancies between otherwise similar components. The FIC, introduced into such video equipment as portable VCRs, eliminates the need for the conventional adjustable resistors normally employed for this task.

Fuzzy Logic. In video, a term applied to a camcorder feature designed to provide better automatic lighting control, especially under difficult conditions. Fuzzy logic uses a special integrated circuit (IC) to "read" the light intensity of more than one area of a scene to be recorded and calculate the average light value. The technique, introduced early in 1990, is said to offer more accuracy and adjust the lens aperture more quickly. The feature appears on relatively few camcorders.

G

Gain. The term used for contrast in video and volume in audio. Also, the whiteness or luminance level of an image. Also, the degree to which a signal is amplified. RF (Radio Frequency) signals are increased by RF AMPLIFIERS while video signals are augmented by VIDEO DISTRIBUTION AMPLIFIERS. In PROJECTION TV screens, gain refers to the measured quantity of light reflected by a screen compared to that reflected by a white matte surface. Therefore, a gain of five translates generally to a screen five times as bright as a flat white surface. See AUTOMATIC GAIN CONTROL, GAIN-UP.

Gain Control. See AUTOMATIC GAIN CONTROL.

Gain Unity. A control on some process amplifiers designed to maintain a neutral position. For example, if the chroma gain control knob or dial is positioned at gain unity, then there is no increase or decrease in the intensity of the color signal. Gain unity applies not only to chroma control, but to luminance level as well. Therefore, if the luminance gain control is set at gain unity, there is neither gain nor attenuation in the brightness level of the signal. See DIFFERENTIAL PHASE AND GAIN, GAIN, PROC AMP.

Gain-Up. A camcorder feature using digital circuitry to increase sensitivity to light so that the camera can be used in low-light surroundings measuring only 1 lux. However, gain-up also has its drawbacks: it is accompanied by an increase in video noise and, often, LAG or IMAGE RETENTION. Gain-up, which is sometimes used deliberately to create special visual effects such as streaking, is accomplished by digitally boosting the brightness of the picture. See DIGITAL GAIN-UP.

Gap. The tiny space in an audio or video head across which the magnetic field is produced during recording and activated during playback. The head is a U-shaped electromagnet, and it is that opening which the videotape must pass and make contact with for quality recording and playback. Video head gap size depends on the speed mode of the VCR. Some VHS machines have four heads, two with a specific micron size for standard speed and the other pair designed for the slowest speed. See VIDEO HEAD.

Gartel, Laurence. VIDEO ARTIST who often uses electronics and video technology to enhance and alter the images of ancient and classical sculpture. A proponent of video still photography, his original way of seeing objects has been recognized in national newspapers and magazines. See VIDEO ART, VIDEO ARTIST.

Gaussian Filter. In electronics, a low-pass filter used for bandwidth limitation and pulse shaping as well as to remove unwanted distortions such as noise, preshoot, overshoot and ringing. Its name derives from the approximation of its amplitude/frequency response to a Gaussian distribution or shape. One application of the filter concerns the extraction of "spikes" or ringing in video line sync pulses that may be caused by high frequency components. Another use of the Gaussian filter is in the field blanking period of television signals. The filter usually comes in modular form for relatively simple insertion into electronic equipment. See FIELD BLANKING, FILTER, LOW-PASS FILTER, SYNC PULSE.

General Purpose Interface Buss (GPIB). A special feature, usually built into some industrial/professional instruments such as COLOR VIDEO NOISE METERS, designed to test various signals automatically.

Generation. In videotaping, one copy or duplicate removed from another tape. The original or master program is known as a first generation. A duplicate from this master is called the second generation tape. Each generation away from the original or master produces increased degradation in the image quality.

Genlock. A feature on a SPECIAL EFFECTS GENERATOR which locks into sync incompatible signals. For example, two cameras, one camera and one prerecorded tape or two VCR signals (but not two prerecorded tape signals) can be fed to a genlock and an SEG for mixing, etc. SEGs, with or without the genlock feature, are professional devices and cost thousands of dollars.

Geometric Linearity. The accurate reproduction of sizes and shapes on a TV screen. The geometric linearity of a TV set or monitor/receiver depends upon the quality of the deflection yoke—that part of the cathode ray tube (picture tube) whose neck contains a complex set of coils. Electron beams emerging from the deflection yoke are bent as they leave the center and must hit the raster or face of the tube accurately. Beams that reach the tube inaccurately result in misshapen circles or straight lines that appear distorted near the edges of the screen. These vertical image distortions, sometimes called a "barrel effect" since they appear in the shape of a barrel, are known as poor geometric linearity. See DEFLECTION YOKE.

Geostationary. Refers to a satellite's orbit around the earth at the same speed as the rotation of our planet, so

Ghost

that the satellite appears stationary. A satellite is "stationed" 22,300 miles above the earth's equator and remains in its geostationary orbit for the ground station to receive optimal transmission. See SATELLITE TV.

Ghost. In video, an unwanted, repeated image that appears almost invariably next to the original. Ghosts are usually caused by reflected signals which travel by longer paths and therefore arrive on screen later than and in a different position from the original signals. Special directional antennas, designed to restrict signals over only one path, may alleviate this problem.

Gigahertz (GHz). A unit of measure used with MICROWAVE, SATELLITE TV, etc. One gigahertz equals a thousand MEGAHERTZ (MHz). Current satellites function in the range of 3.7 to 4.2 GHz, which is the lowest portion of the microwave band. See HERTZ.

Ginsburg, Charles. Engineer who, with RAY DOLBY, developed the first prototype for the video recorder. AMPEX demonstrated the model in 1956.

Glass Delay Line. One of the two kinds of comb filter systems (the other uses a CCD or charge coupled device). The glass delay line technique is less sophisticated, more economical and more popular. Both types perform the same functions. They improve detail and resolution in the TV picture by separating the black-and-white from the color information and thereby prevent VIDEO MOIRE or a rainbow effect. However, it is generally agreed that the CCD comb filter is more efficient. See CHARGE COUPLED DEVICE, COMB FILTER, RESOLUTION, VIDEO MOIRE.

Glitch. The term given to any type of video distortion such as picture BREAKUP or variations of video noise. A glitch usually occurs each time a VCR stops and starts to record. It is less prevalent when going from the Pause mode to Record. More costly VCRs usually incorporate special electronic circuitry that makes editing virtually glitch-free. See AUTOMATIC TRANSITION EDITING, DISTORTION, EDIT CONTROLLER, EDITING, EDITING CONSOLE, VIDEO NOISE.

Glitch-Free Editing. In video, any method or procedure that transfers or duplicates selected recorded segments of tape onto another tape without picture breakup or the familiar "rainbow" effect that appears where two images are joined. Unlike movie film, video tape is difficult to edit because each frame or segment is a diagonal magnetic signal that cannot be seen by the human eye. Precise physical splicing is virtually impossible. Therefore, other methods have had to be devised to accomplish exact editing. At first, home VCRs performed the task rather clumsily, adding picture breakup a rainbow effect, or MOIRE, wherever edits began. Later, some VCR manufacturers introduced improved editing techniques by backing up the tape a few frames each time the Pause and Record modes were pressed. This eliminated much of the visual interference. The FLYING ERASE HEAD was another innovation

Graticule

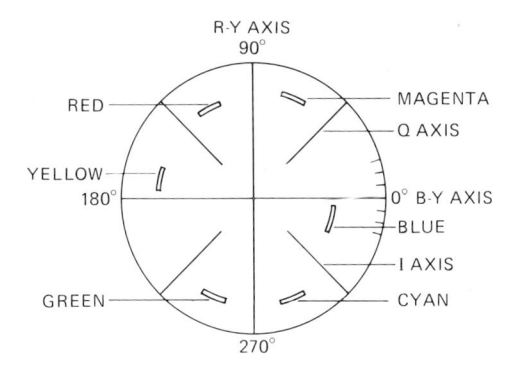

Figure 20. Color vectorscope graticule.

that helped to produce glitch-free edits. Another, known as SYNCHRO EDIT, synchronizes the start/pause modes of both the recording and playback machines during the editing process, thereby eliminating glitches. Other technological advances for glitch-free editing include the use of electronics, digital memory and computers.

Global Satellite. See SATELLITE FOCUS.

Go-To. A VCR feature that, when activated, can automatically locate any point on a videotape. Go-To usually operates with either digital tape counters, which require entering a specific four-digit number, or real-time function in which the hours, minutes and seconds are entered. The latter usually can be activated by using the remote control. See REAL-TIME COUNTER, SEARCH MODE, TIME SEARCH.

GPIB. See GENERAL PURPOSE INTERFACE BUSS.

Graphic Equalizer. See EQUALIZER.

Graphics. In video, a technique involving a computer, a MONITOR and certain SOFTWARE to create designs, drawing and other graphics on a TV screen. Through the use of computer keyboards and video terminals, the electron beam is guided around the screen to produce drawings that can be videotaped. Graphics programs are available for different systems. They can create elaborate graphic designs, abstract art, random patterns; they can change screen color, form block letters, create a figure and move it anywhere on the screen. See VIDEO GRAPHICS GENERATOR.

Graphics Decoder. An electronic accessory, usually built into certain compact disc players, that adds graphics displays to music. Such machines are known as CD+G (for graphics) players. The visuals produced by the graphics decoder are still pictures; the system is not capable of producing continuous motion images. See CD+G.

Graticule. The pattern imprinted on the face of the CATHODE RAY TUBE of a COLOR VECTORSCOPE for mea-

121

Gray Scale

GUARDBANDS

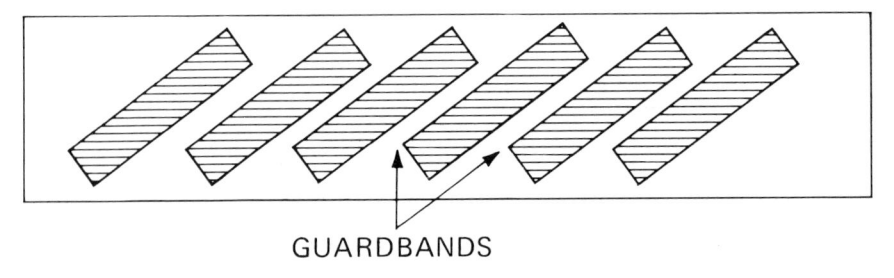

Figure 21. Guardbands between video tracks.

surement purposes. The graticule consists of a 360° circle, a B-Y horizontal axis line, an R-Y vertical axis line, two other lines labeled "I" and "Q" axis and little boxes representing the six colors (blue, red, magenta, green, cyan and yellow). On some graticules, the circle is divided into 5° segments. Signals from cameras, VCRs and other pieces of electronic equipment are superimposed over the graticule to help determine the quality of the color signal and where any problems may lie. For example, if no "I" signal appears, then no red will be visible in the TV image; if there is no "Q" signal, then no green will show up on the TV screen. See "I" SIGNAL, "Q" SIGNAL.

Gray Scale. The range of gray shades in a TV image corresponding to color. A picture with too much contrast has a limited or shortened gray scale. A gray scale, which consists of a series of graded, or regularly spaced, tones ranging from white to black, is one method of judging the quality of a television picture or photograph.

Gray-Scale Chart. See VIDEO TEST CHART.

Grazing. A term used by advertisers, producers and others in the television industry to refer to the practice by viewers of randomly searching through the channels for an interesting program. This practice has alerted the networks, TV producers and programmers to seek new ways to hold the attention of their viewers.

Great Time Machine, The. Quasar's first (now defunct) VCR, model VR-1000, which was incompatible with its two contemporary formats, VHS and Beta. The GTM was introduced in 1976 and used a cassette format called VX200. The machine was replaced in 1978 by Quasar's second VCR, model VH-5000, a more conventional VHS machine. The original recorder had one unique feature for its day—its clock/timer could be set to record more than one event or program during a 24-hour period.

Ground-Loop Hum. Refers to the large horizontal bars that occasionally roll up or down on the screen image. This video hum appears as a result of differences in the grounding of house currents from that of the broadcasting station or TV cable system. Ground-loop hum is sometimes referred to simply as hum.

Guard Band. A space between two video tracks essential in maintaining a correct video signal even with a misaligned video head. The introduction of the AZIMUTH process of electronically printing video tracks diagonally, each with a different pattern, has largely discontinued the need for guard bands. A guard band also refers to a frequency band left empty between two channels to protect against interference from either channel.

Gustafson, Julie. VIDEO ARTIST and producer of VIDEO DOCUMENTARIES. Together with her husband, JOHN REILLY, she made "Home," "Giving Birth: Four Portraits" and other video documentaries, some of which received exceptional recognition. See VIDEO ART.

G-Y Signal. Refers to the green-minus-luminance signal in color television. A primary green signal is produced when the G (green) signal is joined to the Y (luminance) signal either inside or outside the picture tube.

H

HAD (Hole Accumulated Diode) Sensor. A component of some top-of-the-line professional/industrial video cameras that increases horizontal resolution and enhances color quality. The special sensor can produce 700 lines of horizontal resolution.

Half-Inch Format. Refers to the most commonly used tape size for home video. Both Beta and VHS machines use 1/2-inch videotape while profession/industrial/institutional equipment uses other widths such as 3/4-inch and one-inch tape. Newer, smaller VCRs for home use have introduced 1/4-inch videotape, particularly with camcorders.

Half-Load. A VCR feature of a VHS-format unit that keeps the videotape partially in position so that faster searches can be made. Conventional VHS machines usually remove the tape from the VCR mechanism when the machine is in Stop mode. With the half-load system, a desired segment can be reached through either Rewind or Fast Forward mode at a speed of 120 times that of EP speed. Panasonic introduced the half-load system in 1979 with its professional/industrial decks. See FULL LOAD, M-LOAD, U-LOAD.

Hall Effect, The. An audio technique that is not affected by tape speed in the playback mode. Since tape speed influences the quality of audio reproduction, the slow speeds of videotape produce far from good sound. But videotape must travel these relatively slow speeds to accommodate the video signals and to offer reasonable maximum playing time. If a process, such as the Hall Effect, can be developed so that it can be used with recording as well as playback, audio quality will greatly improve.

Halo. An unwanted ring of light circling a spot on the screen of a cathode ray tube.

Hand-Held Television. A small, lightweight, portable TV unit, usually applying LCD (Liquid Crystal Display) technology and other electronic refinements to the video portion. Some manufacturers concentrate on increasing the number of pixels, or picture elements, per square inch to ensure accurate color separation. Others focus on resolution to produce a clear, sharp image. Sony's compact Video Walkman combines an 8mm video recorder with a 3" LCD color TV.

Hand-Held Video Game. See PORTABLE VIDEO GAME.

Hardware. In video, equipment such as VCRs, videodisc players, video games, video cameras, portable VCRs, TV sets, etc. Videotapes, discs and game cartridges are considered software.

Hays, Ron. VIDEO ARTIST, proponent of visual or VIDEO MUSIC, Emmy Award winner, designer of musical synthesizers, creator of the first computer-animated music album "Ron Hays Music Image Odyssey." His work with lasers, computers and other video paraphernalia has been featured in national magazines. His music synthesizer has played a part in films (*Demon Seed*) and such concerts as the "Star Wars" performance by the Los Angeles Philharmonic conducted by Zubin Mehta. Hays won an Emmy (Best Graphic Design) for the opening of a network special and created a 20-minute visualization of Wagner's "Prelude" and "Love and Death."

HBO. See HOME BOX OFFICE.

HDSNA (High Definition System for North America). An experimental method of improving video and audio components for consumer-oriented television that is compatible with existing TV in the United States. HDSNA, which can produce 1,050 lines in contrast to the present standard of 525 lines, features wide-screen capability (16:9 aspect ratio), a high-resolution picture, digital audio sound and the capability of accommodating broadcast, cable and satellite modes of transmission. Unlike HDTV (high definition television), its chief rival, HDSNA does not render the millions of current TV receivers obsolete; however, these sets will not benefit from the improved image.

HDTV. See HIGH DEFINITION TV.

HD VTR. See HIGH DEFINITION VTR.

Head. In video, an instrument that reads, records or erases information by changing the magnetic pattern on the surface of videotape. With many VCRs, the recording and playback of video information are handled by the same head while the audio functions are performed by separate heads. See VIDEO HEAD.

Head Gap. See GAP.

Head-Switching Interval. A sometimes-visible line on the TV screen indicating the completed function of one video head and the start of the second head. If the line appears on the screen, it can be removed in one of two ways. The head switch can be set lower or the vertical size control can be adjusted until the line disappears from the bottom of the image. Head-switching interval is inherent in home video units.

Head-Switching Noise. A line of interference that occasionally appears at the bottom of a picture as a result of video heads being switched on and off. A peculiarity chiefly inherent with helical scan videotape recorders, such as U-Matic units, head-switching noise occurs as a result of the different positions that each format records information on tape. Since the problem is one of recording, not playback, these industrial recorders can be adjusted to produce invisible head-switching.

Head-to-Tape Contact. In video, the extent to which magnetic coating surface of the videotape comes near the surface of the video heads during normal operations of the VCR. High resolution and minimal separation loss occur with proper head-to-tape contact. See VIDEO HEAD.

Helical Scan. The diagonal track across the videotape made by the rotating video heads which spin at the speed of 30 revolutions per second. The diagonal pattern is produced by a tilted drum assembly on which the heads are mounted. Sony's 3/4-inch U-Matic was the first commercially successful VCR to use helical scan, which was invented in the 1960s by Dr. NORIZAKU SAWAZAKI. Helical recording, an important feature for professional video editors, made visual tape searches posssible.

Hemispheric Satellite. See SATELLITE FOCUS.

Hertz. A unit of measurement (expressed in Hz) on the microwave frequency band equivalent to one cycle per second. A thousand Hz is often written as 1 KHz, a million as MHz and a thousand MEGAHERTZ as 1 GHz (GIGAHERTZ). See H.R. HERTZ.

Hertz, H.R. German physicist who first discovered radio waves. It has been internationally accepted that the frequency at which something occurs per second (cycles per second) should be termed Hertz or Hz. See HERTZ.

Heterodyne Color Process. A technique that reduces the color frequency from the normally high requirement of 3.58 MHz (MEGAHERTZ) to the KHz range. Larger format video recorders have no problems processing the high color subcarrier that the color signal needs. The heterodyne process allows smaller format VCRs, such as the 1/2-inch type, to accept the reduced color frequency which is boosted during playback to its original 3.58 MHz. Larger formats process the full color signal directly, which is required to meet the broadcast standards of the FCC, whereas smaller recorders process the color indirectly.

HG. A term given to high grade or premium quality videotape. High grade actually refers to the quality of a ferric-oxide particle used in the production of this upgraded tape. According to some manufacturers, HG tape provides an improved signal-to-noise ratio—which translates into a picture with less VIDEO NOISE, especially at slower tape speeds. In technical jargon, the gain comes to about three decibels. HG tapes offer fewer overall dropouts. See COATING, OXIDE, VIDEO SIGNAL-TO-NOISE RATIO, VIDEOTAPE.

Hi-Band Still Video Standard. See STILL VIDEO.

Hi8. A high-band 8mm video recording format that generates 400 lines of horizontal resolution, improves signal-to-noise ratios and permits up to two hours of broadcast-quality recording on one compact cassette. The format of these VCRs and camcorders raises the luminance carrier from 5 MHz to 7 MHz and expands the frequency deviation from 1.2 MHz to 2 MHz. Hi8, equivalent in many areas to Super-VHS, features separate S-video luminance and chrominance inputs and outputs for excellent video quality with compatible components. The format also offers less generation loss when making duplicate tapes, improved color accuracy, flying erase heads for glitch-free edits and lightweight compact camcorders. However, Hi8 tapes cost more, and the format is incompatible with most existing VCRs.

Hi-Fi. In video, the audio signals recorded across the width of the videotape to produce high-quality sound. VCRs equipped with Hi-Fi can also produce linear recording for the purpose of overdubbing. Hi-Fi does not necessarily produce a stereo effect. The 8mm format, for example, uses similar recording techniques as those of Hi-Fi, but produces mono sound. See HIGH FIDELITY, STEREO.

Hi-Fi Tracking Meter. A VCR feature designed to maximize the tracking of tapes encoded with Hi-Fi signals.

Hi-Z. See IMPEDANCE.

High Definition Broadcasting. The transmission of a broadcast signal which occupies a larger or wider channel than the 6 MHz (megahertz) slot apportioned to conventional TV channels by the Federal Communications Commission. Since space on the ELECTROMAGNETIC FREQUENCY SPECTRUM is at a premium, high definition broadcasting will have to utilize other means of transmitting its signal. Cable systems have the capacity to carry these signals which are approximately five times wider than the typical 6 MHz. Another method is by satellite, but there is none presently available with the capacity to carry high definition broadcasting, which eventually will feature a wider picture, better definition and an image containing more than 1,000 horizontal lines. See HIGH DEFINITION TV.

High Definition TV. An improved broadcasting system utilizing 1,125 lines or more (instead of the NTSC standard of 525 lines) to give a sharper image, more detail and improved color. The higher resolution images rival those from 35mm movie cameras and can be applied to medical, industrial and computer uses as well as future projection TV video theaters. But high definition TV will require special satellite transmission, new video camera technology, different TV receivers, etc. One problem inherent in high definition TV is the need for a wider BANDWIDTH to store the additional information. To accomplish this, HDTV would have to be broadcast over a few existing channels. Although Sony has not solved the question of the expanded bandwidth, the company has made inroads

High Definition VTR

with a non-broadcast system which utilizes advanced recording and processing methods in its own version of HDTV. Sony calls the process High Definition Video System in which videocassettes are capable of storing high-resolution TV for use with special TV receivers. CBS experimented with HDTV as early as 1981.

Other competitive systems have sprung up to rival HDTV. For example, HDSNA (High Definition Standard for North America), introduced by Philips, and a line-doubling technique introduced by Yves FAROUDJA, offer many of the benefits of HDTV without making present TV receivers obsolete. The ACTV, discussed elsewhere, is compatible with the NTSC broadcasting standard. The Del Rey system, also compatible with NTSC and boasting a tripling of the standard resolution, can produce a 14:9 ratio. The Glenn system can be used with present TV sets but requires a larger bandwidth to produce its increased number of lines and wider aspect ratio. The MIT system, with its 19:9 aspect ratio, uses the current 6 MHz bandwidth to produce either an 800-line or 600-line picture, the latter compatible with present TV sets. The MUSE system, incompatible with conventional TV sets, needs 8.1 MHz of bandwidth to produce 1,125 lines with an aspect ratio of 5:3. The NAP system, which uses two signals—one an unmodified NTSC signal, the other an increased signal, can turn out 1,050 lines but depends upon two full 6 MHz channels.

Early in 1990 the FCC chose HDTV over other enhanced TV systems as a national standard. New advances in HDTV include a simulcast system which is compatible with the existing NTSC standard. A simulcast system requires a TV station to use one channel for standard NTSC broadcasts while a portion of an empty adjacent channel is used to transmit information necessary to improve reception for a 1,050-line, 16:9-ratio widescreen image. See BANDWIDTH, HDSNA, IDTV LINE DOUBLER, NATIONAL TELEVISION STANDARDS COMMITTEE.

High Definition VTR. A professional/industrial videotape recorder that makes use of a wideband 30 MHz luminance and 15 MHz chrominance recording system. HD VTRs can record 1,000 horizontal lines as produced by HDTV color video cameras. In addition, the HD tape recorders usually allow multi-generation recordings with a 56 dB signal-to-noise ratio, provide internal editing features and a built-in time code reader/generator.

High Density Videotape. Tape that packs a higher number of magnetic particles per square inch onto the coating. All Beta tapes are considered high density tapes since their OERSTEDS OF COERCIVITY or particle density rating is more than twice that of conventional tape. Some high density tapes claim 230 million particles per square millimeter. The greater the particle density, the better the video image. See COATING, HG.

High Fidelity. In audio, a commonly accepted range of 20–20,000 Hz, plus/minus three decibels (also described as +/-3dB). In most conventional video systems, these measurements are presently not attainable due to the slow speed and the small bandwidth of the videotape. Some of the higher quality VCRs have tested at 40–9,000 Hz (+/-3dB). Videodisc players seem to have the potential to fare better in audio high fidelity. LV discs allegedly are capable of producing a range of from 20 to 20,000 Hz. Sony's BETA HI-FI process, which utilizes the video heads instead of the audio head to place the sound track on the tape, provides improved sound. Sony claims its process attains the high fidelity range. See BETA HI-FI, STEREO VCR.

High Fidelity Video Component. An electronic unit or part of a video system which is designed to reproduce a wider range of picture detail than ordinary video equipment. A major determinant in producing high fidelity video is the frequency range—or bandwidth—of the component. The wider the bandwidth, the greater the image detail. A better-than-average broadcast signal extends over a bandwidth of four million cycles per second or 4 MHz. Since all TV receivers cannot reproduce this range of frequencies, picture resolution often suffers. Indeed, some mediocre TV sets can barely cover a bandwidth of 2 MHz while high fidelity video units come very close to reaching the entire 4 MHz, producing excellent picture detail. See BANDWIDTH, FREQUENCY, HIGH FIDELITY, RESOLUTION.

High Gain. In front projection TV systems, a specially designed screen that increases perceived brightness. High gain helps to alleviate one of the drawbacks of front projection TV—illumination loss.

High Grade Videotape. See HG.

High/Low Sensor. One of the processes used in the Pause circuitry of a VCR. See PAUSE CONTROL.

High-Pass Filter. In video, an electronic circuit that reduces interference on all channels. The filter, sometimes written as hi-pass filter, accomplishes this by allowing passage of frequency components above a predetermined limited frequency while rejecting components below that parameter. See FILTER, LOW-PASS FILTER.

High Power Satellite TV. Refers to DIRECT BROADCAST SATELLITE which transmits its signals at 12.2 to 12.7 GHz within the Ku-Band of the electromagnetic spectrum. High power satellite TV differs from conventional satellite TV which utilizes 4 to 6 GHz low power bandwidth and uses the now familiar large antenna dishes which range from 10 to 15 feet in diameter. Channels broadcast by DBS cannot be received by these large antenna systems. Although high power satellite TV systems are less costly to the consumer than their low power counterparts, they are limited in the number of channels they can hold. DBS can broadcast up to about six channels whereas the typical satellite TV system can receive over 40 cable services. See C-BAND, DIRECT BROADCAST SATELLITE, LOW POWER SATELLITE TV, SATELLITE SIGNAL, SATELLITE TV.

High Resolution

High Resolution. The display of a larger-than-usual number of scanning lines, perhaps 1,000 or more. In HIGH DEFINITION TV, the number of lines exceeds 1,100. Video cameras for home use often have an average of 240 horizontal lines. The higher the number, the better the detail, sharpness and DEFINITION. See HORIZONTAL RESOLUTION, RESOLUTION, VERTICAL RESOLUTION.

High-Speed Picture Search. See VISUAL SCAN.

High-Speed Search Mode. A VCR feature that presents the picture on screen while the tape moves rapidly to a desired point. This is made possible by keeping the tape in contact with the video heads during the fast search. Conventional Fast Forward mode disengages the tape from its contact with the video heads. One shortcoming of this high-speed search mode is that the fast-moving tape, as it passes over the heads, builds up friction and heat, both of which may be harmful to the tape and heads. The high-speed function can usually be adjusted from about 5 to 20 times on some machines and from 10 to 30 times on others in the standard playing speed.

High-Speed Shutter. A video camera feature that uses electronics in the lens shutter to produce slow motion or freeze frame of action scenes without the usual blurring effect that accompanies subject motion during playback of the tape. The high-speed electronic shutter allows exposures as fast as 1/10,000 of a second to "stop" or freeze action. Most camcorders, however, provide shutter speeds of 1/1000 or 1/2000 of a second, which should be sufficient for users to capture major sports activities.

Hiss. The continuous audio noise which emanates from tape playback on the upper end of the frequency band. See DYNAMIC NOISE REDUCTION, VARIABLE WINDOW FILTER.

HiVF. An advanced electronic STILL VIDEO camera that uses still video floppy (VF) disks. The HiVF camera is capable of producing 500 lines of horizontal resolution with a peak recording frequency of 9.7 MHz. It rivals the professional Betacam SP and MII but falls far short of its major competitor, the 35mm film camera, which turns out slides rated as high as 5,000 lines of resolution. In addition, the film format offers better brightness contrast and a higher range of color than anything produced by current electronic still photography (ESP). See STILL VIDEO, STILL VIDEO FLOPPY DISK.

Hold Control. In video, the function knob on a TV receiver or similar unit that controls the stability of the picture. This is accomplished by the capability of the hold control to adjust the frequency of the vertical- or horizontal-scanning pulses. See HORIZONTAL HOLD CONTROL.

Home Box Office (HBO). A PAY TV service carried by a CABLE TV system, telephone wire or MICROWAVE and specializing in recent Hollywood films. HBO also offers some sports programs, entertainment and documentaries. With 17 million subscribers as of 1990, it is the largest pay TV service in the United States. Owned by Time, Inc., HBO currently offers a 24-hour-per-day schedule, although many of the films are repeated throughout the month. The service began operations on a small scale in 1973 in Pennsylvania where six cable TV systems carried it. HBO's decision in 1975 to transmit by satellite was the major contributing factor in the company's rapid growth and success. HBO's first presentation was "The Thrilla From Manila"—the legendary Muhammad Ali-Joe Frazier championship fight. Although Home Box Office is often carried by a cable company, for an additional monthly fee, it is also offered in areas as an individual service by way of microwave. See CABLE TV, PAY TV, TIER.

Home Video. A concept of home entertainment in which a TV set is only one of the components and over which the viewer has some control. Home video provides the viewer with (1) controlled viewing (he or she can slow down, speed up or stop the program with the functions on a VCR or VDP); (2) altered programs (the viewer can edit, dub in a new audio track, etc.); (3) the capability of storing and collecting material; and (4) the freedom to watch a program at any time other than the regularly scheduled one.

Home Video. A chiefly non-technical breezy magazine that featured specialty articles on peripheral areas of video rather than on technology or hardware. The periodical tended to have a monthly theme; e.g., cable TV, science fiction, etc. Heavy coverage of pay TV and cable programs dominated its pages with pseudo-sensational articles about Ted Turner, etc. Very few test reports on equipment appeared in its pages. However, the magazine provided the best and most extended department devoted to technical questions and problems from its readers. The replies were detailed, specific and informative. The design layout was satisfactory. The magazine began operations in June 1980 and by March of 1983 was sold, merging with the more successful *VIDEO REVIEW* magazine. See VIDEO PERIODICALS.

Home Video Game. See VIDEO GAME CARTRIDGE, VIDEO GAME HARDWARE, VIDEO GAME HISTORY, VIDEO GAME SOFTWARE.

Horizontal Blanking. In video, the clipping off of the electron beam between ensuing active horizontal lines during the retrace process. See BLANKING, FIELD BLANKING, LINE BLANKING.

Horizontal Blanking Pulse. In video, a rectangular-shaped flat-topped pulse or pedestal in the composite TV signal. The horizontal blanking pulse exists between active horizontal lines to cut off the electron beam current during the retrace process of the picture tube. See BLANKING, COMPOSITE COLOR SIGNAL, COMPOSITE VIDEO SIGNAL.

Horizontal Centering Function. The control found on television receivers and similar units that allows the viewer to adjust the entire screen image from left to right or vice versa.

Horizontal Hold Control

Horizontal Hold Control. A function, found on virtually all TV receivers and usually located in the rear of the set, designed to adjust the horizontal sweep section that is locked to the horizontal sync pulse. The user can readily and safely use the horizontal hold control for superficial adjustments. Further fine tuning of related controls, such as the width, linearity and frequency, are best left to qualified service technicians.

Horizontal Hum Bar. A wide stationary or moving strip, usually occurring in a series, that appears on the TV screen. These alternating light and dark bars usually are the result of interference at about 60 Hz or a harmonic (multiple wave) of 60 Hz. See NOISE BAR.

Horizontal Image Delineation. An electronic process designed to increase picture contrast and detail at the edges of a TV screen. Horizontal image delineation accomplishes this during the electron beam's horizontal scanning across the screen. The signal level is immediately shortened prior to its reaching a dark-to-light boundary and instantaneously increased before it reaches a light-to-dark border. Horizontal image delineation, sometimes listed as dynamic picture control or contour control, is similar in some respects to DELAY LINE APERTURE CONTROL.

Horizontal Linearity Function. The control built into television receivers and similar units that permits the user to control the width of the screen image.

Horizontal Lines. The TV lines that make up the picture height. The greater the number of lines, the better the vertical resolution or sharpness and detail. The number of lines is derived by multiplying 80 by the peak video frequency of the videocassette recorder or television receiver. For example, if a VCR can record and play back three million cycles per second (3 MHz), then that machine is listed at 240 lines. Black-and-white and color resolution are usually listed separately. See HORIZONTAL RESOLUTION, RESOLUTION, VERTICAL RESOLUTION.

Horizontal Noise Bar. See NOISE BAR.

Horizontal Resolution. A method of judging detail in an image. A test pattern consisting of a series of vertical lines is used in measuring horizontal resolution. The more vertical lines that can be seen clearly, the better the quality of the image. Horizontal resolution should not be confused with vertical resolution. The number of horizontal lines determines the quality of vertical resolution while vertical lines determine the quality of horizontal resolution or the sharpness of the image. However, horizontal resolution is dependent upon the vertical. In other words, if the number of horizontal lines is very high (vertical resolution), then the horizontal resolution will be above average. Some TV receivers can capture a horizontal resolution of from 300 to 400 lines, which covers almost all of the telecast signal. However, the conventional TV set reaches only a horizontal resolution of about 240 to 280 lines. Con-

Hue

ventional VCRs attain a horizontal resolution of about 240 lines, whereas more advanced and costly machines can reach 400 or more lines. See RESOLUTION, VERTICAL RESOLUTION.

Horizontal Retrace. Refers to the line made by the electron spot as it returns from right to left as it creates the picture image on the television screen. The line is not visible on the screen because of the blanking process. See BLANKING.

Horizontal Sync. The sync pulses controlling the electron beam that places horizontal scanning lines on the TARGET AREA of the CATHODE RAY TUBE in a TV set or the CAMERA TUBE in a video camera. In the image-forming process, the horizontal sync pulse follows both the single line scan and the blanking pulse. In other words, the electron beam scans one line, then blanks out before it starts its second line. In this blanking period, the horizontal sync pulse controls the beam so that it performs an orderly job and stays in sync with the entire process. See COMPOSITE VIDEO SIGNAL, HORIZONTAL SYNC PULSE.

Horizontal Sync Pulse. A portion of the last three lines of every video FIELD. This pulse controls or adjusts the electronic "dot" that lays out the zigzag patterns of horizontal scanning lines. Some anti-piracy devices, to prevent duplicating a prerecorded cassette, intentionally alter the horizontal sync signal, causing an unstable picture when the video signal is fed into a second VCR. See COMPOSITE VIDEO SIGNAL, HORIZONTAL SYNC, VIDEO SIGNAL.

Horizontal Time Base. See TIME BASE CORRECTOR, TIME BASE INSTABILITY, TIME BASE STABILITY.

HQ Circuitry. A VCR system of special "high quality" circuits designed to improve video information, reduce interference and present a clearer screen image. Associated chiefly with the VHS format, different circuits and filters perform various functions, such as enhancing detail, reducing luminance signal-to-noise ratio and suppressing color interference. Some of these improvements are accomplished by a technique called DIGITAL VIDEO NOISE REDUCTION SYSTEM. Some special circuits include white clip level extension to improve edge sharpness and contrast, detail enhancement circuitry to enhance image detail, luminance noise reduction to produce whiter whites and darker blacks, and chroma noise reduction to control uniform color in the entire picture. Not all manufacturers utilize all the potential circuits and filters in their models, although they describe their machines as having HQ circuitry. The term itself does not actually imply a particular standard. All this technological advancement, when added to a VCR, reportedly improves picture quality by about 10 percent.

Hue. The general shade of color of a screen image such as blue or red. An ideal picture balances the three primary colors (red, green, blue). If any one of these is emphasized, the hue is then adjusted. Hue is different from

Hue Control

SATURATION, which refers to the intensity of a color. Hue is also known as color phase or tint.

Hue Control. A TV receiver function knob, usually located at the rear of the unit, that adjusts or changes the tint or color of the picture. The hue control is sometimes called the tint control.

Hum. Audio noise in the low frequency range. Hum can result from the AUTOMATIC GAIN CONTROL of a VCR. It is simply picking up and amplifying extraneous sound. Hum can also stem from faulty "ground" connections, from defective internal mics on a camera, etc. See GROUND-LOOP HUM.

Humidity Eliminator Unit. A built-in feature found on some VCRs and videocassette players designed to keep metal surfaces dry when humidity is high. Problems arise when the location of a machine is changed from a cold environment to a warm one. In this situation, the dew indicator will light and the machine will not function as a safety precaution. The user must wait for the dew light to go off before playing or recording. This type of device helps to prevent damage to the videotape. See DEW INDICATOR.

Hyperband. Those frequencies which are located immediately above the superband channels. At 300 to 402 MHz, the hyperband range covers channels W + 1 to W + 17 and is reserved for such special services as aero navigation, the Coast Guard, etc. Other ranges include subband, MIDBAND and SUPERBAND CABLE TV. See BANDWIDTH, FREQUENCY.

I

IBO. See INTEGRATED BROADCAST OPERATION.

"I" Signal. The "in phase" signal generated by a video camera's circuitry. Today's cameras produce "I" and "Q" (quatrature) signals, whose combination and relationship are responsible for all colors. If all or parts of these signals are not in balance or absent, the chrominance will be distorted. The I signal can be checked on a COLOR VECTORSCOPE. See "Q" SIGNAL.

IARC. See INTERNAL ANTI-REFLECTIVE COATING.

IC. See INTEGRATED CIRCUIT.

Iconoscope. An early camera tube invented by VLADIMIR ZWORYKIN in 1923 which led to the first video camera. The iconoscope is no longer in use. This CAMERA TUBE, like others to follow, was capable of converting light values into electronic signals by having a beam of high-velocity electrons scan a photosensitive mosaic capable of retaining a pattern of electrical charges. See CAMERA TUBE, IMAGE DISSECTOR, ORTHICON.

IDTV (Improved Definition TV). A technique that employs digital circuitry to deliver a sharper and clearer picture by manipulating the horizontal lines that appear on the screen. Normally, two sets of lines, even- and odd-numbered, appear on the screen—the even-numbered first, then the odd-numbered. This process is called interlace. The IDTV, or non-interlace, approach shows all the lines at once, resulting in a denser image. This is accomplished by digital memory, which stores the first 262 1/2 lines (one-half of a frame) in memory and then traces the full 525 lines (National Television Standards Committee) of the frame in one sweep. More effective on larger screens where the horizontal lines are usually more noticeable, this method seems to result in a clearer, sharper image for the viewer. IDTV operates within the NTSC, or present, broadcasting standard, thereby not requiring the viewer to buy new equipment. In this way, it has an advantage over the proposed HDTV (high definition television) sys-

IF (Intermediate Frequency)

tem, which transmits more lines and makes present TV receivers obsolete. Some manufacturers have applied IDTV technology to their 27" and 31" television monitor/receivers and projection TVs. These models, which display 480 lines of vertical resolution, are often equipped with other advanced features, such as DIGITAL VIDEO NOISE REDUCTION, a field comb filter, dual tuner and surround sound. IDTV is considered by some to be an interim step between conventional TV and HDTV. However, the Federal Communications Committee's decision early in 1990 to opt for a true HDTV system as a national standard dealt a serious blow to the developers of IDTV. See HIGH DEFINITION TV, INTERLACE, NON-INTERLACED DISPLAY, VERTICAL RESOLUTION.

IF (Intermediate Frequency). A frequency containing a signal wave that has been shifted locally as a halfway stage to reception or transmission. For example, when IF signals pass through an IF AMPLIFIER, they pick up from between 50 to 70 dB (decibels) of amplification and sustain extensive frequency selection.

IF Amplifier. That component of a TV receiver which strengthens the RF (radio frequency) carrier signal fed to it. The signal is a combination of the audio and video signals which the intermediate frequency amplifier increases before it is sent to a video detector for separation. See RF, TV RECEIVER.

Image. A video picture created by electron beams that scan the cathode ray tube. The beams turn on and off to illuminate the groups of phosphorous dots, called pixels (picture elements), with the appropriate intensity required to produce a suitable picture. The image is usually judged by its luminance (brightness) values. Image quality depends on a number of components and factors. For example, the CAMERA TUBE used in the production of the image can determine its luminance, sharpness and resistance to IMAGE RETENTION. HORIZONTAL and VERTICAL RESOLUTION can also affect the detail in the TV image. See HORIZONTAL LINES.

Image Archiving. Refers to electronically storing video images, frames or stills for future reference or use. Such units as STILL VIDEO RECORDER/PLAYERS, which offer high resolution (500 lines) and utilize a 2-inch floppy disk for random access, have been designed for such a purpose.

Image Burn. A temporary loss of picture in an isolated area of the TV IMAGE caused by bright lights or reflection; also, possible permanent damage to the video camera tube as a result of excessive exposure to bright lights such as the sun or studio lighting units. All cameras using vacuum-type tubes are susceptible to image burn. See CAMERA TUBE.

Image Detail. The amount of information and sharpness displayed on a TV screen. With VCRs, image detail can be measured by its response to a 2-MHz signal. Some machines may lose as much as 4 dB (decibels) of the video signal while other models give up as little as 0.13 dB at the 2-MHz frequency. Often, image detail is sub-

Image Enhancer

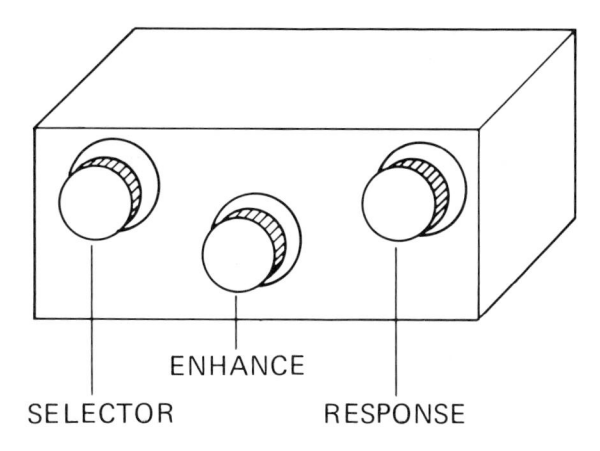

Figure 22. Image enhancer.

jective, although manufacturers keep coming up with various techniques and electronic circuitry, including filters and digital video noise reduction systems, to improve it.

Image Detailer. See IMAGE ENHANCER.

Image Dissector. A vacuum-type image pickup tube developed in the late 1920s by PHILO FARNSWORTH. The image dissector tube operates by having the entire image sweep past an aperture in a series of interlaced lines instead of using a beam to scan the image. It closely resembled the ICONOSCOPE invented during the same period by VLADIMIR ZWORYKIN. Both PICTURE TUBES were able to convert light values into electronic signals. See CAMERA TUBE, ORTHICON.

Image Enhancement. Any improvement of detail, sharpness, color accuracy or reduction of video noise in a TV screen picture. Several techniques have been developed to enhance the screen image, or definition, including HQ circuitry, digital effects, special comb filters, line-doubling and non-interlaced display. Some processes involve increasing the number of horizontal lines. Others use special electronic circuitry to reduce video noise.

Image Enhancer. A signal processing device designed to restore some of the detail which is lost when duplicating a video image. An enhancer operates by increasing or cutting the highest video frequencies. Since this can also increase noise, the unit often has a Response control to restrict this interference. Like most processors, the enhancer works only with a VIDEO SIGNAL, not with RF (radio frequency) SIGNALS such as those broadcast directly to the TV receiver. The unit can be connected between two VCRs as when making a copy of a tape or between a video camera and a VCR. Some enhancers have a selector dial which permits a comparison between the enhanced and unenhanced image. However, the final results of most enhancers are often subjective.

Image Jitter

Some VCRs feature a built-in enhancer, a control dial or knob simply called Picture.

Image Jitter. See JITTER.

Image Lag. See IMAGE RETENTION.

Image Mix. A digital camcorder function that creates a variety of effects by joining live recorded images with a stored still picture. The combinations made by the image mix feature produce a series of special effects, such as simulated dissolves, split-screens or picture-in-picture.

Image Orthicon. See ORTHICON.

Image Pickup Device. See IMAGE SENSOR.

Image Pickup Tube. See CAMERA TUBE.

Image Processor. Special circuitry, usually found in higher-priced TV sets or TV monitor/receivers, designed retain natural colors while continuously adjusting the light and dark values of each scene. The image-processing circuits constantly modify the dynamic range of the cathode ray tube to correspond to the light-to-dark values of consecutive scenes. The image processor is also a general term for one of the many "black boxes" or devices available that can be connected to a VCR, such as an IMAGE STABILIZER, IMAGE ENHANCER and FADER. Some of the more recent image or video processors can accommodate both S-VHS and Hi8 video signals, offer several audio fade controls and provide a power distribution amplifier that permits the user to duplicate several copies of a videotape simultaneously with little or no loss of signal strength. See SIGNAL PROCESSOR.

Image Retention. Unwanted lagging or trailing of previous images or pictures. This effect is usually caused by rapid movements of the subject, quick panning of the video camera or the quality of the CAMERA TUBE. The anomaly often occurs more frequently during low lighting sequences. The term is often mistakenly confused with burn or IMAGE BURN. Image retention is also known as image lag or cometing.

Image Reversal. A process of reversing black-and-white images (dark-to-light, light-to-dark) and inverting color images. The conversion of black-and-white images is called luminance reversal while that pertaining to color requires both luminance and chrominance reversal. Some high-priced video cameras are capable of performing both of these reversal processes by way of a negative/positive image switch.

Image Scanner. A device, supplied with some VCRs, designed to "read" text or a drawing and then superim-

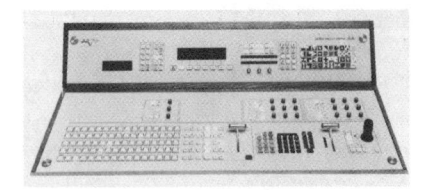

Figure 23. An image processor, designed for a professional studio, that can be used for digital processing. (Courtesy The Grass Valley Group, Inc.)

Image Stabilizer

pose it over a video image. The scanned image can be copied in different colors and sizes.

Image Sensor. The image pickup method used in a video camera to capture the picture. The design may be a charge coupled device (CCD) or metal oxide conductor (MOS), both of which have virtually replaced the conventional image pickup tube. The metal oxide semiconductor chip is a 3/4-inch square solid state sensor containing numerous rows of light-sensing cells upon which the camera lens projects its image. The black-and-white and color patterns of the image activate electrical impulses within each cell which are then generated into a video signal. Although more costly than standard camera tubes, image sensors are not only smaller and lighter, but they minimize IMAGE BURN and IMAGE RETENTION. Some professional camcorder image sensors are composed of two CCDs—one for brightness and the other for color. The two signals are processed separately to produce better resolution and color. Still other more costly professional/industrial components feature an image sensor that utilizes three CCDs—one for each of the primary colors. See CAMERA TUBE, CHARGE COUPLED DEVICE, MOS.

Image Sharpness. In video, a subjective measurement of the amount of detail in a TV screen image. Image sharpness depends upon such factors as RESOLUTION (picture detail as measured by the number of HORIZONTAL LINES), CONTRAST (the relationship between the white and dark portions of an image) and the amount of VIDEO NOISE (unwanted signal interference) present in the screen image. Many TV sets, VCRS, monitors and monitor/receivers have a SHARPNESS CONTROL feature designed to reduce noise and minimize contrast. The control, however, does not basically alter the resolution or number of horizontal lines in the image. See SHARPNESS CONTROL.

Image Shift. A digital television or VCR feature that allows the viewer to exchange the main image and the inset image of the picture-in-picture function. Part of the digital effects process, shift permits swapping pictures while retaining the audio with the main screen image.

Image Smear. See IMAGE BURN.

Image Stabilization. The elimination of "jitter" and vertical rolling of the screen picture. Some Super-VHS camcorders have developed techniques to improve image stabilization. They include electronic and mechanical improvements. Additional electronic correction circuitry ensures stable pictures at all camera speeds while changes in the way tape is transported across the video head drum rectify some types of instability. See AUTOMATIC IMAGE STABILIZATION, JITTER.

Image Stabilizer. A device designed to override some anti-piracy signals of prerecorded tapes and restore the stability of the signal to the TV image. The electronic coding, deliberately placed on tapes to prevent their being copied, changes and weakens the vertical sync signal so that when it is fed

Image Swap

into a VCR, the machine cannot usually lock into it. The stabilizer, therefore, either re-forms the vertical sync pulses or corrects any altered horizontal sync pulses. Some models completely strip the vertical blanking signal (which contains the anti-copying white pulses) and then proceed to rebuild the sync signals to broadcast standards. Without a stabilizer, the picture often rolls and breaks up. However, most late model VCRs have advanced circuitry that defeats this signal without the use of this device. Some image stabilizers have different features, such as loop-through outputs, lock control signal lights and switchable inputs. Other models, such as the copy-protection removers, or digital video stabilizers, use digital filters in their process. See ANTI-PIRACY SIGNAL, COPY-PROTECTION REMOVER, VERTICAL SYNC PULSE.

Image Swap. A feature, found on TV receivers equipped with PIP (picture-in-picture) and split screen capability, that allows the viewer to exchange left and right pictures or main and inset images.

Image Transceiver. An accessory unit, used in conjunction with still video cameras, that permits color photos to be transmitted by way of ordinary telephone lines. Some image transceivers accept a still video floppy disk (VF), which is used in still cameras. However, the loss of image quality produced by this procedure limits its usefulness for professional/industrial purposes despite its several conveniences. See HiVF, STILL VIDEO, STILL VIDEO FLOPPY DISK.

Image Translator. A conversion kit or VCR designed to modify some VHS recorders, using the standard NTSC, to accept European PAL-recorded 625-line color tapes. Developed by INSTANT REPLAY, the videocassette recorder plays back the North American conventional 525-line color tapes as well as PAL and SECAM tapes. Instant Replay claims that their VHS VCR can play and record in 16 world standards.

Imagic. A software company that produced video game cartridges for both Atari and Mattel systems. Founded by four former employees of these two large manufacturers, Imagic was the first company to make games for more than one format.

Impedance. The resistance characteristics of any electronic circuit. The connection of one electronic component to another requires matching their impedances. Video games, VCRs and other units must have the same impedance level as the TV set. Impedance is measured in ohms—a standard unit of electronic resistance. Video involves only two ratings—75 and 300 ohms. An impedance adapter, such as a BALUN or small transformer, is used to convert 300 ohms to 75. Audio units such as speakers, when working as a set, should have matched impedances—usually 4 or 8 ohms. High and low impedances are expressed as hi-Z and low-Z.

Impedance Adapter. A device which matches up the output of one unit with the input of another. For example, to connect the audio from a TV set (one that has an earphone jack, for instance) to the "aux" input of a stereo system, one should purchase the appropriate impedance adapter. This will depend upon the input of the stereo amplifier. Although the connection will work without the adapter, the sound will appear distorted. The impedance adapter is also known as matching transformer.

Improved Definition TV. See IDTV.

In-Camera Recording System. A unit in which the video camera houses the recording components so that a separate VCR/VTR becomes unnecessary. In home videos, some of the various formats include Sony's Betamax and 8mm models and the VHS-C which uses a mini-cassette that is compatible with a standard VHS recorder. In 1982 the Committee on Video Recording and Reproduction Technology, a temporary body of the Society of Motion Picture and Television Engineers, was unable to decide on an industry standard for an in-camera recording system, or camcorder, as the unit is presently called. The marketplace will eventually determine which format or formats will dominate the field. See CAMCORDER.

Index Counter. The VCR feature with three or four digits which rotate with the motion of the tape. The odometer-type mechanical counter, now all but defunct, helped to locate different portions of tape once the numbers were noted down. The numbers are arbitrary, representing neither inches nor time, simply registering rotation. A counter on one machine, in fact, may not match that of another. VCR owners whose machines had mechanical index counters often had to compose their own reference charts, with a list of numbers and matching times; i.e., every five digits equaled 15 or 30 minutes. Today, virtually all VCRs have electronic counters or read-outs, many of which measure real time.

Index Mark. See CUE MARK.

Index Search. A VCR feature that helps the viewer find the beginning of a program automatically and quickly. With the index search method, also known as VISS or VHS Index Search System, an electronic mark is automatically placed on the videotape each time the Record mode is pressed. Later, the viewer just activates the index search to find the beginning of each program until he or she finds the right one. Some machines can mark up to 19 indexes. Index search, which differs from the more sophisticated INTRO SEARCH, can be performed manually as well. However, the VISS system performs indexing only during recording whereas VASS (VHS Address Search System) permits marking scenes during playback.

Indirect Scanning. A technique that utilizes a small beam of light that searches across a subject and then reflects its results to an array of photo tubes. Indirect scanning was employed in early television which was heavily dependent on mechanics. The process is currently used in flying-spot scanning of films. See FLYING SPOT SCANNER.

Industrial VCR

Industrial VCR. A videocassette recorder that is externally similar to a home VCR, but is different in many ways. An industrial machine, whether a 3/4-inch U-Matic model or a 1/2-inch Beta or VHS deck, usually has a rectangular multi-pin jack for use with a TV MONITOR. The machine often plays at only one speed (the fastest) to retain high image resolution. It has special features such as random access and AUTOMATIC TRANSITION EDITING. Many of these features, with modifications, eventually find their way into home VCRs. The costlier industrial machines are more sturdily built for heavier duty and are used more in business and institutional environments than in homes. See VIDEOTAPE FORMAT, VTR.

Infrared Remote Control. On video cameras, a feature that starts and stops the camera recording and controls the zoom lens. This allows the camera user to set up his equipment and operate it from a limited distance without his direct presence affecting the subject. This is a useful feature in capturing shots of animals in the field or children at home or in a studio, subjects that sometimes appear shy before a camera. See REMOTE CONTROL.

Infrared Sensor. The element on TV receivers and videocassette recorders and players that permits wireless remote control to operate various functions from a limited distance. This feature is sometimes called remote control receiver.

Infrared Transmitter. See REMOTE CONTROL, REMOTE CONTROL PANEL.

In-Line Amplifier. See AMPLIFIER.

In-Line Switching. The special circuitry in VCRs that automatically disconnects the tuner signal during the copying process. This entails duplicating by means of audio/video cables hooked up to the VCR.

Input. A jack, receptacle or terminal, or combination of these, designed to accept an electrical input signal into a videocassette recorder or similar unit. Most pieces of equipment have audio, video and related auxiliary inputs.

Input Selector Switch. A VCR feature that permits the user to use the TUNER mode for normal operations of viewing televised programs or prerecorded tapes or the Line mode for viewing a signal from another unit connected to the Audio/Video inputs of the VCR. The input selector switch, sometimes described as a tuner/line switch, is utilized for electronic editing between a VCR or camcorder and a second VCR.

Insert. A general term used in special effects in which a secondary signal in introduced into an already existing, primary image. The insert can be made to appear in any portion of the TV screen. See CORNER INSERT, DISSOLVE, KEY, SPECIAL EFFECTS GENERATOR.

Insert Editing. A technique in video which permits the electronic editing of audio or video or both into any part of a videotape. Its counterpart, ASSEMBLE EDITING, replaces both the old audio and video tracks with new ones. In addition, insert editing preserves the original control track as new material is added. See CRASH EDITING, EDITING.

Insertion Loss. Signal loss caused by such items as cables, connectors, baluns and splitters that connect to or pass through the RF switcher. Insertion loss is expressed in db (decibels). The lower the dB number, the better the signal. Insertion loss is usually corrected by adding an amplifier.

Instant Record. See ONE TOUCH RECORDING.

Instant Replay. A now defunct videocassette magazine founded by Chuck Azar containing entertainment and information that was sold through retail video stores and subscriptions. The video magazine, which first went on sale in May of 1978, consisted of great moments in U.S. and foreign television, new analyses and segments of VIDEO ART, home tapes, concerts, etc.

Instant Replay. Refers to the immediate playback of a recorded event—a technique widely used during sports broadcasts. The term was coined by Ampex in the 1960s.

Instant Review. A feature found on some video cameras which permits the operator to inspect a portion of the previously recorded material. For example, some cameras with instant review automatically rewind the tape several inches while others have a review button on the handgrip. In the latter case, the last few moments of the recording are played back in the viewfinder, first in reverse, then in forward; finally, the VCR is returned to Record/Pause position. Many cameras without this feature permit reviewing has has been shot, but these models use different approaches. In some, the Camera/VCR switch must first be set. This shifts the recorder from Record to Play Mode. The Search function is then utilized (in reverse) to rewind the tape. The Play/Pause control is then pressed to activate tape playback through the electronic viewfinder.

Instant Viewing. The ability to produce on a TV screen any "frame" of a videodisc by using the RANDOM ACCESS feature of an LV videodisc player. Since the LaserVision system provides a contactless or laser beam stylus, one frame or revolution can be played continuously, offering a steady FREEZE FRAME. A typical disc contains as many as 54,000 pictures on each side, with each image having its own designated number. By pressing the numbered keypad, the viewer can locate any one of these frames or pictures. Instant viewing is possible, however, only in the standard play mode (CAV) or 30-minutes-per-side and not in long-play (CLV) or 60-minutes-per-side. See FRAME, FREEZE FRAME, LV VIDEODISC SYSTEM, RANDOM ACCESS, VIDEODISC SPEED.

Integrated Broadcast Operation. Refers to a fully automated, digital television station. As a result of rapid advances in digital technology and

Integrated Circuit (IC)

electronics in general, major manufacturers of professional/industrial audio and video equipment, including Sony and Chyron, have developed an array of sophisticated units. These include graphics and effects devices, editors and complete digital multi-effects systems.

Integrated Circuit (IC). In video, an electronic component that contains several active or passive devices designed to operate all or part of a circuit function. One silicon chip, embedded with interconnecting electrical conductors, may contain diodes, transistors, resisters and photocells. This tiny electronic component is responsible for the miniaturization of various units. For example, most of the electronic make-up of a video camera consists of only a few square inches; the size of the camera is limited by the size of its other components. Since the mid-1970s, manufacturers of VCRs and TV receivers have applied integrated circuits to many different functions and components, including color synchronization, automatic color correction, flesh-tone correction, vertical interval reference adjustments, demodulators and bandpass amplifiers.

Integrated Receiver/Descrambler. An enhanced satellite receiver that applies advanced technology to improve video and audio quality as well as to decode scrambled channels. IRDs, with built-in VideoCipher II circuitry, are capable of unscrambling a high percentage of the coded channels that TV satellite system owners receive. About 57 of the more than 200 available satellite channels are scrambled. Subscribers who want to receive these channels can register for those they prefer. The IRD then automatically opens the signals that have been purchased. Some of these IRD units include additional features, such as menu-based on-screen controls, digital sound, programmable tuners and video noise reduction. See SATELLITE RECEIVER, SATELLITE TV.

Interactive Cable. Cable programming that allows the subscriber-viewer to participate in the televised event. For example, early in 1990 a Canadian cable company offered its subscribers an opportunity to interact with a hockey game. A subscriber, simulating the role of TV director, was able to select any one of three different camera angles. He or she could activate instant replay by pressing a button on the remote control panel. The replay would then appear as an inset and display a backup of the action for ten seconds. In addition, interactive cable holds the promise of other features, including closed captions for the hard-of-hearing, the receiving of videotex information, participation in game shows, shopping at home and the capability of interviewing guests on TV shows. Recent advances in interactive cable eliminate the need for special accessories or adapters by making use of the remote control functions. This is possible because the necessary information is transmitted in the broadcast signal.

Interactive Video. Television programming designed so that the viewer can participate in or influence its method of presentation. One early attempt at interactive video included a children's show in which youngsters "joined" the program by drawing on a vinyl sheet placed on the TV screen. Another program was the "art lesson," which had viewers draw along with a host-artist. Today's popular video games may be considered interactive TV. The Nintendo, Sega and Socrates systems permit the user to affect what occurs on screen. Another sophisticated use is QUBE, in which viewers vote, offer opinions, judge talent shows, etc. Interactive video is being used successfully in schools and private training programs to supplement conventional methods of learning. IV can (1) be cost effective, (2) outperform the individual general instructor by presenting a series of experts in one particular field and (3) offer a larger selection of subject matter. See CHANNEL ONE, INTERACTIVE CABLE, INTERACTIVE VIDEODISC.

Interactive Videodisc. A prerecorded LV disc divided into CHAPTERS and/or FRAMES for easy reference. The disc utilizes two audio channels and permits viewer participation in basic games. The player can locate chapters (segments of a program such as musical numbers) for perusal. Interactive videodiscs allow the user to rapidly locate a single frame (one picture such as a page of a book or catalogue or a painting or photograph). The two audio channels can present different sound tracks such as a foreign language and a translation for home study or alternate narration. "How to Watch Football," produced by NFL Films, was the first interactive videodisc presented for home viewing. It contained 15 chapters, individual frame material, diagrams and a fast-paced football game. Another early entry into the interactive disc field was THE FIRST NATIONAL KIDISC (1981) with its unique programming for children. The interactive disc differs from the LINEAR program which is designed to play from the beginning to its finish (such as a movie).

Intercarrier Interference. Refers to the buzzing noise that is sometimes heard when white titles suddenly appear on the TV screen. Intercarrier interference results when TV cable company amplifiers are incorrectly adjusted. Channels 2 to 13 (the low band channels) tend to be affected more than others.

Interface. A much-used term referring to a circuit which forms the proper hook-up between two components. Originally a computer term applying to the connecting of a computer with another unit, it now refers to the interconnection of any two otherwise incompatible units.

Interference. The appearance of unwanted disturbances joining the signal path that may cause performance degradation, complete malfunction or some other undesired response. Interference may be the result of natural or man-made electromagnetic disturbance, extraneous signals, emissions, etc. Electronic interference is usually eliminated by using an ISOLATOR. Other devices employed to control interference include FILTERS, INTERFERENCE SUPPRESSERS and NOISE REDUCTION SYSTEMS.

Interference Suppresser. A device designed to reduce or eliminate distortion of a TV picture usually caused by household appliances. A suppresser is installed into a wall socket and has apertures for three-prong or two-prong plugs. See INTERFERENCE.

Interlace. The capability of a television receiver to produce exact horizontal lines that are equidistant from each other. The more accurate the interlace, the better the vertical detail of the TV set. A perfect interlace may be expressed in video terms as a 50/50 ratio, indicating that all scan lines have the proper spacing between them. A 60/40 ratio denotes the degree of imperfection. Interlace occurs when one separate set of even-numbered image lines follows another set of odd-numbered lines on screen to make up a total of 525 lines—the conventional NTSC (National Television Standards Committee) standard for U.S. broadcasting. Interlace that was random once plagued low-priced video cameras by placing odd-numbered lines indiscrimately on the screen. Crystal-controlled 2:1 interlace eventually replaced random interlace to give cameras a better scanning technique. Interlace problems of a camera can be discerned by studying the diagonal lines of a resolution chart. Some systems, to improve the image, have introduced a non-interlaced technique. See HORIZONTAL LINES, NON-INTERLACED DISPLAY, VERTICAL RESOLUTION.

Interlaced Scanning. In video, a method of electronic scanning in which the odd- and even-numbered lines of a television picture are transmitted consecutively as two separate fields. The two fields are superimposed to create one frame, which makes up a complete image or picture on the screen. Interlaced scanning is sometimes referred to as line scanning. See FIELD, FRAME.

Intermodulation. Distortion and unwanted signal energy caused by signals being transmitted in other channels. Intermodulation is one of several forms of DEGRADATION that affect NTSC picture quality.

International Satellite. See SATELLITE FOCUS.

International Tape Association. See INTERNATIONAL TAPE/DISC ASSOCIATION.

International Tape/Disc Association. A professional, worldwide body whose members include many of the major companies in the audio and video field. Holding annual seminars since about 1971, the group is concerned with and influential in such areas as audio, video, home video systems and consumers' buying habits. The association has presented its Golden Videodisc Award to discs that topped the one million mark in sales.

International Telecommunications Union (ITU). An organization which designates locations for satellites. With the increase of communications satellites, some order was required to prevent them from interfering with each other or, worse, colliding. The positions extend from 150 degrees to 70 degrees west longitude. See GEO-

STATIONARY, SATELLITE, SATELLITE TV.

Internal Anti-Reflective Coating (IARC). A relatively recent process developed for use with projection TV systems. It is designed to increase brightness, resolution and contrast, three acknowledged problem areas inherent to conventional projection TV.

Interval Timer. A built-in video camera feature designed for time-lapse or animation effects. See INTERVALOMETER.

Intervalometer. A special timing device attached to a camera for automatic single-frame exposure in time lapse photography. It releases the camera shutter at predetermined intervals to condense large portions of time into smaller ones. For example, hours of cloud movements can be recorded to play back with a running time of a few seconds, thus dramatically speeding up the motion. With limitations, the accessory can be used in TIME LAPSE VIDEO when connected to a home video camera. Keeping a portable VCR in Pause mode for extended lengths of time may cause damage to the tape or video heads. When this function is built into a camera, it is usually referred to as an interval timer.

Intro Search. A VCR feature that automatically locates all indexed material on a prerecorded videotape and plays back the opening portion of each section for a few seconds in fast motion. Intro Search helps a viewer quickly find a selection on a tape that he or she is looking for. This feature, sometimes referred to as Intro Scan, differs from INDEX SEARCH, which simply locates the beginning of programs that have been automatically marked electronically each time the Record button is pressed.

Inverted Image. A condition occurring when a small TV set is employed in a budget PROJECTION TV system and no mirror is used. The special lens which projects the image onto the screen also creates an inverted image. This can be corrected in one of two ways. A serviceperson can reverse the vertical connection wires on the yoke of TV picture tube, or he or she can add a picture tube inversion switch which permits either a normal or inverted picture for projection use.

InView. A quarterly magazine aimed at keeping its professional subscribers informed about the latest products and applications in the video arena. InView, which began operations in the spring of 1989, focuses its main articles on both people in the industry and topics related to video. Its departments cover such areas as audio/video, graphics, cable and new products. Also included in each issue are a useful trade show calendar and an informative but short column that answers readers' technical questions and problems.

IRD. See INTEGRATED RECEIVER/DESCRAMBLER.

IRE (Institute of Radio Engineers). A system of measurement used in audio and video. Besides representing the professional organization, IRE is a term utilized for measuring such items as frequencies and video lines. For example, white balance (the amount of color that can be viewed on a neutral object) is measured in IRE. The lower the IRE number, the better the white balance. The figure of 0 IRE indicates perfect white balance, a goal all TV manufacturers aim at but seldom attain. IRE plays an important role in waveform monitors, or oscilloscopes, professional instruments which feature switchable on/off IRE filters.

Iron Oxide. A particle combined with other metallic oxides and ferrite to form a coating on magnetic tape. See COATING, FERRITE.

Iris Control Button. A feature that closes down the iris or aperture of the lens to protect the sensitive video camera tube when the camera is not in operation. Camera tubes that are exposed to overly bright light or sun develop a "burn" which may become permanent. See CAMERA TUBE, IMAGE BURN.

Isolation. The absence of interference in video components. Isolation is measured in decibels (dB). The higher the number, the better the isolation. For example, a video SWITCHER should provide approximately 40 dB of isolation. If the number falls lower than this, interference as well as image degradation are likely to occur. In devices like switchers, isolation refers to the ability of the unit to block signal leakage and interference from disturbing the different signals. Again, the higher the decibel number the better the isolation.

Isolation Transformer. See ISOLATOR.

Isolator. A device designed to prevent electrical interference between units in a video system. The isolator accomplishes this by filtering the line current. The isolator eliminates such problems as hum, crosstalk and voltage differences. Some units permit hooking up any three components to its AC outlets. More sophisticated models, called isolation transformers, offer additional features and uses, such as ensuring signal transmission with more than 120 dB attenuation of interference and applying it to ultra-wide bandwidths and broadcast or remote TV lines.

J

Jack. The opening or receptacle on a VCR, TV MONITOR or other component that accepts a male plug. Features like audio-in and video-out on VCRs and other equipment are also referred to as audio and video jacks while the cables with matching connectors are called audio and video plugs. The terms "jacks" and "plugs" are often used interchangeably.

Jitter. In video, picture distortion resulting from signal instability, synchronization loss caused by equipment mechanism or electrical malfunction, or other problems. Some VCRs have minimized the problem of jitter, or image jitter, jiggle or jiggling, as it is sometimes referred to, by employing several mechanical enhancements, including a twin-projection cylinder, a special impedance roller and a unique digital servo mechanism. These improvements, working in unison, provide better image stability. Other VCR manufacturers have introduced time base correction into their units to stabilize images that have been copied or are played back on other machines. See TIME BASE CORRECTOR, TIME BASE INSTABILITY.

Jog/Shuttle Control. A wheel-and-dial combination, usually found on the front of more costly VCRs, designed to move the video picture one frame at a time and assist in a manual bi-directional search for a particular segment. Used chiefly for editing, the jog/shuttle control, or jog dial shuttle ring as it is sometimes called, allows the user to turn the outer ring for moving the tape from slow motion to search speeds. The inner jog wheel or dial moves the tape frame by frame, helping the user to choose the exact edit point. Some VCR models provide a single control to accomplish both tasks. The feature on these machines does not work as fast or accurately as that of the two-part design. Some videodisc players have a similar jog dial shuttle ring. Located on the remote control, the ring allows the user to view a program on the disc forward or backward and at a variety of speeds, from one frame at a time to 10 times standard play speed.

Joy Stick. In a video game, a lever which controls the movements of the game action on the screen. In relation to a VTR, a joy stick regulates a special effect on the screen or tape movement during editing.

Junction Box

Junction Box. A connector accessory with two wired female connectors and designed so that two male plugs can be joined together. The junction box is most often used by professionals.

JVC. A Japanese electronics firm (Victor Company of Japan) which introduced the first VHS system portable videocassette recorder in 1978. The HR 4100 was a one-speed VCR which could record up to three hours on a battery charge as well as on AC/DC house current. With its built-in RF adapter it could also record and play back through a conventional TV receiver. Including battery pack, it weighed 21 pounds.

K

Kelvin. A term used in video lighting to measure COLOR TEMPERATURE, usually with a degree symbol. For example, an ordinary household light bulb of 100 watts is rated at 2900°k (degrees kelvin).

Key. A feature on some video SWITCHERS which permits creating electronically the illusion of placing one image over another without getting rid of the second image. There are four basic types of keys: external key, chroma key, matte key and self key, each capable of providing a different special effect. For example, a chroma key substitutes a particular color with an image from a different source. A matte key, on the other hand, consists of only one color for keying. See VIDEO KEYER.

Key Light. In video, the main light source of a scene. The key light is usually located near the video camera and above the subject to minimize shadows. This light works best in conjunction with fill lights, etc. The key light can even work with available light, the latter acting as the fill. See LIGHTING.

Key Pad. That portion of a remote control designed to operate specific functions of a VCR, TV set or videodisc player. See NUMERIC KEY PAD.

Keystoning. An effect which results in a narrower or wider projected image at the top of the screen than at the bottom. This is caused when the slide or movie projector is not properly aligned with the screen. In those movie houses where keystoning occurs, the sides of the screen are usually darkened or masked to conceal the effect. Keystoning is also a problem in projection TV systems using mirrors.

KHz (Kilohertz). A unit of measure equal to one thousand cycles per second; or two KHz is equal to 2,000 Hertz or 2,000 cycles per second. See HERTZ.

Kidvid. Refers to television and video aimed at children. TV programming for children, in particular, is big business and has invoked the concern of such diverse groups as educators, parents' organizations and the U.S. Senate Commerce Committee. All have

Kinescope

voiced interest in the number and length of commercials during one children's program, program-length commercials and broadcast licensees' responsibilities in presenting programs that meet the educational and informational needs of preschool and school-age children. The issue of regulating TV programs for children has caused much controversy. President Reagan in 1988 vetoed such a bill passed by the Senate. The National Association of Broadcasters has lobbied against further attempts to restrict the number of commercials that can be inserted during one program. The Justice Department has entered the fray, opposing some restrictions. The debate continues.

Kinescope. An obsolete method of recording television programs. A movie camera would be employed to photograph the images of a TV MONITOR for archiving or for future rebroadcasting. Kinescope was widely used in the early years of television before the introduction of videotape.

Kloss, Henry. A pioneer in the field of PROJECTION TV systems. In 1967 he founded Advent, a company which was one of the earliest to develop large-screen television; he designed projection TV sets unique for their time; he introduced the first consumer-model projection TV in 1973; he started his own company in 1977, introducing the Kloss NovaBeam TV in 1979. While with Advent, he introduced to projection TV the three-tube design—one tube for each of the three primary colors. It produced the largest and brightest image up to that time. His NovaBeam TV was a large-screen system, over six feet in diameter with high-quality sound, remote control, random access tuning and accommodations for videodisc, VCR CAVE TV, etc. Henry Kloss had previously been affiliated with audio companies KLH and AR. See THREE-GUN PROJECTION TV.

Kroma Glass. Colored mirrored glass that reflects and transmits light. It is used in video and photography for special effects.

L

Lag. The temporary retaining of the electrically charged image of a television camera tube. See IMAGE RETENTION.

Lap Dissolve. A film and video transition and special effect in which one scene is faded out while the next scene is faded in, both occurring simultaneously.

Large Screen TV. See PROJECTION TV.

Laser. An instrument that transforms various incoherent light vibrations into a narrow and intense light beam. The word "laser" comes from "Light Amplification by Simulated Emission of Radiation." Laser technology has been successfully applied to such video equipment as videodisc players and projection systems.

Laser-Based Projection System. An experimental projection TV system that features a low dispersion of the beam so that focus is not greatly affected by the angle of the screen. First demonstrated in 1988 at a National Association of Broadcasters convention, the laser-based system, despite some interesting advantages over conventional systems, has suffered several setbacks—chiefly financial. See LCD PROJECTION TV, LIGHT VALVE PROJECTION SYSTEM, PROJECTION TV.

Laser Beam. That part of an optical disc player or system that carries the video signal without making physical contact with the disc. Using only a beam of light, the laser beam stores more video information on the disc than can be packed onto videotape. This results in a clearer, more detailed screen image. Since the laser beam head or arm never touches the laserdisc, the disc is virtually free from deterioration. In addition, the sound from a videodisc player equals that produced by a compact disc system.

Laser-Lock. With an LV videodisc player, a malfunction in which the laser arbitrarily locks into a single frame. This may sometimes be caused by fingerprints on the disc.

Laser Optical Media. Any hard plastic disc of information that has been recorded and can be read by a laser

Laser Optical System

light beam. Discs include a variety of types: the three- and five-inch CD (compact disc), CD-I (CD-interactive), five-inch CD-ROM (CD-read-only-memory), five-inch CD-V (compact disc-video), five-inch CD-Write Once, five-inch DVI (digital video interactive) and eight- and 12-inch LV (laser videodisc).

Laser Optical System. See LV VIDEODISC SYSTEM.

Laserdisc. In video, prerecorded software that resembles a long-playing record and is used in conjunction with a videodisc player. Unlike videocassettes that can both record and play back, laserdiscs can only play back prerecorded programs. Laserdiscs are read by a laser beam that never makes physical contact with the disc, thereby preserving the disc from wear and tear almost indefinitely. CAV laserdiscs, which offer special effects, have a maximum playing time of 30 minutes while CLV discs contain 60 minutes of programming. LV discs come in two sizes, 8 and 12 inches, both utilizing analog video and analog/digital audio. See ANALOG SIGNAL PROCESSING, CAV, CLV, DIGITAL SIGNAL PROCESSING.

Laserdisc Player. See CLV, DOUBLE-SIDE VIDEODISC PLAYER, LASERDISC, LV VIDEODISC HISTORY, VIDEODISC PLAYER.

Last Memory Function. A feature, sometimes found on videodisc or CD Video players, that resumes playing a disc at the same point at which the machine was shut off. This function is sometimes listed as last channel memory.

Latent Image. In video, an image stored in the charged capacitance in an iconoscope. In photography, the latent image remains on the exposed film, invisible to the naked eye, until it is processed or developed.

Lavalier Microphone. A miniature condenser-type mic designed to pin or clip on the clothing of the subject being interviewed and videotaped. The lavalier microphone is OMNIDIRECTIONAL. This unobstructed mic is expensive, provides good fidelity and rejects echoes and other incidental surrounding noises. See MICROPHONE.

LCD Counter. In video, a feature employing liquid crystal display for digital read-outs on more recent VCR timers. LCD provides dark digits against a light background, in contrast to LED, which features red or green digits against a dark background.

LCD Display. A VCR remote control panel using a liquid crystal display to program in recording information which is then transmitted to the videocassette recorder. Normally, with recent VCR models, such information as program number, day of week, start time, stop time, channel and recording speed has to be programmed on screen using the TV set. Remote controls with LCD display bypass this step; programming can be done away from the TV set without the set having to be turned on, and the data sent to the recorder by way of a transmit button on the remote panel. See ON-

SCREEN PROGRAMMING, TRANSMIT BUTTON.

LCD Projection TV. A large-screen television system that uses liquid display panels from one to three inches thick. Retailing for several thousand dollars, or about twice as much as a conventional cathode ray tube-based projection TV system, the LCD projection TV is much lighter—thus more portable—than its counterpart. In addition, it offers screen sizes from 35 to 120 inches with about 350 lines of horizontal resolution. The light source of the LCD projector is projected through a matrix of tiny, semiconductor shutters, whose positions determine the light value of that picture element, or pixel. Some advanced models feature a single beam unit, a relatively light 30-pound projector, exceptionally bright picture quality and improved definition. LCD video projectors usually do not provide any audio circuitry or tuner; these components must be supplied by the user.

LCD Status Panel. A video camera feature that is designed to inform the user about the position of various camera operations. The liquid crystal display status panel is usually built into the camera body.

LCD Television. The use of liquid crystal display panels instead of conventional image tubes in TV sets. LCD TV, although not presently available in all its varieties, promises certain advantages over the ordinary cathode ray tube in every TV set—it is lighter, thinner and uses less power. (However, LCDs need their own light source.) LCDs, which offer hope for the much-touted flat-panel wall television, are currently employed in front and rear projection TV systems. Meanwhile, some manufacturers intend to substitute LCD panels in their popular TV models by the late 1990s. One major drawback is the cost, although LCDs should eventually be cheaper to produce than CRTs. The first LCD TV appeared early in 1981 in an experimental model by Toshiba and featured a two-inch b & w image. See LCD PROJECTION TV.

LCD Video Projector. See LCD PROJECTION TV.

LD 8-Inch Disc. A relatively new laserdisc format designed to play on CD-video machines. One of the uses of the 8-inch disc format is to provide several music videos by one artist. The new multi-function machines can play a variety of LD and CD sizes.

Lead-Acid Battery. A less expensive battery than the NICKEL CADMIUM type. The lead-acid battery has to charge overnight whereas the NiCad can be charged in less than two hours.

Leader. The blank segment found at the beginning of a videotape.

Lead-In. Refers to the antenna cable that is connected to the TV receiver. The lead-in is sometimes called the "download."

Lens. The part of a video camera composed of a transparent substance, usually glass, that transmits light to the CAMERA TUBE. There are two major types of lenses found on video cam-

Lens Extender

eras: the fixed focal length and the zoom. The latter is the more popular and more expensive. It has variable focal lengths. For example, it can be used as a wide angle, normal or telephoto lens with a simple adjustment or can zoom in or out during the recording of a subject or scene. The zoom lens often has a macro feature, which allows the lens to focus on objects as close as an inch or two from the lens barrel. Another feature of a lens is its maximum opening or aperture. The larger the opening, the more light it admits; also, the more expensive the lens. Lens openings are calibrated in f-stops, such as f/11, f/16, etc. Earlier camera models had a fixed or stationary lens. Many of today's cameras have lenses that are interchangeable through the use of a standard C-mount. However, the zoom lens has eliminated the need for changing lenses, at least for most home video users. See EXTENDER LENS, F-STOP, SUPPLEMENTARY LENS.

Lens Extender. See EXTENDER LENS.

Lens Mount. See UNIVERSAL LENS MOUNT.

Lens Stabilization. A unique video camera feature that produces a relatively stable picture by physically adjusting the lens assembly to compensate for camera movement. This is accomplished by the use of sensing devices that scan horizontal and vertical movement. A tiny computer receives these signals and sends them to two miniaturized motors that control the horizontal and vertical motion of the lens, thereby correcting much of the camera movement. Not all cameras offer this feature that provides a steadier image than that usually produced by the conventional video camera. Lens stabilization is also known as auto image stabilizer.

Letterboxing. Refers to the wide aspect ratio or dimensions of theatrical films and their presentation on conventional television screens. Some telecasts, determined not to cut parts of the original film, present the entire wide-screen view, resulting in black borders on the top and bottom of the TV screen. Sometimes television stations add a decorative bezel to these unattractive black borders. The term stems from the shape of the slot in mailboxes.

Lifetime. A CABLE TV advertiser-supported network specializing in health programs, talk shows, family-oriented programs, TV rerun shows, and TV and theatrical films. With its 16-hour daily schedule, Lifetime in 1990 earned ninth place in a list of leading cable networks, with 47 million subscribers.

Light Biasing. A technique employed in some video cameras using a SATICON tube to compensate for image retention. The saticon, claiming improved picture resolution over the VIDICON tube, tends to suffer from lag when the camera pans. Light biasing attempts to correct this by directing light at the back of the sensitive faceplate. See CAMERA TUBE, IMAGE RETENTION.

Lighting

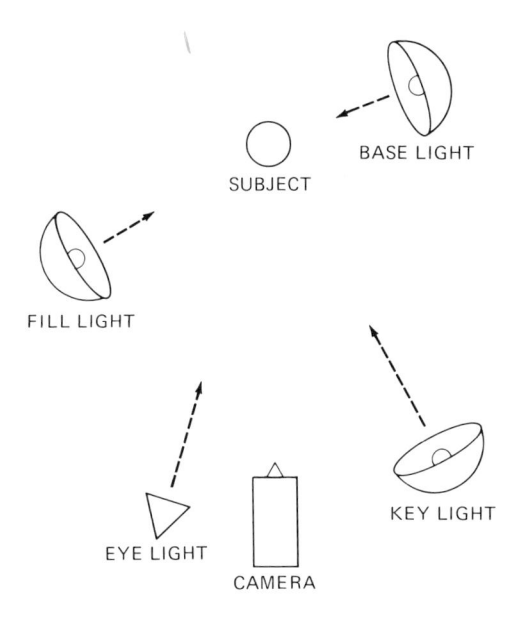

Figure 24. Lighting.

Light Level Meter. An indicator on some video cameras that shows whether the subject has too little or too much light.

Light Spot Scanner. See FLYING SPOT SCANNER.

Light Valve Projection System. A projection TV system that operates by scanning a beam of electrons across reflectors or mirrors coated with an oil film. The electrons distort the surface of the oil, thereby altering how the light reacts when it reflects off the mirrors. The result of this action determines whether the light reaches the screen directly or is transmitted through a diffraction grating.

Lighting. In video, any available, natural or artificial illumination. Standard indoor lighting includes a base light, a key light, a fill light, a black light and an eye light. The base light is usually located over the subject, provides general illumination to the scene area, and assures that the scene is bright enough not to cause any VIDEO NOISE. The key light is the brightest and is aimed at the subject. It may be a spotlight or a floodlight. It accentuates the subject, casting a definite shadow. It is usually positioned 45 degrees from the camera and higher than the subject. The fill light, often the same as the base light, is soft and lights up the dark areas of the scene. It is not as bright as the key light. The back light, usually a rear spotlight, provides definition when aimed at the subject. It separates the subject from the background. The eye light is a tiny spotlight, which, when aimed at the subject's eyes, causes highlights in them, making them appear more lifelike. See BACK LIGHT, BASE LIGHT, FILL LIGHT, KEY LIGHT.

Lightning Arrester

Lightning Arrester. A commercial protective device designed to reduce the danger of damage to TV and related units caused by lightning. The accessory provides a bypass directly to the ground for lightning discharges that reach the antenna. Lightning arresters are usually installed in conjunction with outdoor antennas mounted on the roofs of homes. See ANTENNA.

Limiting Resolution. In video, the measurement of the resolution as determined by the maximum number of lines per picture height as registered on a test chart. See HORIZONTAL RESOLUTION, RESOLUTION, VERTICAL RESOLUTION.

Line. In video, a single tracking of the electron beam or "dot" as it travels from left to right across the TV screen. See LINE SCAN.

Line Blanking. Refers to the period of time that the scanning dot or spot takes to return from the end of one line scan to the beginning of the next. The dot moves from left to right as it scans each of 525 lines which form the NTSC standard. As it moves from right to left, the video camera emits no signal. This line blanking, or horizontal blanking, as it is often called (since the scanning dot moves in a horizontal direction), permits only the left-to-right scanned information to be traced for a clear video image. Line blanking differs from FIELD BLANKING or VERTICAL BLANKING in which the scanning spot moves vertically from the bottom to the top of the TARGET AREA. See FIELD BLANKING, LINE SCAN, NATIONAL TELEVISION STANDARDS COMMITTEE.

Line Control. A feature on a TV set which permits enlarging or reducing the picture. In the case of horizontal lines appearing on the top and bottom portions of the TV screen, the linear control is adjusted so that the screen image is enlarged enough to eliminate the lines. If the control is turned too far, the picture will look elongated. Also, certain titles will be cut from the picture. See OVERSCAN, UNDERSCAN.

Line Doubling. An image enhancement technique, used in video recording and applied to broadcasting, that improves picture quality. When projected, the image almost equals that of 35mm film projection. Developed by French engineer Yves Faroudja, who calls his technique Super-NTSC, the line doubling system is a strong competitor of high definition TV. The former does not make present TV receivers obsolete, whereas the latter does. If line doubling is adapted as the new American standard, owners of conventional TV sets will receive the standard picture they are familiar with. To enjoy the benefits of the improved picture, they will need an adapter. With HDTV, current TV equipment will be useless. The line doubling process, which increases the number of line scans on the screen from the NTSC standard of 525 to 1,050, works by displaying each original line twice. This results in an image of higher density and greater stability. See YVES FAROUDJA.

Line Filter. In video, an electronic component, containing one or more inductors and capacitors, that is placed between a transmitter or receiver and the power line to prevent noise signals and other interferences.

Line Input Terminals. The audio/video input and output jacks, usually found on the rear of VCRs, that are used for copying and editing. The direct line input terminals are preferred for these operations over the antenna connections, which may produce grain and color changes to the copied or edited videotape.

Line Level Impedance. A low IMPEDANCE signal of 600 ohms. Line matching transformers are used for matching the impedance of various components, such as a microphone to the input of a mixer. See TRANSFORMER.

Line Matching Transformer. See TRANSFORMER.

Line Resolution Chart. See RESOLUTION CHART.

Line Scan. The rapid movement of the electronic beam across the TV screen of the CATHODE RAY TUBE. Different TV broadcasting systems have different numbers of line scans per picture FRAME. The NTSC (American) standard requires one frame of 525 lines (actually two FIELDS of 262 1/2 lines each). These line scans are not to be confused with the lines of HORIZONTAL RESOLUTION. See FIELD, FRAME, INTERLACED SCANNING, NATIONAL

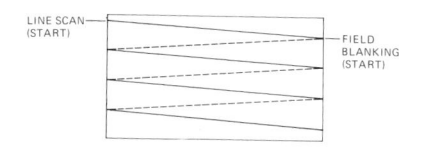

Figure 25. Line scan.

TELEVISION STANDARDS COMMITTEE.

Linear Audio. A method of placing the sound track on videotape. Linear audio may be mono or stereo. Another method of recording sound on videotape is diagonal recording—placing the audio track along with the diagonal video track for better quality. See LINEAR STEREO.

Linear Editing. Refers to a restrictive process of editing tape by recording predetermined scenes in sequence on another tape. This meant that after the second tape was completed, any additional editing required a third tape—and another generation loss of detail. This method has been replaced by non-linear editing, in which information about different sequences is stored in memory until the final tape is made. If additional changes are required, another tape can be produced without any generation loss by referring to the stored memory rather than the edited tape. See EDITING, NON-LINEAR EDITING.

Linear Program. Program material on tape or disc which the viewer plays through from the beginning to end. Linear programming, such as films, plays, etc., is usually contrasted with INTERACTIVE TV or INTERACTIVE VIDEODISC in which segments of a

Linear Stereo

program are encoded with a signal for easy access.

Linear Stereo. The use of conventional, low-quality mono audio tracks, located near the edge of the videotape, for the stereo audio signal. Linear stereo splits the audio track into two, separating the pair with a narrow guard band. In contrast to linear stereo, or linear track stereo as it is sometimes called, the superior Beta or VHS Hi-Fi technique records the audio signal along with the diagonal video signal tracks for better sound reproduction.

Linear Time Counter. See REAL-TIME COUNTER.

Linearity. A testing procedure which measures the ability of a VCR to reproduce a series of shades of gray in a uniform (linear) pattern. The more linear the pattern of shades which range from black to white, the better the VCR's ability to reproduce the original picture. In addition, linearity refers to the horizontal and vertical controls that affect the "size" of the TV image. For example, the picture is enlarged so that it fills the screen without exhibiting lines above or below the image. See GEOMETRIC LINEARITY.

Linearity Chart. See VIDEO TEST CHART.

Liquid Crystal Display (LCD). See LCD COUNTER, LCD PROJECTION TV, LCD TELEVISION, LCD STATUS PANEL.

Loading System. See BETA, VHS, VCR.

Local Pickup. A condition in a TV receiver or a VCR in which the internal tuner substitutes as an antenna, producing ghosts on the TV screen. Proper shielding of the tuner minimizes local pickup. VCRs are usually free from this anomaly.

Longitudinal Recording. Tape recording by means of the tape moving past stationary heads. Audio tape recorders use the longitudinal process, as do VCRs whose stationary heads lay down audio tracks across the top portion of the videotape. Early VTRs recorded information longitudinally until the helical scan technique with its rotating video heads set at an angle was developed. This process permits more information to be stored diagonally on narrow tape or shorter lengths. See AZIMUTH, BETA HI-FI, HELICAL SCAN.

Longitudinal Time Code. See LTC.

Longitudinal Video Recording. See LVR VIDEO RECORDING SYSTEM.

Loop. See PROCESSOR LOOP.

Loop-Through Jack. A feature found in TV monitors which permit several monitors and VCRs to be hooked up to the same signal source. A panel switch on the rear of the monitor selects either high input impedance or 75-ohm impedance. The first is used when the set transmits its signal to other units while the 75-ohm setting is used when the monitor is the final set in the series. Some more expensive industrial-model CHARACTER GENERATORS offer this feature.

Low Light Sensitivity. A video camera feature that helps to produce clear, detailed images. The LUX rating affects the low light sensitivity of a camera. The lower the number, the less light that is needed. Advanced video cameras with high-speed shutters require low lux numbers or low light sensitivity to ensure good screen pictures.

Low Noise Amplifier. The component of a SATELLITE TV system that is mounted inside the FEEDER HORN assembly of an antenna and is designed to amplify the signal it receives from the dish before it reaches the SATELLITE RECEIVER. The effectiveness of an LNA is measured by how much GAIN it gives to the incoming signal and by its noise-temperature rating. Although the antenna itself increases the signal sent to it, the LNA should boost it more by 50 decibels (dB) of gain. A lower noise-temperature rating means less noise; some amplifiers provide a number of 120 degrees, which is considered good.

Low-Pass Filter. A device often employed on a two-way cable system to restrict the flow of high frequency information while permitting the passage of low frequency information. The filter is sometimes written as lo-pass filter. See FILTER.

Low Power Satellite TV. Refers to satellite TV systems which broadcast within the 4–6 GHz C-band. To receive this low power signal, the EARTH STATION, or receiving base, requires a large dish (parabolic or spherical antenna), usually 10 to 15 feet in diameter. Although these dishes cost thousands of dollars, they are capable of receiving dozens of cable services. The electromagnetic spectrum or microwave frequency range of a satellite TV signal usually falls between 5.9 and 6.4 GHz when transmitted to a satellite and ranges from 3.7 to 4.2 GHz when returned to earth. Medium power satellite TV refers to a bandwidth of 11.7 to 12.2 GHz and requires a four-foot antenna. Although this system is less costly than lower power satellite TV, it is severely limited to the number of channels it can handle (currently, about four). Medium power satellite TV is similar to DIRECT BROADCAST SATELLITE which utilizes a different bandwidth and a smaller-diameter antenna. See C-BAND, DOWNLINK, EARTH STATION, HIGH POWER SATELLITE TV, SATELLITE TV, UPLINK.

Low Power Television. A system of broadcasting that permits thousands of new local stations to broadcast within a radius of 10 to 20 miles. Low power television, or LPTV, is accomplished by limiting VHF stations to ten watts of power output and UHF stations to 1,000 watts. These channels will be subject to fewer regulations than conventional ones and in part will serve local communities, minority groups, colleges, etc. Basically a line-of-sight medium (the flatter the terrain, the larger the radius of the low power signal), LPTV was given a boost in March of 1982 when the Federal Communications Commission approved a set of final rules governing the 4,000 anticipated new stations. Although chiefly supported by local businesses, LPTV stations in 1990 received about 15 percent of their advertising revenues from national advertisers.

Low-Z. See IMPEDANCE.

LP Speed. The middle speed (Long Play) of a three-speed VHS format VCR. With a standard T-120 videocassette, the LP mode records and plays back for four hours. The other two speeds are SP (Standard Play) which records for two hours and EP (Extended Play) or SLP (Super Long Play) which provides up to six hours of recording time. Some machines no longer record in the LP mode but do offer it in playback only. These VCRs, usually containing four heads, optimize two for SP speed and the remaining two heads for EP mode.

LPTV. See LOW POWER TELEVISION.

LTC (Longitudinal Time Code). Refers to the time code recorded on an audio track of the videotape. Time codes are digital addresses that distinguish each frame, thus permitting access to it. LTCs are written longitudinally, as opposed to video information and some audio information, which are recorded diagonally. See TIME CODE.

LTR (Linear Time Readout). See REAL-TIME COUNTER.

Lumen. A measurement of light used chiefly in reference to the light output of front projection TV systems. For instance, front projection TVs whose light output measures 300 lumens or better are considered excellent. Rear projection TV uses the term "peak brightness level" instead of "light output" and is measured in footlamberts,

the number based on a surface that emits one lumen per square foot.

Luminance. In video, the brightness of an image. Also, the amount of light given off or reflected by a surface. Luminance is important in judging projection TV systems, TV receivers, etc. See AMBIENCE, AMBIENT LIGHT, PROJECTION TV.

Luminance Noise. Refers to a type of video interference which influences both black-and-white and color signals. Luminance noise is listed as a number in specification sheets of components and in tests reports. It differs from CHROMINANCE NOISE, which affects only color.

Luminance Noise Reduction. A special electronic circuit designed to reduce unwanted noise or interference in the brightness signals, thereby producing brighter whites and more intense blacks. Luminance noise reduction usually is part of the broader HQ (high quality) circuitry of many VCRs. See HQ CIRCUITRY.

Luminance Reversal. See IMAGE REVERSAL.

Luminance Signal. A signal whose amplitude depends on the light values of a televised scene. The luminance signal, although part of a composite color signal, can generate its own, complete monochrome picture. Luminance (brightness) and chrominance (color) components are recorded separately. The former is recorded on an FM (frequency modulation) carrier while the latter is recorded as an amplitude modulation. See CHROMINANCE SIGNAL.

Lux. A measurement of light in relation to the sensitivity of video cameras. One FOOTCANDLE equals ten lux. Thus the sensitivity (the minimum amount of light needed to produce a usable image) of a camera may be rated at 50 lux (five footcandles). The lower the lux number, the lower the lighting conditions the camera can handle. Lux is the measurement recommended by the International System of Standards.

LV Videodisc System. One of two major types of machines that play back records containing pictures as well as sound on a standard TV receiver. The LaserVision player uses a highly reflective grooveless disc which is "read" by a small laser beam. In its standard speed (30 minutes per side) the player provides such various sophisticated functions as random access to CHAPTERS and FRAMES, FREEZE FRAME, VISUAL SCAN, etc. Because the system employs a laser, the disc is virtually indestructible. The laser tracks from the inside of the disc to the outside (in contrast to conventional audio recording), but never makes contact with the surface of the disc. The LV player has two speed modes. CAV, which is its standard speed (30 minutes), and CLV with an extended play of one hour. See CAV, CED SYSTEM, CLV VIDEODISC, VIDEODISC PLAYER.

LV Videodisc System History. After a false start in 1935 when Phonovision introduced "recorded television records" to London consumers, the innovative system gained a more firm foothold in 1978 by way of Magnavox. The company offered Magnavision, a laserdisc player, using the same technology that is currently employed in laser videodiscs. But the new system was soon challenged by RCA, which brought out its own CED (Capacitance Electronic Disc) format (using a conventional stylus to read the disc), and VHD (Video High Density), a Japanese system that never made it to American shores. Faced with a choice of three possible incompatible systems, a confused public was less than enthusiastic. Instead, most consumers opted for videocassette recorders, which offered both recording and playback advantages. Meanwhile, Pioneer, one of the leaders in the field of laser video technology, added CX noise reduction in 1982. Several important events occurred in 1984: RCA dropped its CED system, digital audio was added to laserdisc, and the first LV/CD combi player reached the marketplace. By the end of the decade, LV players provided additional innovations, including digital special effects, built-in time base correction and continuous playback.

LVR Video Recording System. A now defunct video recording system which passed tape at a high speed over a fixed recording/playback head. Introduced in 1979 by Toshiba and BASF, the Longitudinal Video Recording process played 220 parallel tracks of audio and video signals on tape that was magnetized along its length, hence its name— Longitudinal Video Recording. Because of its stationary head design, the machine was less costly. It accommodated 1/2-inch tape in a special cartridge. The LVR system, the first attempt at video recording, was simply an accelerated version of an audio recorder.

M

Macro Lens. A special lens designed to focus very close to the subject. Macro lenses, some capable of functioning as close as an inch or two, are particularly helpful in nature work and hobbies involving stamps, coins, models, etc. Most macro lenses also serve as a normal lens when not in the macro mode. See LENS.

Macro Mode. An alternate function of a dual-purpose lens that can take extreme close-ups of such tiny objects as coins, stamps and insects as well as operate as a normal lens. See MACRO LENS, MACRO VIDEO.

Macro Video. The use of extreme close-ups with the macro part of a video camera zoom lens. Standard on most new, quality video cameras, the macro feature permits focusing as close as an inch or two from the subject which fills the TV screen. Macro shooting provides a very narrow DEPTH OF FIELD (that which is in focus in front of and behind the subject) so that focusing becomes extremely critical. Also, any slight movement becomes highly visible on the screen. Therefore, a good tripod is recommended. If a camera is not equipped with a macro lens, or the zoom feature (which disengages in the macro position) is desired, the video camera user may add a special series of close-up lenses. These are measured in diopters, such as +1, +2, etc. Kits are available with various diopter lenses.

Magnetic Coercivity. See COERCIVITY.

Magnetic Recording. Capturing audio and video frequencies by magnetizing areas of tape which can be played back by moving it past a head where the magnetized areas are reconverted into electrical energy.

Magnetic Tape. The medium used for recording and playback on tape recorders. The most popular width of consumer tape is 1/4-inch while industrial tape may be 1/2-inch, 3/4-inch, 1-inch or 2-inch. See VIDEOTAPE, VIDEOTAPE FORMAT.

Manual Focus. A video camera function that allows the user to override the autofocus feature. Manual focus provides several uses. Camera owners may prefer this mode as a means of extending battery life, which is affected by continuous use of autofocus and other automated features. In addition, manual focus permits the user to add individual creativity to his or her work by producing special effects, such as out-of-focus fades or scene transitions. See AUTOFOCUS.

Manual Interval Time Lapse Mode. A video camera feature that permits the camera user to add animation effects. The operator accomplishes this by pressing the pause mode, then slightly moving the object that is being recorded, and finally pressing the record button. The process is repeated until the entire cycle of desired movement has been completed. Many video cameras provide an automatic TIME LAPSE feature. See INTERVAL TIMER.

Manual Iris Control. A video camera function that permits the user to manually control the amount of light that enters the camera lens. Manual iris control is sometimes described as "exposure" control. See AUTO/MANUAL APERTURE CONTROL.

Manual White Balance. A video camera function that permits the user to control the way the camera views different colors. This is important in maintaining correct color when various light sources produce changes in the color mix, or color temperature. See AUTOMATIC WHITE BALANCE, WHITE BALANCE, WHITE BALANCE CONTROL.

Mask. In video, refers to the device mounted in front of a television picture tube to limit the viewing area of the screen. The mask is sometimes referred to as a frame.

Masking. A term employed in the Dynamic Noise Reduction system and referring to the capability of a program to conceal its background noise. DNR utilizes a special dynamic filter which eliminates high frequencies (mostly in the form of hiss or noise) whenever the signal is not strong enough to cover the hiss. But when the signal does "mask" or cover this noise, the filter permits the high frequencies to pass through. See NOISE REDUCTION SYSTEM.

Master. In video, an original recording on disc or tape from which copies may be made.

Master VCR. The VCR deck or machine that plays the tape during the duplicating process onto one or more SLAVE machines (or VCRs doing the recording). See COPYING.

Matching Transformer. See IMPEDANCE ADAPTER, TRANSFORMER.

Matrix. In video, that part of a color TV circuit that joins the I, Q and Y signals and converts them into red, green and blue signals which are then applied to the picture tube grid.

Matrix Surround. A surround-sound system similar to, but not as sophisticated as, Dolby Surround.

Matrix Wipe

Matrix Wipe. A special effect designed to tessellate a video image. A MIX/EFFECTS SWITCHER is used to produce this effect as well as to change the picture in each square in a seemingly random pattern. See WIPE.

Matrixing. The conversion of a master videotape into a glass laser videodisc master. A heavy-duty laser beam etches microscopic pits into the surface of the disc which is then used as a master to produce the videodisc stamper.

Matte Key. See KEY.

Mattel. See INTELLIVISION.

MCA/Disney vs. Sony Lawsuit. The famous case in which Universal City and Walt Disney Studios filed suit in 1976, charging that Sony and others, by selling videocassette recorders, damaged the studios financially and infringed upon copyright laws. The first major decision concerning this case occurred on October 1, 1979, ruling in favor of Sony. Then in October 1981, an appellate court reversed the decision in favor of the plaintiffs. Sony, however, appealed this ruling and eventually won. The suit had many ramifications for studios, equipment and tape manufacturers as well as the general public. There have been occasional suggestions of a "royalty fee" placed on VCR and tape sales. See THE FERGUSON DECISION.

McLuhan, Marshall (1911–1981). Media expert, writer, social commentator. His terms and phrases such as "global village" and "the medium is the message" have become household expressions. He perceived television transforming us into a global community of world citizens. On media in general he once wrote: "The way we acquire information affects us more than the information itself." Also, "By knowing how technology shapes our environment, we can transcend its absolutely determining power." He firmly believed that the electronic media would bring about international social and cultural growth.

MDS. See MULTIPOINT DISTRIBUTION SERVICE.

Mean Time Between Failures (MTBF). The estimated life of an electronic component. Although it is impossible to determine the exact working life span of a general unit such as a VCR or a video camera because of the number of components within each piece of equipment, it is possible to estimate the life of each individual component. Engineers, for example, have placed an average life span of 100,000 hours on VCR components. However, other parts of the machine, such as video heads and rubber belts, are much more vulnerable to wear and deteriorate first. VCRs may have an MTBF of up to 10 years, but servicing of these vulnerable parts may be required within two to three years.

Megahertz (MHz). A unit of measure used in video to rate frequency. One MHz is equal to one million cycles per second. MHz is often used to determine RESOLUTION in VCRs, TV sets, etc. See BANDWIDTH, HERTZ, RESOLUTION.

Memory. In video, a digital VCR feature that permits the viewer to lock in a still picture from a TV broadcast. In addition, the picture, which usually appears in a corner of the TV screen, can be stored in memory until the VCR power is turned off. This feature is sometimes listed as TV memory or TV memo.

Memory Backup. Refers to the capability of a VCR to retain its programmed instructions and other timer functions in the event of a power failure. The first technique VCR manufacturers employed for this purpose was a built-in nickel cadmium battery which lasted a relatively long time (several hours)—covering the length of most electrical outages. The battery has recently been replaced by a smaller and less costly super capacitor, a device that can store an electrical charge powerful enough to keep a VCR timer active long enough to cover some power failures. However, the average capacitor has only enough storage to last from about five seconds to approximately 30 minutes. For those VCR users who intend to be away from home for long periods of time and want a stronger assurance that their machine will record an important program, a special external accessory, known as an uninterrupted power supply unit, is available. This item, which usually sells for several hundred dollars, comes in several models, depending upon the number of watts.

Memory Bank. A VCR feature designed to permanently store such programming information as day, time, channel and an identifying code, all of which can be recalled later for future scheduling without re-entering each item separately. The short code may consist of a few recognizable letters, such as "six" for *Sixty Minutes*. This will help the user to recall the data when he or she wants to repeat the recording process at a future date.

Memory Controlled Effects. A special feature usually built into a professional PRODUCTION SWITCHER and designed to store dozens of complete set-ups. In addition, the MCE remembers transitions previously put into production.

Memory Pause. Refers to a videodisc player that has the capability of stopping a program at an exact point and then continuing to play the disc from that same point. This feature is helpful to those viewers who are often interrupted by telephone calls and other similar disturbances.

Memory Rewind. A feature on some older-model videocassette recorders which, when pressed, stops the tape during Rewind or Fast Forward when the INDEX COUNTER reaches 000 (or 0000 on some machines). Memory rewind works in conjunction with the tape counter and is useful in locating a preselected portion of the tape for replay. This feature is different from the ELECTRONIC PROGRAM INDEXING.

Menu. A feature, usually found on some VCRs or TV receivers, that displays on screen a vast choice of operating options that the viewer activates by way of the remote control. There are programming menus to make it easier for the user to set the time, day and channel of programs to be recorded. Setup menus help the new owner of a TV set make the proper wire and cable connections. Other menus include audio functions, such as adjusting bass and treble, and video functions, such as controlling sharpness, contrast and color.

Metal Evaporative Tape. See VAPOR DEPOSITION.

Metal-Particle Tape. A high-grade videotape composed of needle-like metal particles and a roughly textured base to hold the particles. The tape, which differs from standard tape that uses metal oxide instead of metal particles, reportedly is free of dropouts while producing a higher frequency response than other high-grade tapes. By replacing the oxide with metal, manufacturers have more than doubled the strength of the magnetic field.

Metal Oxide Semiconductor Chip See MOS.

Metamorphosing Animation. A special effect that creates changing shapes and color for specific needs. The technique is particularly useful for television meteorologists who preprogram much of their weather animation. By incorporating metamorphosing animation, they can show the movements both of numbers and storm fronts across the TV screen.

Metering System. Refers to the technique used by a video camera or camcorder to measure the light necessary for the proper exposure of a scene or subject. One simple system produces a single value, with the emphasis placed on the center portion of a given scene. Another, sometimes referred to as the two-field metering system, takes one reading of the entire field and another reading of the central zone, thereby assuring a correct exposure.

MHz. See MEGAHERTZ.

Micro-Monitor. A small TV monitor with a 1 1/2-inch screen and a one-inch speaker designed to be used in the field. The micro-monitor also permits the viewing of a tape while recording.

Micron. A measurement equal to one-millionth of a meter and used to determine video head gaps. Beta II uses a head gap of 30 microns while Beta III has a 19.5 micron gap. On some six-hour VHS machines, the gap originally was 58 microns, but presently is 30 microns. With other models, one pair of heads may have a gap of 58.5 microns for SP speed while a second pair may have a gap of 19.5 microns for the EP mode. See GAP, VIDEO HEAD.

Microphone. A device used with video cameras, portable VCRs and home models to record sound onto videotape. A mic converts sound to electrical energy. Mics have different patterns. Some basic types are the OMNIDIRECTIONAL, BI-DIRECTIONAL, DIRECTIONAL AND CARDOID. There are also mics for different purposes such as the LAVALIER, BOOM, etc. Other types are the CONDENSER and DYNAMIC micro-

Microphone Splitter

phones. All mics have some degree of COLORATION which alters its flat response. Basically, the less the coloration, the better the microphone.

Microphone Combiner. See MICROPHONE MIXER.

Microphone Frequency Response. The measurement of the amount of coloration in a microphone. Since virtually all VCRs have a response that is less than that of high fidelity quality, it is easier to match a microphone to the machine. Many ANALOG recorders register an audio frequency response of up to 9 through 12,000 Hz. Therefore, a mic with a range of 80 to 12,000 Hz will provide a smooth response.

Microphone Impedance. The resistance a microphone offers to the sound signal it is picking up. Each mic has its own IMPEDANCE which must be matched to the input impedance of the VCR or other similar unit. This is a relatively simple task involving a matching transformer, available at most electronic stores. The addition of this accessory will assure that the mic will operate at its peak frequency response. See BALUN, MATCHING TRANSFORMER.

Microphone Jack. A receptacle or opening which permits the connection of a microphone plug to the video camera or videocassette recorder. There are chiefly three kinds of mic jacks: the RCA phono jack, the 1/4-inch jack and the most frequently used with home video components, the 1/8-inch mini-jack. Appropriate adapters are readily available for connecting any mic jack with any other unit. See JACK, PHONO JACK, PLUG.

Microphone Mixer. An accessory which accepts several microphones and controls the volume of each mic separately. Mixers for home video range in price roughly from $125 to $1,000. They are usually limited to four inputs and offer some degree of portability. The mic mixer permits the use of only one mic when required, even while three other mics are connected to the console. The other major purpose of the mixer is to blend the sounds of several mics into one signal while balancing the output of each in relation to the others. There are active and passive mixers. See ACTIVE MIXER, PASSIVE MIXER.

Microphone Mixing. A video camera feature that can combine the soundtrack of the tape with an external source. The process can occur either during recording or editing.

Microphone Pickup Response. See POLAR RESPONSE.

Microphone Splitter. An accessory designed to split a single microphone line into multiple outputs. The mic splitter allows the divided mic level signals to feed various mic inputs on such components as video recorders (both VCR/VTR formats), audio recorders, monitors, speakers, etc. This device provides proper isolation between outputs. The number of outputs depends on the particular splitter. There are basic units described as 1X3 mic splitters which split one line into three outputs, while more complex models can divide each of four

Microphonics

mic lines into three outputs. There are also more sophisticated types such as the mic splitter/combiner which splits and/or mixes mic signals in a variety of combinations. Mic splitters/combiners can be either passive or active. Active units provide a gain of +/−6 dB maximum and are designed for use with equipment that does not have its own output transformer. Passive splitters are usually of low impedance to match that of the mic outputs while active ones are of line level. See LINE LEVEL IMPEDANCE, MICROPHONE MIXER.

Microphonics. Interference in the form of a series of horizontal lines on a TV screen caused by extreme surges from loudspeakers, applause or certain musical instruments. These loud bursts affect the picture tube in the video camera but cause no permanent damage to the equipment. This effect can be avoided or minimized by keeping the video camera out of the direct range of these instruments and not standing too close to the loudspeakers.

Microreflection. In video, one of several forms of degradation that affects NTSC picture quality. Microreflections are caused by waves that strike a medium of different characteristics and are then returned to the original medium. See DEGRADATION.

Microwave. A very high frequency range in which the transmitted wave lengths are extremely small. The microwave band contains a variety of information and is used for different purposes. For example, its lower portion, between 3.7 and 4.2 GHz, contains the band of the satellite channels. Other segments of the band are allocated to amateur radio operators, police radar, telephone companies, etc. DBS (DIRECT BROADCAST SATELLITE) systems operate in the 12 GHz range of the microwave band.

Microwave Interference. In satellite television, interference from generators, transformers and other like devices, usually installed by utility companies in the vicinity of a PARABOLIC ANTENNA. If these objects fall in the line of view between the antenna and the transmitting satellite, they can adversely affect reception. The owner of a satellite TV system can obtain an FCC license which will assure him or her of interference-free reception. See SATELLITE TV.

Microwave Transmission. A method used by some PAY TV systems to transmit over-the-air, point-to-point video signals. The encoded programs are beamed to subscribers who are equipped with decoders. Besides microwave, PAY TV can also transmit programs by way of telephone wires and cable. See CABLE TV, PAY TV.

Midband Cable TV. Channels that occupy frequencies not used for TV broadcasting. Midband channels, like superband CABLE TV channels, are television channels. The most popular midband channels are channels A through I. The frequencies from 120 to 126 MHz are occupied by A, while those of B-1 continue on until just under channel 7 at 174 MHz. Channels A-1 are called midbands because they fall between channels 6 and 7 (which is the lower end of what is known as the high band). There are also sub-band channels which fall below channel 2; these are utilized for special transmissions.

Mid-Range Switcher. A video switching device, falling somewhere between a low-cost consumer SWITCHER and an expensive professional/industrial DIGITAL or PRODUCTION SWITCHER. Mid-range switchers may provide up to eight primary inputs, black and color backgrounds, all-linear keying, and a variety of wipe effects.

Mid-Side Principle. A technique employed in stereo microphones (especially SINGLE-POINT STEREO MICS) in which a single internal component "listens" to both its right and left while another element picks up information from a forward position. Many recording patterns are possible by electronically mixing the various combinations of outputs of the two components. This technique is also known as the MS principle.

Mil. The equivalent of .001 inch. The mil is used in measuring the thickness of a video tape.

Millimeter. A monthly trade magazine for those involved in the motion picture industry and television production. Providing news and information in these areas for more than a decade, the magazine offers more departments than articles to its readers. Subject matter for feature articles ranges from animation and technical stories about specific hardware to interviews with well-known film directors and producers. Regular monthly items include such topics as teleproduction, commercials, motion pictures, graphics and effects, audio technology and news of corporate television. Top-heavy with trade advertisements, the magazine provides plenty of illustrations and graphics to support its written contents.

Mini-Enhancer. A device designed to improve the video signal of portable VCRs, video cameras, etc. The mini-enhancer attaches between camera and recorder and is meant to be used in the field. It is also useful when camera extension cables are used. The accessory usually contains a bypass switch which allows a comparison of an enhanced and unenhanced image. See IMAGE ENHANCER.

Mini-Jack. A phono JACK or PLUG used in the audio inputs and outputs of a Beta VCR. The 1/8-inch jack is smaller than the more popular RCA JACK generally used in the VHS format. Sony uses the RCA-type in its video input and output together with the mini-jack in its audio lines. The size of the jacks becomes important when copying tapes from one machine to another and using the audio/video rather than the RF connections. See COPYING.

Mini-VCR. Refers to 1/4-inch-size videotape in a compact cassette which operates inside a smaller than usual portable VCR. The first such mini-VCR was Technicolor's model 212 which weighed seven pounds, measured 10 inches square and 3 inches deep and used a cassette just slightly larger than an audiocassette. The mini-VCR format is incompatible

Mini Video

with others such as Beta and VHS in terms of videocassettes. But the machine can be connected to any model for duplicating tapes and can be hooked up with any camera and other components as long as the proper cables are obtained.

Mini Video. An alternate VCR format using 1/4-inch tape. The 8mm video system is designed mainly for a one-piece video camera/recorder, or camcorder, with one- or two-hour maximum recording time. The advantage of a mini video include lighter equipment, smaller components and relatively less expensive tape. The conventional Beta and VHS home video formats will maintain their market strength and sophisticated superiority, especially with new developments in video noise reduction, STEREO, special editing techniques, full function remote control, longer playing tapes and other features not yet technically possible in the mini format.

Minimum Illumination. The least amount of light necessary to produce a noise-free image with a video camera. Minimum illumination may depend upon the F-STOP of the LENS, the sensitivity of the CAMERA TUBE, etc. See FOOTCANDLE, LUX, SENSITIVITY RANGE, SENSITIVITY SWITCH, VIDEO CAMERA SENSITIVITY.

Misalignment. A condition in which one of the primary colors in video (red, green, blue) appears on the side of a subject or object, as if that particular color is "bleeding" or not registering properly. Misalignment, sometimes referred to as misregistration, is the result of improper convergence—the inability of electron beams to strike the face of the picture tube precisely. With a video camera, the problem stems from a faulty camera pickup tube. See CONVERGENCE, REGISTRATION.

Misregistration. See MISALIGNMENT.

Mistracking. See TRACKING, TRACKING CONTROL.

Mix. The superimposition of one image over another. With the use of a mixer/fader device, the image of one source can be increased or decreased on the screen by the movement of a simple control. A mix differs from an effect which keeps the entire image even during superimposition.

Mix/Effects Switcher. A multi-buttoned, box-like electronic console used by professionals to create a variety of special video effects. The instrument can produce a number of preset patterns, including wipes, keys and digital effects. Each row of buttons is called a "bus." See KEY, WIPE, DIGITAL EFFECTS VIDEO PALETTE.

Mixer. See MICROPHONE MIXER.

M-Load. The loading system used by VHS videocassette recorders and first invented by Sony. The machine wraps the tape around the head drum in an "M" design, hence the name. To do this, 13 inches of tape are removed from the cassette, in contrast to the Beta system in which 24 inches are taken out. Two TWIST PINS angle the

Monitor

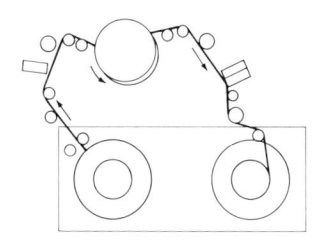

Figure 26. M-load of VHS machines.

tape to its path around the video head drum. Each time the VCR enters the Stop mode, the tape is loaded back into the cassette. Some critics complained about early VHS machines, suggesting that the loading system placed too much stress on the tape. They believed the strain would cause the tape to break or stretch. See FULL LOAD, HALF LOAD, U-LOAD.

Mode Switch. A feature on some VCRs for selecting Mono, Stereo or SAP broadcast mode. Some TV receivers have these modes on the remote control unit.

Modern Satellite Network. A cable service offering programs aimed at women. The channel presents cooking shows, exercise classes, films, etc. Owned by Modern Talking Picture Service, Inc., and originating in 1979, the service presents its almost four million subscribers a schedule from 10:00 A.M. to 1:00 P.M. Monday to Friday.

Modification Kit. A kit of parts and instructions for adding one or more special effects to older VCRs. Conversion kits are available for VHS systems. The kit provides VISUAL SCAN, SLOW MOTION and FREEZE FRAME. One company provides a videocassette of instructions along with its parts, wire and printed instructions.

Modulation. A technique which adds audio or video signals to a preselected signal. See DEMODULATED.

Modulation Transfer Function. See DEPTH OF MODULATION CHART.

Modulator. A miniature transmitter whose circuitry carries the raw audio and video signals from microphones and video cameras and puts them in a designated bandwidth channel or range of frequencies. TV modulators register the video and audio signals into separate carriers. A demodulator, or tuner, then strips the video signal from its carrier to reproduce the original video signal. See RF MODULATOR, SATELLITE TV MODULATOR.

Moire. A tremulous spectrum of color caused during editing when a VCR is backspaced and the beginning of a new scene is superimposed over the end of another scene. The wavy, satin-like optical effect occurs when converging lines in the picture are nearly parallel to the scanning lines. Video moire, sometimes called a rainbow effect, occurs also when the VCR is in Record mode and Pause is pressed. Recent machines have introduced special circuitry which all but eliminates moire caused by the record/pause functions.

Monitor. A television set minus receiving circuitry. Monitors have no VHF/UHF tuners, IF (intermediate frequency) AMPLIFIERS or video detectors. Some monitors do not have any audio system. A monitor is used basically to display video signals. Often more costly than a TV receiver, the monitor, with its more advanced and sophisticated circuitry, produces a superior picture. It also contains audio/video input jacks and other features such as picture focus, internal sync and horizontal and vertical scanning. The major advantage of the monitor over the TV set is that the composite video signal travels directly to the video amplifiers and control circuits while the audio is fed directly to the audio amplifiers. These direct processes retain the original quality of the signals. A monitor differs from a monitor/receiver. One of the more recent developments in monitors is the AUTO-SETUP monitor that can line itself up automatically. See AMPLIFIER, COMPOSITE VIDEO SIGNAL, MONITOR/RECEIVER.

Monitor Analyzer. A device, introduced in 1978 to the home video field, which helped adjust the brightness and color of a TV set. The plastic monitor analyzer, which came with a ready-reference chart, sold for about $25 and worked only with color bar test patterns on the TV screen.

Monitor/Receiver. A component that looks like a conventional TV set but has direct audio and video inputs as well as additional features usually not found in TV receivers. Like all monitors, these sets accept direct hook-ups of VCRs, videodisc players, video games, etc. By avoiding RF inputs, the monitor/receiver can retain the original signal. More expensive than the ordinary TV receivers, monitor/receivers are often sold by audio/video dealers rather than by conventional retail stores. Monitor/receivers began to appear in the marketplace in the fall of 1981. See MONITOR.

Monitor/Receiver Reference Tape. A specially prepared videocassette of tape containing miscellaneous video test patterns for the proper adjustment of monitor/receivers. Often as short as ten minutes, the tape features tests for CONVERGENCE, flesh tones, COLOR BARS, GRAY SCALE, etc. The reference tape permits adjusting equipment for color purity, ALIGNMENT, TINT, brightness, contrast, CHROMINANCE, etc. The reference tape is also known as setup tape.

Monitor Speaker. Refers to a small built-in video camera speaker that permits the user to check the soundtrack. The monitor speaker eliminates the need for headphones.

Monochrome Signal. That part of a color TV transmission signal wave which controls the luminance, whether it is displayed in black and white or color.

Monopad. In video, a one-legged support for a video camera. It is a temporary or limited alternative to a video tripod. Although monopads are less costly, lighter in weight and easier to carry than tripods, they require continuous handling to maintain steadiness. See TRIPOD.

Morita, Akio. Co-founded Sony in 1946 with an initial capital of $500; chairman at Sony in 1964 and promoter of the first consumer video recorder. A reel-to reel VTR, it incorporated HELICAL SCAN and recorded and played back in monochrome only. Although its $1,000 price tag was prohibitive for its time, it became the forerunner of Sony's now legendary Betamax introduced ten years later. See BETAMAX.

MOS. A metal oxide semiconductor chip that replaces the conventional SATICON and VIDICON camera tube. While the HORIZONTAL RESOLUTION of the MOS chip doesn't match other high quality cameras, the advantages of this technology include a lighter and smaller camera, no waiting for tube warm-up, less power consumption and no IMAGE BURN or drag. Hitachi, with its model VK-C1000, was the first company to feature a camera using the MOS image sensor. See IMAGE SENSOR.

Mosaic. A special-effects feature that breaks up a video image into hundreds of little squares or rectangles. The effect is often uses in commercial broadcasting to hide the face of witnesses during interviews and hearings or to prevent nudity from showing up on TV screens. This feature is usually found on certain digitally equipped units such as EDITING CONSOLES.

Motion Adaptive Interpolation Circuit. An advanced electronic technique used in large-screen Improved Definition Television to give a three-dimensional effect to the screen image. The special circuitry, sometimes described as a motion adaptive comb filter, first utilizes inter-line and interfield information and then processes it by applying two separate types of digital memory buffers. The entire process reportedly produces a 3-D effect, especially in action sequences. See IDTV.

Motion Analysis Camera. A professional/industrial video camera designed for stop-motion videotaping. Used for industrial, medical, laboratory and military applications, the camera usually contains a variable speed shutter with speeds capable of from 1/500 to 1/10,000 of a second. Motion analysis cameras perform such highly technical tasks as color spectral analysis, spray-flow study, high-speed microscopy, speeding-bullet analysis, etc. Many models are compatible with Beta, VHS, 3/4-inch and 1-inch formats.

Movie Channel, The. A PAY TV movie service offering recent Hollywood films, foreign films, classics, lesser-known works, etc. Owned by Warner Amex, The Movie Channel offers its over one million subscribers movies 24 hours a day. It began operations in 1980.

MP-1000. A video game system by APF once popular in the early 1980s. The system included two numeric keypad/joystick controllers and a console. An optional computer keyboard was available for approximately 160 dollars, converting the system into a home computer. The combined units made up The Imagination Machine. Dozens of games were available for the system.

MSO. The abbreviation for a multiple system operator or CABLE TV owner. Larger cable companies like Tele-

MS Principle

prompter and Warner Amex actually own and control many cable systems.

MS Principle. See MID-SIDE PRINCIPLE.

MTBF. An industry-consumer term for mean time between failures, or the average length of time between successive failures of a piece of electronic equipment. See MEAN TIME BETWEEN FAILURES.

MTTF. The industry-consumer term for mean time to failure, or the measured operating time for a piece of electronic equipment divided by its total number of failures during that time.

MTS (Multichannel Television Sound). Refers to stereo TV that is broadcast along with a secondary audio program (SAP), both of which require a special decoder. The quality of MTS depends on the effectiveness of the decoder and the addition of dbx—an audio noise reduction system.

MTS Decoder. That part of a VCR or stereo TV receiver that produces stereo separation, measured in decibels. In addition, the quality of a decoder depends on its signal-to-noise ratio and frequency response. Because they are difficult to align at the factory, many MTS decoders leave something to be desired in stereo separation. Separate units, sometimes called MTS/SAP (Secondary Audio Program) decoders, are available for TV sets and monitor/receivers not equipped to receive stereo broadcasts.

MTV (Music Television). A 24-hour video music channel distributed to more than 500 cable affiliates in 48 states. MTV, which was the first CABLE TV stereo music channel, coming on the scene in summer of 1981, specializes in rock music, live performances, in-studio interviews, news of concerts, etc. Its success has sprouted other music endeavors such as Nashville Network, another cable station, featuring country music. Owned by Warner Communications and the American Express Company, MTV started with 2.5 million subscribers in 1981 and now boasts more than 49 million and in 1990 was listed as the sixth most popular advertiser-supported cable TV network.

MII Component System. A professional/industrial 1/2-inch tape-recording format that provides full NTSC bandwidth. Features include 90-minute recording time, field color playback, a built-in digital time base corrector, time-code reader/generators, four audio channels, composite and component video inputs and outputs, and several advanced editing capabilities. The MII VTR can be integrated with other formats, including S-VHS, 1-inch, U-Matic, Beta and Beta-SP. Working in conjunction with a professional digital video camera, the MII system permits the 1/2-inch cassette recorded in the field to be loaded directly into a studio recorder for studio-level results that compete favorably with the one-inch C format.

Multi-Brand Remote Control. See MULTI-PURPOSE REMOTE CONTROL.

Multi-Burst Chart. See VIDEO TEST CHART.

Multi-Dimensional Autofocus. A camcorder feature that provides uninterrupted focus from the lens surface to infinity. This does away with the necessity for the conventional macro setting for extreme close-ups.

Multiformat. In video, a VCR that can play back videotapes recorded in foreign countries that have different broadcast signals than those in the United States. Usually, such machines, which compensate for American and foreign differences in the number of scan lines and house-current cycles, can process signals from PAL or SECAM tapes.

Multi-Frame Edit Viewer. A method of showing a series of frames in sequence. The unit displays fixed multiple images which can then be considered as edit points according to their sequence. The major benefit of multi-frame edit viewing is its doing away with the need to continually move the tape.

Multi-Gun Tube. A cathode ray tube containing more than one electron gun. Color television receivers use multi-gun tubes, as do multiple-presentation oscilloscopes. See SINGLE GUN COLOR TV TUBE, THREE-GUN PROJECTION TV.

Multi-Pin Connector. A component with a predetermined number of metal PINS attached between a video camera and a VCR. Each pin is assigned a special function; e.g., audio output, video ground, video signal input, etc. Early Beta recorders first used a 10-pin connector but soon switched to and presently use a 14-pin system while VHS machines adopted the 10-pin configuration. Akai, although a VHS system, at one time developed its own 7-pin connector. Not only are these systems incompatible, but all the pin functions of a 10-pin VHS camera connector do not necessarily match every VHS recorder. (Beta 14-pin connectors are standardized.) Some VCR manufacturers and independent companies sell ADAPTERS which permit attaching 14-pin connectors of Beta cameras to VHS recorders which, of course, accept only 10-pin connectors.

Multi-Standard Switchable Encoder. A feature generally found on a high-priced professional video camera designed with special outputs to accommodate other formats, such as Beta, S-VHS and MII.

Multiple-Effects Generator. A relatively costly editing unit for post-production work that provides an array of professional-type functions. The machine features a special-effects generator for fades and wipes in dozens of patterns, a colorizer, an audio/video processor, a color processor and a genlock/power supply. The last item permits the user to dissolve from one video source to another. See CHARACTER GENERATOR, SPECIAL EFFECTS GENERATOR.

Multiple Picture Inset. See MULTI-SCREEN DIGITAL FREEZE.

Multiple-Player Video Game System. A process using telephone lines or cable TV systems to permit 2 to 10 players to participate in one video game. One such system, Game Line, by the Control Video Corporation, allowed owners of the Atari 5200 console to receive games by telephone. The Ohio-based Compuserve Information Service supplies games to computer owners, as does the Plato computer system introduced by the University of Illinois. Multi-player video game systems can be relatively costly. One such computer service charges $5 an hour. See VIDEO GAME SYSTEM.

Multiplex Output. A television or monitor/receiver connector that permits adding a special audio adapter unit for receiving stereo TV programs.

Multiplexer. A relatively expensive multimedia device which permits dissolving or cutting from films or slides or any combination of these. Utilizing various lenses, prism beam splitters and mirrors, the multiplexer can handle many input sources simultaneously.

Multipoint Distribution Service. The utilization of a microwave system of over-the-air, line-of-sight broadcasting of video programs over a single channel. Subscribers use specially equipped receivers to pick up MDS programs. This system differs from that of CABLE TV or SUBSCRIPTION TV. Cable uses coaxial cable while STV makes use of conventional VHF and UHF channels. MDS also differs from other microwave systems in that it transmits a weaker signal in all different directions. The service is regulated by the Federal Communications Commission. See CABLE TV, LINE-OF-SIGHT, MICROWAVE, SUBSCRIPTION TV.

Multi-Purpose Remote Control. A preprogrammed remote control pad that can "communicate" with a large variety of VCRs, cable converter boxes and TV sets. Unlike other advanced remote controls, such as universal remotes, these multi-brand units, as they are sometimes called, cannot be "taught" the codes of other machines. Instead, they have been preprogrammed to work with many VCRs, TV sets and cable converters. See REMOTE CONTROL, UNIFIED REMOTE CONTROL, UNIVERSAL REMOTE CONTROL.

Multi-Purpose Zoom Lens. A video camera lens that offers several sophisticated features along with the conventional zoom function. Usually found on higher-priced cameras, the multi-purpose zoom lens provides automation for previously manual functions. For example, focus memory maintains the sharpness of the subject as he or she moves toward or away from the camera. Other features may include zoom memory, autofocus macro and AUTO FRAMING. See ZOOM LENS.

Multi-Screen Digital Freeze. A digital VCR feature that allows the viewer to fill the TV screen with several stationary images simultaneously. Bringing up this feature on the TV set does not affect the audio portion of the broadcast.

Multi-Stage Noise Shaping. A relatively recent enhancement built into some videodisc players that transfers unwanted noise into inaudible portions of the frequency spectrum. This is accomplished by a newly developed digital-to-analog conversion process, sometimes referred to by its acronym, MASH.

Multi-Standard TV. A digitally equipped television receiver capable of handling any video standard. Some of these TV sets have reduced all of the complex, sophisticated electronic circuitry to three chips.

Multi-Strobe. A digital VCR feature that continually and successively updates a series of strobes or still pictures generated on the TV screen by the strobe function. See STROBE.

Multivision. A system designed to feature simultaneously more than one image on the television screen. Early experiments in the 1960s with multivision TV made no impact on the public. Sony offered a model with three black-and-white nine-inch picture tubes placed side by side; RCA followed with a 25-inch color set and three 10-inch black-and-whites in one cabinet. The concept was dropped until 1979, when Sharp introduced dualvision, a TV set which presented a second smaller black-and-white image within its regular picture. In 1980 Sharp introduced a model in which nine different color channels could be viewed simultaneously, rotated or placed into Freeze Frame. See PIP.

Municipal Cable TV. A CABLE TV station owned by a local government. Fewer than three dozen exist of more than the 4,000 cable systems in the United States. Municipal cable differs from cable TV cooperatives which are subscriber-owned. See PUBLIC ACCESS TV.

Munsel Color Chip Chart. A video test chart used to check the color-producing ability of a video camera. Used more frequently by professionals, the Munsel chart contains specially prepared strips of color similar to the familiar color bar signal which is generated electronically. The chart can be used in conjunction with a COLOR VECTORSCOPE. The user first points the camera at the vectorscope and them aims it at the chart, noting any differences. See VIDEO TEST CHART.

Music Television. See MTV.

Music/Voice Switch. A tone control feature on some TV monitor/receivers. The music/voice switch is designed to provide the full frequency range of the audio system in the Music mode. When the switch is placed in the Voice or Speech position, the bass and treble responses are modified to minimize hiss and to sharpen voices.

Muting Circuit. An electronic feature built into virtually all VCRs to minimize instability or noise in the video signal. Some older machines go to a blank screen rather than display signal deterioration such as picture breakup. However, in newer VCRs, this muting lasts for a fraction of a second. Muting is sometimes called blanking. See also AUDIO/VIDEO MUTE.

N

Narration Microphone. A second microphone built into a video camera. This auxiliary mic allows the videographer to record narration while shooting on location.

Narrowcasting. A term applied to special- or limited-interest programming targeted at a small portion of viewers. For example, the Christian Network and other religious channels engage in narrowcasting, as opposed to broadcasting as carried by the three major networks, ABC, CBS and NBC.

National Captioning Institute. See CLOSED-CAPTIONED DECODER.

National Home Entertainment Show. An annual exhibition of video equipment, both hardware and software, aimed particularly at the consumer and hobbyist. Begun in 1981, the National Show usually draws tens of thousands of curious spectators to its displays of VCRs, TVs, video games and related paraphernalia. Besides the latest in video technology, the show features video art in the form of demonstrations and workshops in different areas. Another aspect of the show is the presentation of CASSIE AWARDS—a competition for those who are creative with their video cameras.

National Television Standards Committee (NTSC). The group responsible for setting standards for commercial TV broadcasting. Organized in the United States in 1940, the NTSC, made up of engineers of all the major and some minor companies, conceived in 1941 the black and white television standards and by 1953 the color television standards. It decided upon 525 line scans made up of two FIELDS of 262 1/2 lines each and at the rate of one frame (two fields) every 1/30 per second or 30 frames per second. Since TV broadcasting on a wide scale began in the United States, the NTSC standards for North America were the first to be adopted, but, unfortunately, were not the best in quality. Europe, which entered commercial TV a few years later, had the benefit of more technological developments and adopted different and higher standards with its PAL and SECAM systems, both incompatible with NTSC. The U.S. standard has been adopted by Canada, Japan and several South American countries. At one time NTSC was facetiously referred to as Never Twice the Same Color. See FIELD, FRAME, NTSC COMPOSITE VIDEO SIGNAL, PAL, SECAM.

Neck. The narrow cylindrical section of the cathode ray tube used in television sets and monitor/receivers. The neck extends from the funnel part of the tube to its base and contains the electron gun.

Negative Modulation. A technique employed to attenuate the energy transmission of bright images. Overly bright scenes, otherwise, tend to cause buzzing. Negative modulation is used by virtually all current broadcasters.

Negative/Positive Image Switch. See IMAGE REVERSAL.

Nesmith, Michael. Creator of the first prerecorded STEREO videocassette, "Elephant Parts." Produced by Pacific Arts, it was released in 1981. Ex-Monkee Nesmith's made-for-video tape provided comedy skits and musical numbers.

Neutral Density Filter. A transparent piece of smoked glass (or similar substance) placed on the front of a camera lens to restrict the amount of light entering it, thus affecting the SENSITIVITY RANGE of the video camera. For instance, without a neutral density filter, the sensitivity range may be from 5 to 6,500 footcandles; with the filter in place, the range may drop to 40–50,000 footcandles. A neutral density filter is sometimes built into a camera, in which case it is activated by a switch. It is used in very bright light such as snow scenes to give more flexibility to the iris which otherwise would be almost closed. Another use of the filter is to shorten the DEPTH OF FIELD. Since the iris must be opened wider when the filter is in place, only the subject focused upon will be sharp while objects in the foreground and background will be deliberately thrown out of focus. Neutral density filters are sold in a variety of ranges which require the lens to increase its f-stop one or more openings. See FOOTCANDLE, SENSITIVITY RANGE.

New York Expo of Short Film and Video. The oldest annual festival for short subjects in the United States. Founded in 1966, the N.Y. Expo presents awards for outstanding entries in fiction, documentaries, experimental work and animation. Directors whose early works were displayed at past Expos include film director Spike Lee and character actor Danny DeVito. See TOKYO VIDEO FESTIVAL, VIDEO EXPO NEW YORK.

Newvicon. An image pickup tube designed by RCA (who also invented the VIDICON camera tube). It helps to reduce SMEAR, BLOOMING and IMAGE RETENTION. Other tube designs include the SATICON and TRINICON. These image pickup tubes have largely been replaced in home camcorders which have adopted the CCD. See CAMERA TUBE, CCD.

Next Function. A feature, found on some VCRs, that allows the user to switch directly from Rewind to another mode. When Rewind is pressed, followed by another selected mode such as Eject or Play, the "next memory function," as it is sometimes called, will activate the unit to rewind the tape and then proceed to execute automatically the second function.

Nickel-Cadmium Battery. A battery used with portable VCRs that can be charged in less than two hours. The NiCad battery is more costly than its alternative, the lead acid type, also found in some portable systems.

Nickelodeon. A CABLE TV advertiser-supported network oriented toward children. Approved by the National Education Association, the channel presents talk shows, science programs, documentaries and cartoons to its millions of subscribers. Owned by Warner Amex, the service began in 1979. By 1990 it became the fifth most popular cable TV service with almost 50 million subscribers.

Nintendo Entertainment System. A home video game consisting of a control deck push-button controller that utilizes different game cartridges from the company and third-party manufacturers. By the late 1980s, Nintendo became the most popular video game on the market, selling more than three million systems in 1988 alone. By 1989 Nintendo and its licensees grossed $2.7 billion, capturing 80 percent of the video game market. This was accomplished by offering better design and technology than those provided by other systems and games that became popular in the early 1980s. In addition, the company maintains a tight rein over the 54 companies licensed to market Nintendo games, making certain that quality control is maintained. Nintendo also introduced the first home video game with a "save-game" feature, a concept heretofore found only on computer software games. In 1989 the company brought out Game Boy, a portable model of its game system that uses miniature cartridges. One million units were sold that year. See VIDEO GAME CARTRIDGE, VIDEO GAME HARDWARE, VIDEO GAME HISTORY, VIDEO GAME SALES, VIDEO GAME SOFTWARE.

Nipkow, Paul. Inventor of a scanning device in 1884 that was capable of producing about 4,000 pixels (picture elements) per second. Nipkow's scanner was the first apparatus that could analyze a scene for generating electrical signals applicable for transmission. Made up of a rotating disc with 18 small apertures laid out in a spiral pattern and stationed in front of a photo-electric cell, the scanner was capable of reproducing a scene by projecting it. The image passed through a similar, synchronized rotating disc that cast the image onto a screen. Although other experimenters tried to elaborate upon Nipkow's device in later years, it contained severe limitations, including an inadequate light source, too many mechanical parts and a weak optical system.

Noise. See VIDEO NOISE.

Noise Bar. A horizontal line across the TV screen, usually caused by misalignment of video heads and tape. Noise bars often occur in special effects modes such as FAST SEARCH or FREEZE FRAME. Some VCRs are relatively free from these noise bars, also known as INTERFERENCE, noise or video noise. These machines accomplish this in one of two ways. Some have four video heads, two for recording and playing back a program in one of the normal tape speeds and the other two heads for playing back special effects. Other VCRs have adopted digital recording, as opposed to analog recording, to eliminate the annoying noise bars and produce a better image generally. See DIGITAL VIDEO.

Noise Reduction System. A process designed to improve the sound quality of an audio or video unit by cutting down on the amount of hiss, thereby removing most of the noise without affecting high frequency response. There are various kinds of audio noise reduction systems: DOLBY, CX, dbx, Dynamic Noise Filter (DNF), Dynamic Noise Reduction, etc. Most noise reduction systems operate on a compression/expansion principle. Each system, however, goes about this in a different way. DNR and DNF are unlike the others listed above which mainly expand audio signals during recording and condense them during playback. Instead, these two systems employ a dynamic filter which eliminates high frequencies (mostly noise or hiss) whenever the signal is not strong enough to cover this noise; when the signal does cover the hiss, the filter lets the high frequencies through. Another way in which DNR and DNF differ is that they function only in playback to minimize existing noise while Dolby and most of the others cannot reduce noise already

in the signal. They have to be utilized during recording and playback, encoding and decoding the signal. See DYNAMIC NOISE FILTER, HISS.

Non-Interlaced Display. An enhanced scanning system utilized by some manufacturers of IDTV (improved definition television). This system doubles the number of scan lines from 262 1/2 to 525 every 1/60 per second. The result reportedly improves vertical resolution by eliminating the visible spaces between scan lines. Viewers of conventional TV images are often distracted by the horizontal lines that make up a screen image, particularly on larger sets. Instead of presenting two separate groups of lines (each composed of 262 1/2 lines), one following the other, non-interlaced display produces both sets of lines at one time. This creates a picture of greater density. See IDTV.

Non-Linear Electronic Editing. A process that does not limit the editor to follow a particular order or length of events in assembling a finished tape. Advanced editing techniques and digital editing consoles allow designated sequences to be stored in memory until the final sequence is ready. The editing console then reproduces the selected sequence from the original material onto a different tape on another deck. An early non-linear editing system used a computer disk that stored the scenes to be edited. The disk provided almost instant access to any point. These systems are both time-savers and creative devices. Some non-linear editing systems, sometimes referred to as tape-free editing systems, use a storage system of

Non-Transitional Digital Video Effects

several laserdiscs. Selected dialogue, effects and music can be recorded directly onto the discs from videotape without external facilities. Each video frame and sound byte is then instantly available to the editor by way of a keyboard and a control wheel.

Non-Transitional Digital Video Effects. Refers to correcting visual errors that appear on original videotape by using special sophisticated equipment such as a DIGITAL VIDEO EFFECTS SYSTEM during POST-PRODUCTION editing. A video camera operator may inadvertently pick up on tape an unwanted shadow, window glare or part of an overhead microphone. The DVE system allows the editor to eliminate these intrusions by enlarging the picture until the intrusive image is no longer in view. This type of function differs from TRANSITIONAL DIGITAL VIDEO EFFECTS.

Notch Filter. A TV monitor/receiver circuit that helps to eliminate horizontal interference or video noise that occasionally shows up on the edge of the screen. The notch filter suppresses an annoying frequency segment of the chrominance signal that often affects color contrast. Since these filters may also affect the sharpness of the screen image, some manufacturers add a manual notch filter switch to the monitor.

NTSC. See NATIONAL TELEVISION STANDARDS COMMITTEE.

NTSC Compatible. Any proposed, experimental or future high definition television system that will operate with TV sets built for the NTSC (National Television Standards Committee) standard. The problem with many of these advanced and improved HDTV systems is that they make the present crop of TV sets obsolete. See HIGH DEFINITION TV.

NTSC Composite Video Signal. A video signal composed of various types of information as designated by the National Television Standards Committee. This signal, known also as simply the composite video signal, contains the primary color (red, green, blue) signals which drive the respective color guns in the CATHODE RAY TUBE; sync pulses; and black-and-white picture signals.

Numeric Key Pad. That part of a remote control containing the ten numeric keys designed to operate various functions. With a VCR or TV set, for instance, the chief function of the numeric key pad is to change channels. With a videodisc player, on the other hand, the key pad serves to control direct chapter and track search.

O

OEM (Original Equipment Manufacturer). A plant that produces units for other companies. For example, Hitachi supplies Pioneer with its VCRs whereas Matsushita makes VCRs for General Electric, Magnavox, Panasonic and other companies. See VCR MANUFACTURERS.

Oersteds of Coercivity. Refers to the density of particles of a videotape; thus, it is a method of measuring the ability of tape to store information.

Off-Line Editing. Refers to the re-organizing of material on a videotape in preparation for editing but without actually recording the new arrangement. This may be done by electronically marking different segments of the tape or by composing a list of them. See ON-LINE EDITING.

Off-Tape Monitoring. A technique present in some professional video machines that permits viewing the videotape as it is being recorded. This is accomplished by employing two additional video heads utilized expressly for playing back the information as it is recorded. In copying tapes on home VCRs using two monitors, the picture played back from the set connected to the recording VCR is not the one being recorded, but the signals the VCR is receiving and transmitting to the TV set. Conventional home video recorders have two or four heads that either record or play back; they cannot do both simultaneously.

Omnidirectional Microphone. A mic designed to respond to sound from all directions. It is particularly effective when recording more than one person in a scene. Used by most video cameras, the omnidirectional mic is recommended for close recording. See MICROPHONE.

On-Camera Remote Control. A feature on or added to a VIDEO CAMERA for starting or stopping the VCR, switching between Record and Playback modes, using Forward and Reverse Visual Scan and operating Freeze Frame and Single-Frame Advance. Some video cameras offer this unit as optional, in which case it is usually connected to the bottom of the

On-Line Editing

camera or to a removable shoulder pod.

On-Line Editing. The final recording onto a videotape based on rearranged material from an original tape. See OFF-LINE EDITING.

On/Off Switch. See POWER SWITCH.

On-Screen Clock. A feature, found on some TV sets and virtually all TV monitor/receivers, that presents the time and day in large figures and is usually operated by a remote control button.

On-Screen Display. A menu or set of menus of a TV or VCR that appears on the screen and is designed to help perform certain operations. A TV monitor/receiver display, for example, may include menus for current time set, on/off timer; video adjustments of hue, color, brightness and sharpness; or audio adjustments of treble, bass and balance—all tuned by remote control. A VCR display may include prompts for programming the timer or checking the day and time functions. These items and more are usually found on the remote control. Some of these menus have submenus for even more subtle adjustments.

On-Screen Graphics. A picture-in-picture (PIP) display menu for the adjustment of the inset picture in addition to other on-screen displays.

On-Screen Programming. A VCR feature that permits the machine to be set up for recording by menu displays on the TV screen. Usually operated by remote control, on-screen programming provides either summary-style menus or individual program menus. Other special features include flashing prompts that ask for the next item to be entered or a changing line that lists the possible options. Once a feature only of the more costly VCRs, the on-screen programming function has trickled down to less expensive models.

On-Screen Prompt. A flashing or blinking sign or word that appears on the TV screen and is designed to cue the viewer to enter information. Usually part of a VCR's on-screen programming feature, the prompt shows up in conjunction with certain menus that make programming easier and virtually error-proof. Information may be entered from the remote control buttons.

One Touch Recording (OTR). A convenience feature on many VCRs designed to provide a simple procedure for instant recording without the normal programming steps. With either the ordinary power or the timer on and the proper channel tuned in, each touch of the OTR button activates the Record mode for either 15 or 30 minutes. With some VCRs, the first touch of the OTR button activates the Record mode while each successive touch gives 30 minutes of recording time up to a total of four hours. These times vary, depending on the VCR. When OTR is pressed, all other functions cease to operate, as in the typical timer position of other machines.

Open Architecture Television. A concept designed to do away with or limit TV obsolescence by allowing new circuit modules to replace old ones. The idea is to make the TV receiver so accessible that further advances in technology and equipment could be assigned to plug-in boards. These accessories, when inserted into TV sets, would provide the receivers with the necessary improvements so that they would not have to be discarded. Open architecture TV requires the use of computer chips and digital processing.

Open Channel. A channel on a TV receiver or a VCR that is not occupied by a local broadcaster. In most communities channel 3 or 4 is considered an open channel and can be used for transmitting a signal from a VCR without INTERFERENCE or BLEED THROUGH. Virtually all VCRs have a switch in the rear for selecting either of the two open channels.

Optical Animation. A professional video technique using image processing and computer-generated programs to create animated screen images. The process usually provides sophisticated swooping and spinning enhancements along complex motion paths that provide professional results.

Optical Recording. A videodisc whose audio and video information is encoded by a laser beam and whose signal is read by a beam of laser light. These discs have a longer life than videotapes since nothing but the weightless beam of light ever touches their surface. See VIDEODISC, VIDEODISC PLAYER.

Optical Scanner. In video, a pen-like device used in conjunction with a programming card to transmit information to a VCR. Part of the bar code programming process, the optical scanner is used by the VCR owner to mark the day, time and channel to be recorded. The information is then sent by infrared beam to the VCR and can be confirmed on the TV screen. See BAR CODE PROGRAMMING.

Optical Viewfinder. A mechanical device on a video camera that gives an approximation of what the lens sees. Similar to the viewfinder on less costly still cameras, it is usually found on the more inexpensive cameras. It provides no data about focusing or lighting. But it could be supplemented by a TV monitor connected to the camera; however, this procedure limits mobility. Each year fewer cameras use the optical finder system. Other types of finders include the THROUGH-THE-LENS and ELECTRONIC VIEWFINDER, the latter being the most prevalent.

Orthicon. An early camera tube of the 1930s. The orthicon improved upon the primitive ICONOSCOPE and IMAGE DISSECTOR tubes, both of which required extremely bright scenes to reproduce an image. The orthicon was replaced by the image orthicon tube in 1945 and was considered the first dependable and sensitive pickup tube. By 1952 it, too, succumbed to an improved version, the VIDICON, a smaller and less bulky tube. See CAMERA TUBE.

Oscilloscope. A testing instrument that measures electronic signals. It is utilized to align video components.

OTR

The oscilloscope, also known as a waveform monitor by video professionals, may be a portable field model that offers limited features such as vertical centering for adjusting blanking to zero IRE, rotation, sweep rate, IRE filter and gain boost. The fully featured studio waveform monitor is designed to provide more sophisticated signal analysis. See VECTORSCOPE, WAVEFORM.

OTR. See ONE TOUCH RECORDING.

Output. The terminals where the circuit or unit may supply the current, power, voltage or driving force. Such pieces of equipment as monitor/receivers and VCRs, for example, have audio and video outputs.

Overenhancement. A result of rotating the enhance control of an IMAGE ENHANCER too far. Overenhancement causes thin outlines to appear around the edges of the TV picture. The ideal adjustment involves moving the control to a point where the picture appears sharp but slightly grainy. Then the response control takes over, reducing the noise as much as possible without affecting the sharpness or detail in the picture. See ENHANCE CONTROL, RESPONSE CONTROL.

Oversampling. A digital video feature that uses a frequency several times higher than the standard rate for filtering purposes. The term refers to the number of times the special circuitry of a unit, such as a CD player, scans the digital information that is "read" by the laser. Oversampling transfers the noise band to a higher frequency where a special analog output filter eliminates much of the higher frequency distortion. The standard oversampling rate is 44.1 KHz. More costly CD and laserdisc players use advanced filters that measure the code at a speedier rate, usually in multiples of two or four. The higher the oversampling number, the more desirable the result. Some CD video combination players include an 8x oversampling digital filter, which places these units far above other models, generally rated at 4x oversampling.

Overscan. A term used in reference to a video camera's optical viewfinder which usually views more than the lens captures on the tape. In television, overscan refers to a larger image, of approximately 10 to 15 percent, projected by the electron gun than the one seen on the face of the TV tube. Manufacturers of television receivers often deliberately employ overscan since a TV image shrinks as the set ages. See UNDERSCAN.

Oxidation. A defect, chiefly associated with videodiscs, that shows up on a screen image as colored dropouts. Oxidation, sometimes called color flash, has been attributed to various factors, including a chemical reaction between the glue that binds the two sides of a videodisc and the reflective aluminum layer. One unfortunate characteristic of oxidation is its rapid acceleration, resulting in the eventual loss of a large part of the video signal.

Oxide. A general term used to describe the accumulation of magnetic particles which form the coating of videotape. Audio and video signals are electronically printed on the oxide COATING. The term "oxide" also applies to the flaking particles which eventually fall from the tape and clog the video heads, causing white specks to flash across on the screen during playback. This problem is known as DROPOUT. See COATING, VIDEOTAPE, VIDEOTAPE HISTORY.

P

Pac-Man. Originally a highly successful arcade video game consisting of a disc-eating creature that travels through a maze. Atari adapted its coin-operated game for its VCS home system in 1982 but with some modifications. Some of the graphics were missing as was the original sound track.

Paik, Nam June (1932–). VIDEO ARTIST, co-developer of the VIDEO SYNTHESIZER. Born in Korea, Paik has become the most widely recognized practitioner of VIDEO ART. The Whitney Museum of American Art in New York City showed 60 of his works in the spring of 1982. Using stills, moving images, technology, myriads of TV sets and other paraphernalia, Paik demonstrated the immense creative potential of the art form. Such works as the abstract, sculptural "Magnet and TV" (1965); the satirical "TV Chair" (1968) which positions a TV set face up under an ordinary kitchen chair; the humorous "TV Bra for Living Sculpture" (1969); the architectural "TV Garden" (1974), consisting of 30 TV sets placed among 30 palm trees—all reveal the visionary genius of Paik. Perhaps his most thought-provoking work is his "TV Buddha" (1974), which suggests that technology is draining the human spirit. In it an 18th-century statue stares eternally into a TV screen which is telecasting the figure's own image. Paik and his colleague SHUYA ABE are credited with developing the video synthesizer.

PAL (Phase Alternating Line System). A system of TV broadcasting used in England and many other countries. Not compatible with the North American NTSC (National Television Standards Committee) system, PAL, invented in 1961, has a 625-line scan picture delivered in 25 frames per second. Besides providing a better image generally, PAL has an improved color transmission over NTSC. This discrepancy results from America's entry into commercial TV in 1948 while Europe's PAL system, having the advantage of technological advancements, entered TV broadcasting in the 1950s. PAL is used in China, Australia, India, Brazil and Argentina as well as in most Western European countries. SECAM, another system used in Europe, is incompatible with either PAL or NTSC. See NTSC, SECAM.

Paley, William S. Legendary head of Columbia Broadcasting System (CBS). In 1927, at age 26, he bought a 16-station radio network, United Independent Broadcasters. In less than two years it grew to 47 stations and was renamed CBS.

Pan. To move a video or motion picture camera slowly sideways or up and down while shooting to create a panoramic result. The pan shot is also used to follow a moving subject.

Panning and Scanning. The selection by studio technicians of portions of a wide-screen theatrical film so that they fit into the conventional TV screen format. Digital transfer is another, more sophisticated, technique of presenting wide-screen films on the video screen. See ASPECT RATIO, DIGITAL TRANSFER, SCANNING.

Parabolic Antenna. A 10- to 15-foot-diameter dish antenna at whose focal point is a permanently attached FEED HORN. The antenna, usually made of aluminum or Fiberglas, is mounted on a swivel base so that it can be realigned for each satellite. The critical factors of a parabolic antenna are its size and its location. A 10-foot dish may work well in some areas of the country, but not in others, sometimes only 10 or 20 miles away. Dishes placed in mountainous or tree-lined locations limit the effectiveness of satellite broadcasts. The antenna receives the MICROWAVE signals from the satellite and changes them to electrical impulses. Next, it increases the signals so that they can operate a TV set. Finally, it transmits these modified signals through an RF CONVERTER which uses any open channel on the TV receiver. See SATELLITE, SATELLITE TV, SPHERICAL ANTENNA.

Parallax Stereogram. A technique used in three-dimensional television in which vertical slats are employed to enhance the effect of 3-D. Although no special glasses are required by the viewer, the process is not compatible with 2-D (conventional television). See THREE-DIMENSIONAL TELEVISION.

Parental Lock. A VCR or TV set feature that allows the owner to prohibit one or more channels from appearing on screen. A boon for parents who want to prevent their children from seeing certain objectionable programs or channels, the parental lock is sometimes known as lock down or parental channel lockout.

Passive Mixer. An accessory used with audio to combine and control the level of various signals, without the addition of electronic circuitry or components. Since signals entering this mixer are not amplified, they undergo some loss in strength. Passive mixers do not affect the nature or quality of the signal other than its level of attenuation or boost. In addition, they do not need any electrical power. See ACTIVE MIXER, MICROPHONE MIXER.

Passive Phase Direction Autofocus

Passive Phase Direction Autofocus. An alternative focusing system found on some camcorders. The system, which operates through the lens, does not use any conventional infrared transmitter, thereby eliminating possible focusing problems if an object passes between the lens and the subject being shot. See AUTOFOCUS.

Passive Switcher. See SWITCHER.

Patch Box. An accessory unit designed to accept different video components into its inputs, simplifying the selection of these sources to be played on the TV receiver. A patch box is a relatively inexpensive method of interconnecting antenna, cable, pay TV and VCR inputs and directing them through TV sets, a VCR, etc. It is not as permanent a connection as that of a SWITCHER, although it is not as costly. The patch box is also known as a patch panel.

Patch Cable. A wire (usually in pairs) with the proper connections used to hook up one VCR to another for copying or editing purposes. One cable connects from the video output of the playback machine to the video input of the recording VCR, while the second cable plugs into the audio output of the first unit and the audio input of the second machine. VCRs of different formats may require special connections such as RCA plugs or miniplugs. See COPYING, EDITING.

Patch Panel. See PATCH BOX.

Pause. A feature on all VCRs which stops the movement of videotape for a short time during the recording process while holding the tape against the video heads. The Pause mode is used instead of Stop to produce less picture breakup when the machine finally returns to Record. Virtually all VCRs warn against extended use of Pause. As a safety precaution, machines provide an automatic shut-off if the Pause mode is activated for more than five or six minutes. Pause circuits differ from one VCR to another. Some machines use a high/low sensor while others employ a pulse on/pulse off circuit. In the Pause mode the PRESSURE ROLLER which normally holds the tape firmly against the CAPSTAN becomes disengaged so that the tape stops and remains pressed against the VIDEO HEADS. Pause differs from Still or Freeze Frame, a mode which operates by freezing a picture on screen during playback, but not during record. See FREEZE FRAME.

Pay-Per-View. A CABLE TV/PAY TV service which allows viewers to pay for special programs of their choice. A special box, called an addressable decoder supplied by a local cable company, is attached to the TV set and releases access to paying customers. The viewer is then billed for the service, along with the regular monthly cable TV fee. PPV movies usually appear from one to two months after their release on videocassette. PPV programs are transmitted to cable companies by way of satellite on scrambled channels. Viewers can order these programs in one of two ways— by telephoning the cable service or by pressing an appropriate button on the addressable decoder box. Some consumer groups see pay-per-view as a mixed blessing, believing that the public will eventually be paying for "specials" that it currently receives for free. According to reliable estimates, pay-per-view, which has been around since the 1950s, has produced billions of dollars in revenue. See ADDRESSABLE BOX.

Pay TV. A programming service offered to subscribers for a monthly fee. It can be carried by MICROWAVE, as part of a CABLE TV system or by telephone wire. It can be supplied independently or as part of a larger system, in which case the service is called a TIER. The programming is varied. Some pay TV systems offer only feature films while others feature sports, cultural programs, R-rated films, children's programs, etc. Some popular pay TV channels are BRAVO, CINEMAX, THE DISNEY CHANNEL, HOME BOX OFFICE, THE MOVIE CHANNEL and SHOWTIME.

PCM. See PULSE CODE MODULATION.

Peak. In video, a signal strength's highest point. Peaks can be measured with video level or VU meters.

Peak Brightness Level. A light output measurement used with rear projection TV systems. Peak brightness, or relative brilliance, is measured in FOOTLAMBERTS. Many current rear projection TVs produce enough brightness to compete with conventional, direct-view television sets. Peak brightness level differs from light output, the latter used in reference to front projection TV systems and measured in lumens.

Peak Level Meter. A feature found on some videocassette recorders designed to indicate the peak volume level of the recording and playback. Similar in function to a VU METER, the level meter displays a string of horizontal lights instead of the conventional dial and indicator. See PEAK.

Pedestal Level Control. A function of industrial-type video cameras which affects the contrast levels as well as the shades of gray of the screen image. The pedestal level control is also designed to balance the images of multiple cameras used simultaneously. See BLACK LEVEL CONTROL.

Personal Video. See PORTABLE VIDEO.

Phase. Refers to the timing of the voltage highs and lows of an AC (alternating current) signal. Two signals are in phase when they have simultaneous highs and lows, the same frequency and produce two times the peak-to-peak voltage. Phase plays an important role in video, audio and RF signals.

Phase Alternating Line System. See PAL.

Phase Cancellation. See PHASING.

Phase Lock Loop (PLL). One of several automatic-fine-tuning methods designed to keep a broadcast frequency aligned, or in phase, with a constant, internally generated reference frequency. PLL, which uses a special electronic circuit, functions in a more advanced manner than Automatic Fine Tuning (AFT). Quartz-synthesized tuning, which employs the constant vibration of a quartz crystal, is even more sophisticated than the phase lock loop system.

Phase Noise. Refers to the random phase instability of a signal. Phase noise is one of several types of degradation that affects the NTSC (National Television Standards Committee) picture quality. Continuous research and studies have been made to correct these influences so that transmission signals can be improved to match the improved quality TV monitors and television in general. See DEGRADATION.

Phase Reference Switch. On a COLOR VECTORSCOPE, a control designed to bring two video signals "in phase" or together simultaneously at the same point. For example, the switch is activated to test whether the signals from two video cameras are in phase when using a genlocked system with an external sync generator. The two vectors produced by the cameras will appear superimposed on the GRATICULE of the vectorscope if they are in phase. If they are not, the signals are out of phase and the phase adjustment of the cameras or the cable lengths should be checked.

Phasing. Refers to the cancellation of two sound waves that move in opposite directions. The problem occurs with the improper mixing of two microphones or stereo tracks to mono. Separate stereo and mono mixes usually solve phasing or phase cancellation problems.

Philips VCR Format. A 1/2-inch cassette format not compatible with either BETA or VHS. The Philips V-2000 VCR, marketed in the early 1980s by Germany's Grundig company, had one speed and could record for four hours on each side of its videocassette for a total of eight hours. The cassette size was similar to that of the VHS. Although the tape was 1/2-inch wide, two 1/4-inch tracks were recorded on it. This permitted the cassette to be flipped over for extended play without lowering tape speed which usually reduces picture quality. Other features included noise-free automatic scanning, FREEZE FRAME and SLOW MOTION, an exceptionally FAST FORWARD and Rewind and programmability for 100 days and five different events.

Phono Jack. A popular-size receptacle on VCRs and other audio and video components. Known as the RCA phono jack, it accepts the RCA phono plug. On VHS machines phono jacks are used for both audio and video inputs and outputs. See CONNECTOR.

Phosphor. The chemical used for coating the inside face of a CATHODE RAY TUBE. The phosphor glows when it is struck by electrons. The intensity of the glow depends on the strength of the electrons. The stronger the beam, the brighter the phosphor. However, brighter pictures do not necessarily guarantee better pictures, since they tend to sacrifice color fidelity. Some manufacturers have experimented with modifying the green phosphors to improve color accuracy.

Phosphor Color. In projection TV, the precise color at which the material at the front of the screen glows when it is struck by the electron scanning beam. Cathode ray tube projectors often use phosphors that fall between high light output and exact color rendition. Other projection TV systems face the same dilemma as a result of trying to produce maximum light output while competing with ambient light usually present when projection TV is operating. See PROJECTION TV.

Photo-Diode. In a laser-optical videodisc player, the component which conducts electricity proportionate to the amount of light that strikes its surface. The signal formed by this process carries two audio tracks as well as the video material. LaserVision players do not use a stylus which physically touches the videodisc. Instead, a beam of light "reads" the information encoded as microscopic pits on the disc. The fluctuations in the light intensity reflected from these pits are focused on the light-sensitive surface of the photo-diode. See LV VIDEODISC SYSTEM.

Pickup Device. See IMAGE SENSOR.

Pickup Response. See POLAR RESPONSE.

Picture Carrier. The carrier frequency above the lower frequency limit of a standard NTSC television signal. Sometimes referred to as the luminance carrier, the picture carrier is used in color television to transmit luminance information. Its frequency is rated at 1.25 MHz above the lower limit.

Picture Contour. See HORIZONTAL IMAGE DELINEATION.

Picture in Picture (PIP). A TV receiver feature that sequentially scans several channels and displays their programs in the form of inset pictures in the main screen image. This permits the viewer to survey the contents of other programs while simultaneously watching a main program. PIP, which depends on digital circuitry for its special effects, allows the main picture to be swapped for the inset. In addition, stereo sound can be exchanged between the two pictures. Some TV monitor/receivers can display up to nine images at a time, freeze the PIP image, store the main image for later recall or view a strobe effect of nine images simultaneously. The first viable use of PIP appeared in 1980 with Sharp's DualVision TV set which was capable of producing a 4-inch b & w picture within a 17-inch color image.

Picture Indexing. See ELECTRONIC PROGRAM INDEXING, INDEX SEARCH.

Picture Search. See INDEX SEARCH, VISUAL SEARCH.

Picture Search/Lock. A VCR feature that permits the user to view the contents of a videotape in forward or backward mode at up to about 7 times the normal speed in Standard Play mode or about 21 times normal speed in Extended Play. See SEARCH MODE.

Picture-Tube Brightener. A device added to an old picture tube to augment the brightness of the image. The accessory increases the electron emission from the cathode by raising the filament voltage of cathode emissions that fall below normal.

Piezoelectric Microphone. A microphone containing a ceramic or crystal element that produces a signal voltage whenever the movement of the diaphragm puts stress on the element. these inexpensive mics, also known as ceramic or crystal mics, have a limited frequency range and are easily affected by changes in temperature. See MICROPHONE.

Pilot Light. Refers to a special light that indicates whether a unit or one of its parts is functioning. The pilot light usually maintains a particular position or color to differentiate it from other similar lights. Some VCRs, for example, may have a yellow light to signify that the machine is in Record mode, an amber light for the Pause mode and a green light which designates that the power is on.

Pin. A part of a multi-pin connector used between a video camera and a VCR. Each pin of a 10- or 14-pin connector is designed for a special function. For example, one may be for video signal output, another for video input. The VHS format uses the 10-pin connector while the BETA cameras utilize the 14-pin configuration. Of course, with the introduction of camcorders, which combine the camera and recording components into one unit, many of these pins have become unnecessary. See MULTI-PIN CONNECTOR.

Pinch Roller. An integral part of a VCR designed, along with the capstan, to help control the tape as it moves along its path. Usually composed of rubber, the pinch roller presses the videotape to the capstan. See CAPSTAN.

PIP. See PICTURE IN PICTURE.

PIP Accessory. A video unit that takes advantage of digital technology by providing insert pictures so that the viewer can watch two live broadcasts at the same time. VCRs, videodisc players or video cameras can be connected to provide the second source image. The accessory, which may come equipped with one or two built-in tuners, produces pictures instantaneously. In addition, the insert can be changed in size and positioned anywhere on the screen. Other features include freezing the action of the inset, exchanging the main picture with the inset and monitoring another location through the use of a video camera.

PIP Source Mode. A feature, appearing only on some TV receivers, that allows the viewer to select the source, such as another television channel or a VCR, for a picture-in-picture. The PIP source mode usually is displayed as one of the features of a special screen menu and is activated from a remote control.

Pipeline Architecture. A special hardware process, used with some character generators, designed to regenerate an entire video display for each video field. This technique adds flexibility when animating characters in real time. Pipeline architecture differs from the conventional frame-buffer method used with many character generators.

Pirate Box. A picturesque term for an illegal decoding device that allows unauthorized viewers to watch otherwise scrambled satellite transmissions of cable services. These "black boxes," which may cost as much as $1,000, chiefly interest the more than two million owners of satellite TV systems. Title 47, Section 605, of the United States Code, warns that the penalty for unauthorized interception of satellite signals is two years in prison and a $50,000 fine. Various agencies, including the F.B.I. and the Motion Picture Association of America, have vowed to go after and prosecute the manufacturers and distributors of these decoders.

Pixel. In video, a term referring to a picture element. The number of pixels helps to determine the sharpness of the picture. In high definition TV, with its 1,125 horizontal scanning lines, the pixel number is said to be five times greater than that of the NTSC 525-line standard, resulting in a sharper picture than that of 35mm movie film.

Pixilation. A rudimentary method of creating video animation. The Record mode is pressed to record a subject, then stopped. The subject is moved slightly while the camera is in the off position. The Record-Stop cycle is then repeated to create the animated effect. See INTERVALOMETER, TIME LAPSE VIDEO.

Play. A control button on all VCRs which brings the information on the videotape to the TV screen in the form of a picture. In more advanced machines, the Play mode when activated permits the use of certain special effects. If REWIND is pressed halfway, reverse VISUAL SCAN may be activated; if FAST FORWARD is pressed in a similar manner, VISUAL SCAN may be engaged. Other effects such as SLOW MOTION and FREEZE FRAME operate when the machine is in Play position.

Play Speed Indicator. A feature on many VCRs which tells the tape speed mode when the machine is in Play. Most VCRs simply play back a cassette automatically in the speed in which it has been recorded, regardless of the selector button. The indicator presents no problem in Record or Stop mode, since the machine will operate in the function that it is set for. But since the VCR speed is activated automatically by electronics when it is in Playback, it is difficult to determine the mode without the play speed indicator.

Playback. A function that permits a VCR to convert the recorded magnetic tracks on the videotape into a signal for display on a TV screen, usually with the help of an amplifier and speakers. The same video heads used for recording also "read" the information on tape during playback.

Playback Amplifier. In video, a circuit that increases the VIDEO SIGNAL in the VCR before it is supplied to a TV receiver.

PLL. See PHASE LOCK LOOP.

Pludge Pulse. A signal, found on some color bars, that helps in adjusting TV screen brightness. The best luminance occurs when the brightness control is turned so that the pludge pulse, or gradations on the color-bar pattern, is barely visible. Color bars often appear before a channel begins broadcasting or after it goes off the air. A professional test-signal generator produces an official color bar, but the unit sells for more than $1,000 dollars.

Plug. The male connector or counterpart to the JACK. The audio and video inputs and outputs on components are also called audio and video jacks into which compatible plugs are connected. However, plugs are often referred to as jacks and vice versa. The plug, such as the RCA phono plug, is usually the end part of a cable or wire. See AUDIO INPUT, AUDIO OUTPUT, VIDEO INPUT, VIDEO OUTPUT.

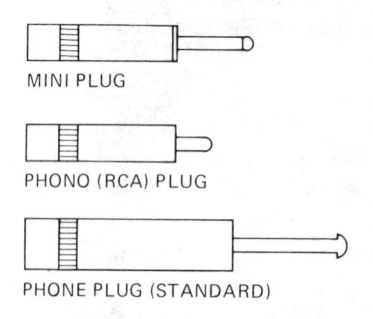

MINI PLUG

PHONO (RCA) PLUG

PHONE PLUG (STANDARD)

Figure 27. Plugs.

Plumbicon. A high-quality, costly image pickup tube once used in professional broadcasting video cameras. The Plumbicon tube was partially responsible for the development of color TV broadcasting in 1964. The tube is alleged to feature greater sensitivity, higher resolution and less lag characteristics than the SATICON camera tube.

Polacoat. A special material used in making REAR PROJECTION SCREENS for FILM-TO-TAPE TRANSFERS. A sheet of Polacoat is usually stapled to a home-made wooden frame and a projected image is cast on the rear of the screen. A video camera captures the bright image from the front. Polacoat is preferred over front projection beaded screening which diffuses the light instead of concentrating it. See FILM-TO-TAPE TRANSFER.

Polar Mount. In a SATELLITE TV system, the base of an antenna designed to be aligned on true north. The polar mount permits the antenna to be moved by remote to the appropriate angle so that it can receive signals from any SATELLITE.

Polar Response. The pattern that illustrates the direction or directions from which sound waves reach a microphone. Polar response patterns are usually circular. For instance, an OMNIDIRECTIONAL microphone responds to sound from all directions—in front, at the rear and at the sides of the mic. Polar response is also known as microphone pickup response, pickup response and polar response pattern.

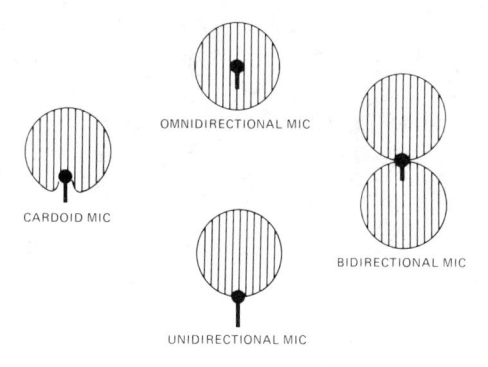

Figure 28. Polar response patterns.

Polarity. The form in which light waves vibrate.

Polymorphic Tweening. A dramatic effect used by professional videographers that changes the shape of a screen object into another shape by controlling motion, timing and perspective. This special effect is accomplished by blending image processing and animation with the help of a computer-generated program.

Pong. The first successful electronic video game. Invented by Nolan Bushnel and introduced by Atari in 1972, "Pong" now seems primitive when contrasted with today's sophisticated visual and audio effects incorporated into video games. Although it was a simple black-and-white nonprogrammable game, its success led to other companies invading the video game field with similar products. See VIDEO GAME, VIDEO GAME CARTRIDGE, VIDEO GAME SYSTEM.

Portable Video. A hand-held portable television or VCR unit. The history of personal video ranges from small black-and-white TV sets introduced by Philco in 1960 to a combination VCR and color TV monitor. The compactness of these advanced systems owes its success in large part to the Sony's introduction of the 8mm videotape format in 1985. These portable videos usually weigh less than three pounds and have been compared in size to a large sandwich. The flat screen LCD (liquid crystal display) measures about three inches diagonally. Portable video is sometime described as personal video.

Portable Video Game. An electronic game that can be hand-held, taken anywhere and operates on either LED, LCD or fluorescent display. These games were popular for a while in the early 1980s but have since failed to capture the consumers' imagination. The LED (light emitting diode) with the familiar red display was used in many portable games by Mattel and Entex. The LCD (liquid crystal display) resembled a mini-TV screen, was less costly to produce, used less power than the LED and provided a more accurate video image. The last type of hand-held game, the fluorescent display, was the only system to

199

Portable VCR

use color and feature quality graphics. Portable video games lacked the sophisticated programs, visuals and sound effects of conventional video games, but they sold for much less. See VIDEO GAME, VIDEO GAME SYSTEM.

Portable VCR. A small and light version of a table-top or home model VCR. The portable can be taken into the field and operates on batteries as well as with AC power. First introduced in 1978 (a VHS model by JVC and a BETA by Sony), the portable, designed for the consumer, climbed in popularity in the early 1980s. The unit was used both in the home and in the field with a video camera that usually required a separate recording unit. Originally a stripped-down VCR with few features, one speed and only two video heads, the portable VCRs soon grew in sophistication to rival their big brothers, the table-model VCRs. The portable VCRs usually consisted of two units, a VCR DECK and a TUNER/TIMER, which could be purchased separately or as a complete system. The portable was powered several ways, depending on the model: (1) the internal battery which was able to handle both the recorder and camera for up to one hour, (2) the AC adapter, (3) an external battery pack for two additional hours of recording and (4) a cable connected to a car battery. See PERSONAL PORTABLE VIDEO, TUNER/TIMER, VCR DECK.

Portholing. An effect created by using lenses at full wide angle or keeping the iris open to its maximum. Either of these conditions may cause the video image to "roll off" at the corners, resulting in a distorted background. Portholing, especially critical when doing CHROMAKEYING, can be avoided by checking the video signal in a WAVEFORM MONITOR.

Position Identification. A professional tape-editing procedure that first uses a cue tone, then a time code. Considered an early version of VIRTUAL EDITING, position identification editing made edit reviewing possible for the first time.

Positive Ghost. A ghost signal displayed on a television screen and containing the same tonal variations as those of the original image. See GHOST.

Positive/Negative Switch. See IMAGE REVERSAL.

Positive Trapping System. A technique employed by some CABLE TV operators to prevent non-subscribers from receiving a PAY TV service. A signal is transmitted in the center of a pay TV channel similar to SHOWTIME, causing picture breakup and sound instability for unauthorized receivers. The interfering signal is eliminated for subscribers by the use of a special "trap."

Post-Production. The stage of producing a TV program requiring editing. Post-production work almost invariably requires more time than the actual shooting. There are studios with professional equipment available, especially in larger cities, where this phase of video production can be done.

Power Fluctuation

Figure 29. A post-production system with graphics, switching and other features designed for field news operations and industrial use. (Courtesy The Grass Valley Group, Inc.)

Posterization. A VCR feature that permits changing the gradation of images during the editing process. Posterization, sometimes known as polarization, often provides several gradations as it exaggerates colors for a cosmic sunspot effect. This feature appears only with digital videocassette recorders. Professional editing consoles also offer this capability among their many special effects.

Potentiometer. In video, a device on the control unit of a VCR modification kit. It is designed to decrease the speed of the slow/still mode. When turned on fully, the potentiometer creates a FREEZE FRAME. The modification kits are aimed at adding special effects to older model VCRs.

Power Adapter. See AC ADAPTER.

Power Belt. A portable power source for a video recorder. Worn around the waist like a conventional belt, the accessory is usually made of leather, contains nickel-cadmium cell pockets and includes a built-in overnight charger. The power belt is basically used in industrial/educational applications with broadcast-type video cameras.

Power Fluctuation. A condition in video caused by a VCR or TV set connected to the same AC line as a major appliance such as a refrigerator or air conditioner. Power fluctuation can cause vertical distortion on the top third of a VCR picture. See INTERFERENCE, INTERFERENCE SUPPRESSER.

Power/Intensity

Power/Intensity. A control usually found on the front panel of a vectorscope. This knob functions both as an on/off switch and a brightness control for the vectors which are displayed on the CATHODE RAY TUBE screen. SEE COLOR VECTORSCOPE.

Power Saver. A video camera feature that automatically shuts off the power of the camera after a certain time to preserve battery life.

Power Switch. The electrical component that connects or disconnects one or more pieces of equipment from the power line. The power switch is also referred to as an on/off switch.

Power Zoom. An electronic trigger-type control that permits the video camera lens to be moved in and out, thereby increasing or decreasing the screen image size of the subject. It is the most sophisticated of zoom controls, the others being the mechanical zoom ring and the zoom ring lever. See ZOOM LENS.

Pre-Echo Effect. See PRINT-THROUGH.

Premiere. The ill-fated PAY TV network that was formed in 1981 by major film studios and designed to compete with HBO. In that same year a federal court ruled against Premiere on anti-trust grounds. It was owned in part by Paramount, 20th-Century Fox, Columbia, Universal and Getty Oil.

Premium Channel. A pay-TV cable channel that the subscriber must order separately from the basic cable charge. Premium channels, such as sports, foreign programming and those offering recently released films, are scrambled so that other viewers cannot gain access to the programming.

Premium Videotape. See HG.

Pre-Production. The preparations required before the actual video camera recording begins. This may include script writing, rehearsing, lighting, etc. See POST-PRODUCTION.

Prerecorded Tape. A commercial videotape that is recorded with a specific program, usually copyrighted and packaged for sale or rental by retail stores or through mail order. These tapes come in both BETA or VHS formats and are usually recorded in the fast speed of each format (Beta II and VHS SP). Many prerecorded tapes are encoded with an ANTI-PIRACY SIGNAL to prevent unauthorized duplication. In the late 1970s and early 1980s, when several film studios refused to release prerecorded tapes of their blockbuster features, many illegal copies of these coveted movies became available—for a premium price. The studios read the writing on the wall and grudgingly began to release their films on videotape. The sudden explosion of numerous video rental shops that began in the early 1980s—an unexpected national phenomenon—reaped huge rewards for those same studios and ushered in a new industry. Various catalogs list thousands of titles of prerecorded programs. Film critics occasionally highlight the best of the new videotape releases, and newspapers offer regular columns devoted to prerecorded tapes. See THE VIDEO SOURCE BOOK.

Pre-Roll. A VCR feature, found chiefly on S-VHS decks, that runs the playback video for a few seconds—to attain the appropriate head and tape velocity and sync—before entering the recording mode. The purpose of this procedure is to provide greater accuracy in recording and editing while reducing picture breakup.

Presentation Products Magazine. A monthly trade publication covering slides, video, graphics and other forms of projection and presentation, all aimed at corporate, education, government and advertising institutions. Articles and reviews deal with systems hardware and software as well as supplies. Much of the subject matter focuses on how presentation users are finding new ways of applying the latest technology. The magazine has several departments, including one that describes new products and another containing news notes about personnel changes in specific companies. Plenty of attractive graphics accompany the written material.

Pressure Roller. A flat, horizontal metal arm with a small vertical rubber cylinder at the end to hold the videotape firmly against the CAPSTAN which controls the speed of the videotape. The pressure roller is disengaged when the VCR is in Pause mode, causing the tape to remain stationary but in contact with the VIDEO HEADS. See CAPSTAN, PAUSE CONTROL.

Previsualization. A relatively recent technique used in filmmaking to transfer and assemble still images of scenes and artists' conceptions onto videotape accompanied by the appropriate sound effects and dialogue. Previsualization gives the director a better opportunity to perceive an early version of his films. The director FRANCIS FORD COPPOLA is a proponent of this technique.

Primary Colors. In video: red, green and blue. Green is the reference color to which the others are registered and adjusted; it is the most stable. Blue is usually used as a background in special effects since it is the opposite of flesh colors. Blue is also used with white titles and credits because video is a high-contrast medium. Red and orange are less stable and apt to streak. Yellow is considered a difficult color to capture on television. Yellow is achieved by subtracting the blue from the green. White is an equal mix of the three primary colors.

Print-Through. The superimposing of the audio and video signals of one part of the tape onto another. This problem may arise if the tape is very tightly wound and stored for long periods of time. This causes one layer to "print through" part of its signals onto adjacent layers of tape. However, this appears to be a minimal problem with videotape. Print-through is also known as pre-echo effect.

Pro Logic. In audio, a term that refers to the directing of surround sound signals to different channels with greater effectiveness.

Proc Amp

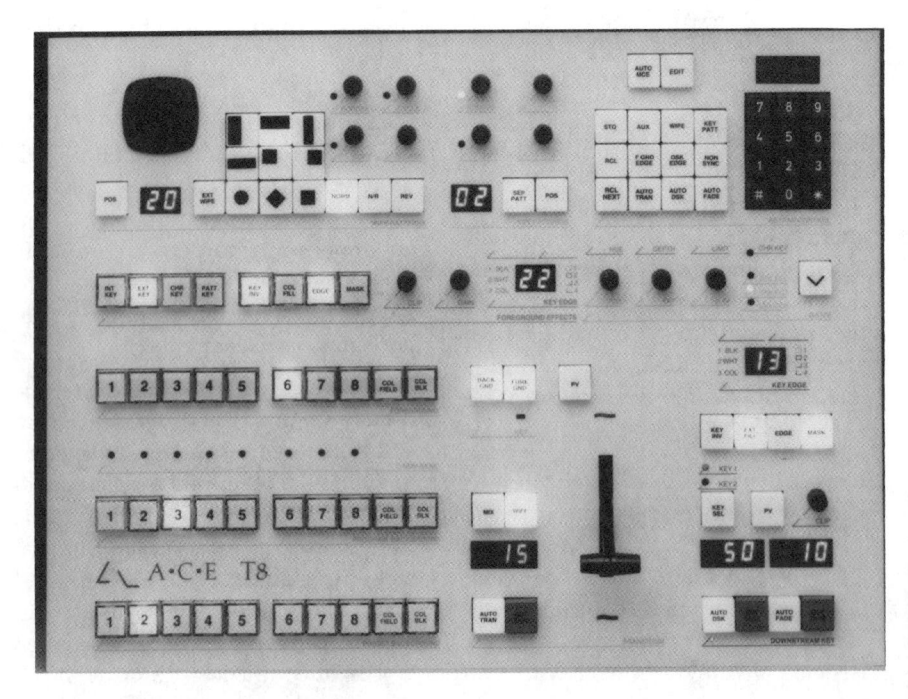

Figure 30. A production switcher that features flexible keying capabilities, split-screen patterns, memory-controlled effects and 10 inputs. (Courtesy Midwest Communications Corp.)

Proc Amp. A processing amplifier that permits individual control of certain aspects of a video signal, such as color hue and saturation as well as fading to and from black. Sometimes mistaken for a COLOR PROCESSOR, the proc amp increases the number of corrections and special effects of the processor. A proc amp can be used between VCRs to correct color or prevent color-fringing during black-and-white dubbing. Professional proc amp units stabilize the picture by eliminating the old synchronizing pulses and creating new ones. Sync pulses control the precise timing of video signals inside a VCR or video camera. Some proc amps employ a joy stick to control overall color. Most models incorporate a distribution amplifier as part of the unit. See FRINGING.

Processor Loop. An electronic connection on various units designed to link up other units such as a color corrector or special effects generator. A processor loop (sometime referred to as "loop through") can accommodate several video signals, including terminals for audio, composite video and S-Video signals. In addition, the connection can increase the number of components attached to an audio/video switcher.

Production Switcher. A professional switcher/fader used chiefly in broadcast, production and post-production

Programmable Timer

work. The sophisticated unit offers such features as 24 video inputs, a control panel divided into groups, informative displays, dual generators capable of producing scores of transition wipes, and pattern keys for masking, etc. In addition, the production switcher usually provides fade-to-black, a split-screen generator, a memory-controlled effects function that can store up to 64 predetermined set-ups and full event auto-sequencing. Other models may offer a range of digital effects, including image compression and rotation, creation of three-dimensional effects and modification of screen image perspectives. See DIGITAL SWITCHER, SWITCHER.

Program. As a noun: any complete show, concert or other event that appears on a videotape, disc or TV set. As a verb: to set up, schedule or organize the recording capabilities of a VCR or the playback functions of a videodisc player. A VCR can be programmed to record one or more events from one or more TV channels for future viewing. Many videodisc players can be programmed to play a series of chapters or tracks in any order the user wishes to arrange them.

Program Indexing. See ELECTRONIC PROGRAM INDEXING.

Program Overlap Warning. A VCR feature that automatically detects overlapping conflicts in programming times. If one selected time range spills over into another time span already preset, the program overlap warning mode blinks an alert so that an adjustment can quickly be made.

Program Start Locator. A VCR feature that is part of the INDEX SEARCH. Usually operating from a jog/shuttle dial, the program start locator is activated by moving the dial or search function to a desired starting point of an edit and pressing a button to store the location. VCRs which offer this feature often provide for up to 99 individual index searches.

Programmable Scan. See AUTOMATIC CHANNEL SCAN.

Programmable Speed. A VCR feature that permits the presetting of the videotape speed in which a program is to be recorded when the viewer is not present. Since TV programs to be set for recording may differ in content (talk shows, dramas, musicals), the viewer may wish to use a different speed for each show. Talk shows, for example, do not need standard play (SP mode), so the viewer can set the six-hour SLP mode while using the SP mode for musical to ensure the best audio reproduction capable on the VCR.

Programmable Timer. A feature on VCRs which permits setting the timer to record more than one event on different days and on various channels. The timer consists of a digital clock, control panel, indicator lights and a microprocessor capable of programming the VCR. The original timers on early machines could program only one event in a 24-hour time period. The serial timer was next: it could activate the VCR every 24 hours, but the channel remained the same. The programmable timer, introduced in 1978

Programming Card

by RCA, allows programming multiple events on different channels for several weeks in advance. Machines in the lower price bracket usually permit four events to be programmed in advance while more costly models often feature eight events. Of course, the switching of channels depends on the type of tuner. The original mechanical/rotary tuner allowed only for one channel per programming session. It was not until the electronic tuner was introduced that VCR owners could program more than one channel. Programming has become simpler with the introduction of the bar code, available on several VCR models. See BAR CODE PROGRAMMING, EVENT, TUNER.

Programming Card. A chart containing bar-line patterns that are "read" by a pen-like optical scanner to transmit information to a VCR. Part of the bar code programming process featured on several VCR models, the programming card lists days-of-the-week, a full range of time slots and numbers for channel selection. Along with this data are corresponding bar-line patterns. The VCR owner can set up his machine for recording by using the scanner to mark the appropriate information on the card. An infrared beam transmits the data to the VCR which then can display the recording information on the TV screen for confirmation. See BAR CODE PROGRAMMING, OPTICAL SCANNER.

Projection TV. A video image magnified many times onto a screen by the use of lenses and/or mirrors. Projection TV, which offers images larger than those of the CATHODE RAY TUBE, was introduced by HENRY KLOSS to consumers in 1973. Two basic projection TV systems are the single-tube and the three-tube, each offered in a one-piece or two-piece unit. The single-tube, one-piece system uses a conventional TV set, a special lens system, a mirror and a screen, all in one cabinet. The single-tube, two-piece system employs the TV set and lens in one unit and a separate special screen. In the three-tube, one-piece system, each tube carries a different primary color (red, green, blue). The tubes project a bright picture onto a reflective screen. In the two-piece unit, the three tubes project their images directly onto a separate screen. These systems are more sophisticated in their audio and video components—and more costly. The latest and most successful system, however, employs a one-piece unit, three tubes, a mirror and a rear projection hard plastic screen. This rear projection system combines a bright, large image and a compact cabinet. A steady stream of improvements has seen a renewed interest in projection TV. One of these is a new cooling system that allows the tubes to function at higher levels of brightness while improving contrast. Another improvement is an automatic contrast correction that brings out details in extremely light or dark picture areas. In addition, automatic focus compensation adjusts for the disparities in distance between the projector and the screen center and projector and the screen edges. These enhancements, as well as others in the area of sharpness, have increased the screen size of some models to 70 inches or more. See LASER-BASED PROJECTION TV, LCD

PROJECTION TV, LIGHT VALVE PROJECTION SYSTEM, REAR PROJECTION TV, THREE GUN PROJECTION TV.

Projection TV Remote Control Panel. An accessory that permits the user to perform an array of functions while remaining away from the projector. Features available on the remote control pad include the usual registration and alignment and automatic test sequences.

Protection Tab. See RECORD-PROTECT TAB.

Psychoacoustics. The study of how the human brain records and deciphers the signals that indicate the direction from which a sound originates. This science has been helpful in producing audio/video equipment that attempts to duplicate the sound often heard in movie theaters. Recently known as surround sound, home systems employing this technique can simulate sound sources that may not actually exist. Special speakers built into a TV set, although installed close to each other, can give the impression of multiple speakers in several locations. See SURROUND SOUND.

Public Access TV. A local channel (or part of one) of CABLE TV which is free to individuals or groups of a local community, based on a first-come, first-served schedule. Whether as a result of pressure from commercial TV to frustrate cable growth and competition, from political activists seeking an outlet for their voices or from idealistic or socially motivated governmental agencies, the FEDERAL COMMUNICATIONS COMMISSION (FCC) passed its first public access regulations in 1972. Originally ordering four stations (public access, government, education and leased access), the regulations over the years were eventually eased to one channel. Even this ruling was overturned in 1979 by the Supreme Court which claimed that, since these channels were operating locally, the FCC was overstepping its authority. The court did, however note that local authorities and states can require public access channels. Despite the above decision, many cable companies, aware either of their civic responsibilities or the public relations benefits, still provide some public access time. Of the more than 4,000 cable systems in the United States, more than two dozen of these are CABLE TV COOPERATIVES and MUNICIPAL CABLE TV stations. Ralph Nader's National Citizens Committee for Broadcasting is one of the few agencies that can help communities gain public access to television.

Pulfrich Illusion. A process that is part of a three-dimensional television system. Utilized in Japan but originally developed in California, the technique employs a gray filter over one eye to create an illusion that objects moving sideways are coming toward or moving away from the viewer. The Pulfrich illusion process is compatible with 2-D, conventional television. See THREE-DIMENSIONAL TELEVISION.

Pulse Code Modulation. A process that divides the light and dark portions of an image and reassembles

Pulse Cross Display

them into a flow of data or digital information. In audio, a digital recording technique that converts videocassette recorders into high fidelity audio recorders. Pulse Code Modulation utilizes digital encoding in place of the conventional analog technique to reproduce music on tape. PCM presents a distortion-free, truer and cleaner sound whose dynamic range is extended 20 decibels. In addition, PCM provides stereo recording capability on 8mm videotape. Although the recording (limited to a frequency response of only 15 KHz) does not capture the full range of human hearing, it can capture the full range of all sounds broadcast over the air; it produces more dynamic range capacity than is otherwise available from either FM or TV audio transmissions. Adapters required for the PCM system originally retailed for thousands of dollars.

Pulse Cross Display. An electronic unit which repositions a monitor picture so that sync and blanking pulses become visible. This permits videotape recorders to be adjusted for proper skew, tension and tracking. The pulse cross display also checks vertical interval test signals and VCR/VTR head switching transients. Other applications of the device include monitoring servo lock, tape tension and proper sync of editing decks as well as testing edits before the actual dubbing and distribution. A typical display unit can sell for a few hundred dollars. See FIELD BLANKING, LINE BLANKING, SYNC PULSE, VERTICAL INTERVAL REFERENCE SIGNAL.

Pulse On/Pulse Off Circuit. One of the methods employed to activate the Pause circuitry of a VCR. See PAUSE CONTROL.

Q

Q Channel. A designated band used in the NTSC color television system to send green-magenta color information. The American National Television Standards Committee has set the 0.5 MHz-wide band for the Q channel.

"Q" Signal. In video, one of two signals (the other known as the "I" signal) that modulates the chrominance subcarrier of color television receivers. Also, the quatrature signal generated by a video camera's circuitry. This signal is normally 90 degrees out of phase, in contrast to the "I" (in phase) signal. Today's cameras produce both "Q" and "I" signals, whose relationship affects the reproduction of all the colors. If any part or all of the signals are not in proper proportion or are absent, the chrominance will be distorted. The "Q" and/or "I" signal can be measured on a COLOR VECTORSCOPE. See "I" SIGNAL.

Q-Phase. A color TV signal carrier that has a phase difference of 147° from the color subcarrier. The Q-Phase is sometimes referred to as the quadrature carrier.

Quadraplex. A system of videotape recording using four rotating heads on a drum perpendicular to the tape. Original experiments in videotape recording used stationary heads similar to audio tape recording. But since this method required an unusually high tape speed of approximately 100 inches per second, it took thousands of feet of tape to record only a few minutes of video information. Obviously, this was impractical and too costly. Quadraplex, with its rotating heads, was developed by Ampex in the 1950s. Using two-inch tape traveling at only 15 inches per second, the improved technique increased the quality of the video signal placed on the tape and enabled an hour of information to be stored on convenient-sized reels. Quadraplex is still used today in broadcasting studios. See AMPEX, VTR.

Quadrascan. A technique used in arcade video games permitting high resolution images to enter the screen at different speeds and from all directions. The images, appearing as targets, reveal exceptional detail in their drawings. Quadrascan, not presently

Quadrature Carrier

applicable to home video games, was developed by Electrohome Electronics for Atari's arcade game "Asteroids."

Quadrature Carrier. See Q-PHASE.

Quarter Size. A special feature, usually found on professional/industrial equipment such as time base correctors, that uses digital circuitry to produce an image one-fourth the size of the TV screen. The quarter-size picture can usually be made to appear in any corner of the screen.

Quartz. A mineral found in nature and whose crystals are highly useful in radio and carrier communication. Quartz crystals that are electrically charged vibrate and keep extremely accurate and stable frequencies.

Quartz Light. A special source of illumination for use with video cameras. Preferred in low light situations, the quartz lamp is balanced for the proper color temperature. Although more costly, quartz is preferred over the incandescent light because it produces a bluer light which is closer to daylight, gives more light and throws off less heat. Some models feature an air-cooled housing and a distinct reflector for maximum efficiency. The light can be hand-held or attached to the video camera. See COLOR TEMPERATURE, COLOR TEMPERATURE SWITCH.

Quartz-Synthesized Tuning. One of several automatic-fine-tuning methods designed to capture a strong broadcast signal and keep it in phase. Quartz tuning utilizes the dependability of the constant vibration of a quartz crystal as its reference point. The crystal, which vibrates at the exact frequency of the selected channel, attains very precise tuning regardless of frequency variations that may occur between channels and broadcast signals. Phase lock loop is another method used in TV monitor/receivers for automatic fine tuning. See AUTOMATIC FINE TUNING, FREQUENCY-SYNTHESIS TUNER, PHASE LOCK LOOP.

Qube. A two-way cable system based in Columbus, Ohio, that started operations in 1977. QUBE allowed subscribers to reply to video events via a small computerized console. Warner-Amex's QUBE cable system, introduced in Columbus, was the first major experiment in INTERACTIVE TV. Viewers have judged talent shows, given their opinions on political issues, voted, decided the fate of characters in soap operas, etc. Other QUBE systems sprang up across the country, but they all eventually ceased operations for one or more reasons, including high expenses and poor marketing of the service.

Quick-Start. A mechanical procedure incorporated into some VCRs designed to speed up such functions as play, fast forward and rewind. To perform some of these commands, several seconds are normally required as the videotape engages or disengages itself around the rotary head drum of conventional VCRs, particularly those of the VHS format. Some machines have modified the internal mechanism so that the tape can be wound or unwound in less than two seconds.

R

Radiated Interference. Television picture disturbance, often from such sources as automotive ignitions, power transmitter lines and radio stations. Radiated interference is usually the result of a poorly shielded TV tuner. See INTERFERENCE, INTERFERENCE SUPPRESSER.

Rainbow Effect. See MOIRE.

RAM (Random Access Memory). In video games, that part of the electronic console that determines the quality or complexity of the graphics. The more the RAM is boosted, the higher the screen graphics or resolution. With some game systems, for example, their limited power inhibits the sophistication of the games designed for them. But if the RAM is increased, the systems take on a game-playing power equivalent to more costly consoles.

Random Access. In viewing and editing, the automatic, rapid cuing to any point, particularly with laserdiscs, digital discs or solid-state memory components. On a videodisc player, random access permits the viewer to locate instantly any FRAME on the disc. The frame number usually is displayed digitally on the screen by punching in the number of the frame on a keypad.

Random Access Tuning. A feature on some TV sets, VCRs or videodisc players permitting the selection of any programmed channel by means of touch sensor buttons. This is in contrast to older rotary tuners in which other numbers must first be passed by. On a remote control, random access is achieved by a set of separate keys for each number from 0 to 9. Random access channel selection is sometimes referred to as direct access.

Random Interlace. A video camera problem, usually associated with older, less costly models, that inaccurately registered scan lines. The normal procedure for producing quality images requires that even-numbered scan lines appear first, followed by odd-numbered lines placed carefully between the even-numbered lines. Lower-priced cameras tended to drop the odd-numbered scan lines arbitrarily. Eventually, improved circuitry incorporated into virtually all cameras eliminated random interlace, although it occurs occasionally in some TV monitors and projectors.

Rapid Access. A feature on videodisc players designed to locate a desired scene or segment on the disc. The picture is not visible during the rapid access search, but the time delay, operating in conjunction with the stylus, acts as a cue to help find the particular excerpt on the disc.

Raster. That portion of the CATHODE RAY TUBE upon which the image or zigzag pattern of horizontal scanning lines is reproduced. That rectangular image is called a raster and contains light and dark-light points as seen by the video camera.

RCA (Radio Corporation of America). Developed and presented the first color video tape when it showed a two-minute experimental demonstration on October 23, 1955, on NBC's "Jonathan Winters Show." The company, of course, pioneered in video by presenting the first public demonstration of television at the 1939 World's Fair and by first demonstrating the present TV color format in 1953. See DAVID SARNOFF.

RCA Jack. A popular phono jack (sometimes called plug) used in VHS video recorders for both audio and video inputs and outputs. It is used in Beta machines for video inputs and outputs only. To copy tapes from a VHS to a Beta machine (or vice versa) the user will need mini-plugs and RCA plugs. Beta uses the 1/8-inch mini-plug for its audio input and output. The RCA phono plug has become the industry standard for home video systems. RF connections use the F-type connector. See CONNECTOR.

Read/Write Head. In video, the part of a videocassette recorder that reads and writes information on tape. See AUDIO HEAD, VIDEO HEAD.

Real-Time Counter. A VCR feature that reveals tape time precisely in hours, minutes and seconds. This is accomplished by keeping the tape in contact with the control head. Thus, the counter can calculate the control pulses. This handy feature, sometimes called real-time tape counter, linear time counter or linear time readout, can measure accurately the length of programs and can help the viewer to locate any point in a tape by simply entering how far along in "time" that point is. See FULL LOAD.

Real-Time Digital Storage System. Professional/industrial equipment, designed chiefly for post-production work, that uses random access memory (RAM). These systems can store several seconds of digital video information and play it back instantly as a digital videotape recording. Individual video frames can be retouched and repaired. Graphics can be added to any portion of the image, and the color composition of the frame can be changed. Film can be transferred to digital videotape, after which shadows can be added or deleted, lettering enhanced, colors manipulated and objects highlighted and textures recreated. All changes are stored in memory and can be transferred back onto videotape which can then be edited

Rear Projection TV

Figure 31. A rear projection TV with a 46-inch screen, surround sound audio system and remote control. (Courtesy Zenith Electronics Corp.)

with a conventional editor. One of the major benefits of using these professional workstations is that scenes do not have to be reshot.

Real-Time Tape Counter. See REAL-TIME COUNTER.

Rear Projection Screen. In video, a screen used for projection from behind the screen with a video camera in front to make FILM-TO-TAPE TRANSFERS. Rear projection screens are recommended over conventional ones which diffuse the light instead of concentrating it. A rear projection screen can be made at home by constructing a wooden frame and stapling onto it a special screening material called POLACOAT.

Rear Projection TV. A one-piece console TV system that produces a large projected image on the rear portion of a special screen. Rear projection TVs usually consist of three color tubes (red, green, blue) and a highly reflective mirror within the projector-cabinet. The picture is projected onto a plastic screen containing a series of etched concentric rings. This specially constructed screen helps to deliver evenly distributed light over its entire surface as well as a sharp pic-

Receiver

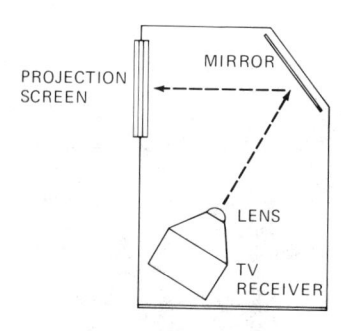

Figure 32. One-piece rear projection TV system.

ture. In addition, the screen provides an image bright enough to be viewed with ordinary room light. Rear projection TV systems have gained in popularity because they are better suited than two-piece front projection TVs where space is limited. Some rear TV systems utilize a black matrix lenticular screen or black stripe projection TV process to increase picture contrast. Improvements in contrast, brightness and detail have pushed the screen size of rear projection TV to 70 inches or more. See BLACK STRIPE PROJECTION TV, PROJECTION TV.

Receiver. A component of a communications system that converts electrical waves into audible and visible forms. Digital TV receivers, integrated receiver/descramblers, monitor/receivers, satellite receivers, transmitter/receivers and TV receivers are some of the units that play important roles in video.

Record. A control button on all VCRs which, when pressed, permits the machine to record information onto videotape. The VCR can record off the air, from CABLE TV, from a video camera or from another VCR. This last method is known as copying or duplicating a tape. Some machines provide a safety feature to prevent accidental recording: two buttons, Play and Record, must be engaged simultaneously to activate the Record mode. If Pause is pressed while the VCR is in Record, the machine will stop taping until either the Stop button is pressed or Pause is disengaged, in which case the machine will continue to record.

Record Lock. A feature found on some portable VCRs designed to keep the tape wrapped around the drum when the machine is shut off. By not having to be unthreaded and threaded again, the tape remains accurately in position, ready for the next glitch-free recording. Record lock puts the VCR into Record mode when the machine is turned on. In addition, the VCR uses no power in Off while the camera continues to receive power, thereby preserving the battery. If Record lock is pressed down after Off is pressed, the VCR will enter the Pause mode when the machine is turned on. In both cases, by having the machine begin in either Record or Pause, it provides for glitch-free edits between scenes.

Record-Protect Tab. The small square tab on a videocassette that, when removed, prevents accidental erasure of the recorded material on the tape. Placing a piece of vinyl tape over the opening once again prepares the tape for recording. Virtually all VCRs have a mechanical sensor that detects the obstruction and treats the videocassette as though it has the protection tab in place.

Refractive Projection TV

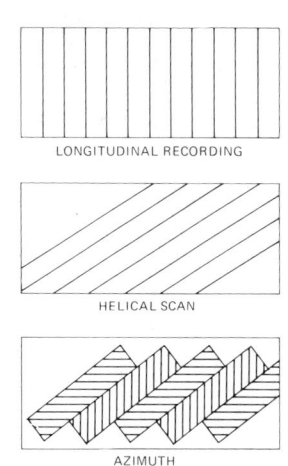

Figure 33. Recording methods.
(LONGITUDINAL RECORDING / HELICAL SCAN / AZIMUTH)

Record Review. A feature on a video camera permitting automatic playback of the last few seconds of a recorded tape. The tape is then placed in position for the next scene. With some cameras, this feature works only when they are connected to certain VCRs.

Recording. See COPYING, RECORD, VIDEO GAME RECORDING.

Recording Amplifier. A device used in a VCR to adjust the level of the video signal before it goes to the video heads. The recording amplifier serves a different function from that of the PLAYBACK AMPLIFIER.

Red/Blue Balance Control. A feature on a video camera to permit adjusting for maximum blue and red reproduction. The effect is either read on an accompanying meter or judged on a nearby color monitor or receiver. The balance control may also be part of a separate accessory, the COLOR CONTROL UNIT, or in more sophisticated camera models, may work automatically through special circuitry. See WHITE BALANCE CONTROL.

Red Eye. A term used by professionals to describe the red recording light on the front of video cameras. The light goes on when the camera is in operation as a cue to the actor or subject. Several home video cameras provide this feature.

Reel Rocking. Refers to finding an exact editing point by physically moving back and forth the two reels of an open-reel tape recorder. By listening or looking carefully for the cue, one can find the edit point. This method was introduced to professional tape editors in 1974 by Television Research International (TRI). Prior to reel rocking, technicians used a more cumbersome numerical procedure for editing. Today's more sophisticated professional editing techniques include, among others, magneto-optical non-linear disc-based systems or non-linear editing used in conjunction with solid-state video recorders.

Reference Black Level. Refers to the picture-signal level that complies with a designated maximum limit for black peaks. See BLACK LEVEL, BLACK LEVEL CONTROL.

Reference Tape. See MONITOR/RECEIVER REFERENCE TAPE.

Refractive Projection TV. A simple, relatively inexpensive one-tube, two-piece projection TV system consisting of a TV set, a special magnifying lens

Registration

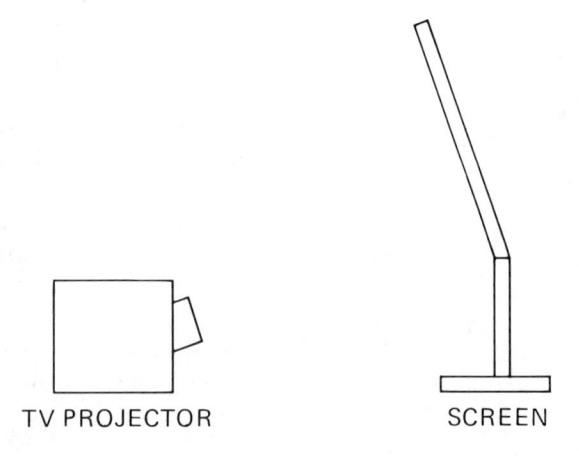

Figure 34. Refractive (two-piece) projection TV system.

and a large projection screen. The small TV receiver (usually 12 or 13 inches measured diagonally) is often enclosed in some type of box or cabinet with the lens at the opposite end. The enclosure prevents ambient light from dimming the image. If the system is the result of a do-it-yourself project, the projected image will be backward, upside down and rather washed out because of the nature and quality of the magnifying lens. To correct these shortcomings, manufacturers have produced systems with special TVs that have inverted images and high quality lenses. However, these refractive projection TVs cannot compete with the image quality and brightness of the more costly THREE-GUN PROJECTION TV systems. These limitations no doubt have accounted for the general demise of the refractive system and the success of other large-screen techniques.

Registration. The exact superimposition of the primary colors (the red, green and blue signals) to form a correctly colored picture. When the color signals do not overlap properly, misregistration, or MISALIGNMENT, occurs, causing colors to form on the side of the objects. See MISALIGNMENT, SHADOW MASK.

Registration Chart. See VIDEO TEST CHART.

Reilly, John. Promoter of independent videomakers; video documentarian; winner of several awards for his video productions; co-founder of GLOBAL VILLAGE VIDEO RESOURCE CENTER, INC. Together with his wife, JULIE GUSTAFSON, he produced several VIDEO DOCUMENTARIES which received critical acclaim, among them "Home" and "Giving Birth; Four Portraits."

Remote Channel Change. A feature found on the remote control panel of most VCRs. It permits changing the channel number of the recorder's tuner. It also converts the TV receiver into a remote control set by using the

VCR's tuner and the TV's open channel (3 or 4). Older VCRs equipped with a mechanical rotary tuner were incapable of utilizing the remote channel change. See REMOTE CONTROL.

Remote Control. The distant operation of various functions of a VCR, TV set, videodisc player, Projection TV or other similar component, from a wired or wireless device. Early model VCRs had a basic, one-function remote control panel that permitted activating the Pause mode so that commercials and other material could be edited out during recording. Remote control panels may be divided into four categories: (1) full function, which usually includes the capability to enter Record mode and channel selection; (2) playback-only functions such as FREEZE FRAME, SLOW MOTION and VISUAL SCAN; (3) wired; and (4) wireless. Of course, to offer remote channel selection, the VCR must have an electronic tuner rather than the mechanical, rotary type (which are no longer manufactured). Each TV set, VCR or VDP offers different features on its remote control panel. More recent developments have spawned a new generation of remote controls, including a MULTI-PURPOSE CONTROL, UNIFIED CONTROL, UNIVERSAL CONTROL.

Remote Control Directivity. Refers to the useful angle range of a remote control pad in relation to a TV receiver, videocassette recorder or videodisc player. The effective angle of remote control directivity depends on the distance between the control pad and the unit. In other words, the farther away from the infrared sensor on the unit, the greater the angle or degrees. Directivity to some extent depends on the condition of the batteries in the remote control pad. See REMOTE CONTROL, REMOTE CONTROL RANGE.

Remote Control LCD Display. See LCD DISPLAY.

Remote Control Panel. An electronic component containing all the function buttons necessary to operate a VCR, VCP, VDP and similar units that are some distance away from the viewer. Remote control panels can be portable which fit into the palm of a hand or a desktop or console model. See REMOTE CONTROL.

Remote Control Range. Refers to the effective distance between a remote control pad and a VCR, TV or videodisc player. The range varies with different units and often depends on the freshness of the batteries in the remote control pad. See REMOTE CONTROL, REMOTE CONTROL DIRECTIVITY.

Remote Control Unit. See REMOTE CONTROL, REMOTE CONTROL PANEL.

Remote Pause Button. A control on a video camera that permits operating the Pause mode of a videocassette recorder from the camera. This feature, which gained prevalence in 1979, gives the camera operator more flexibility by centralizing more functions on the camera. However, some older model VCRs are not designed to operate in conjunction with this feature.

Remote Pause/Still. A button on a VCR remote control panel which (1) stops the tape in Record mode to eliminate undesired material, such as commercials, from being recorded and (2) stops the tape in Play mode to provide a still picture, or FREEZE FRAME, on the screen. This feature was first introduced by JVC.

Remote Sensor. A feature on all remote-controlled VCRs that senses the infrared signal transmitted from the remote control pad. To operate correctly, the remote sensor depends chiefly on two factors. First, the sensor, located on the front of all videocassette recorders, should not be blocked by external objects. Second, the batteries in the remote control panel should be in good condition.

Remote Telephone Programming. A VCR feature that permits the owner to program the unit by telephoning in the required instructions. A conventional modular connector links the VCR to the telephone. When the phone rings, the machine, which contains a blank videocassette, answers with a beep, signaling that it is ready to receive instructions. One enhancement to remote telephone programming is the electronic voice which responds to the telephone call and asks, in sequence, for information concerning the day, time and channel for the intended recording.

Remote VCR On/Off Switch. A function on the remote panel that controls the VCR button on the videocassette recorder. In the VCR position, the tuner of the machine supplies the channel to the TV set which now acts as a monitor. In the TV position, the tuner of the television set becomes the source for the screen image, and the VCR remains inactive unless it is recording another program from a different channel.

Repeat Play. A VCR feature that beckons the machine to play a videotape up to an index signal, stop and rewind to a previous index signal and continue to play that portion of the tape indefinitely until the Repeat Play mode is pressed again. The length of the section to be replayed depends upon the minimum time between two indexes. The Repeat feature also appears on several videodisc players. Some offer two Repeat modes—one that replays the the present chapter or track, and another, sometimes called Repeat All, that plays back the entire disc. Repeat Play differs from Auto Repeat, a feature on some VCRs and VCPs that replay the entire videotape. See AUTO REPEAT.

Residual Elongation. Refers to the capabilities of a videotape to return to its original length after it has undergone tension for a long period of time. Most tapes tend to stretch to some degree. However, beyond a certain point, usually listed at about 5 percent, elongated tapes become unplayable. Residual elongation is only one of many factors that determine the overall quality of a tape. Manufacturers rarely list this measurement on their packages.

Resolution. The sharpness or detail of the video image. Resolution is measured in LINES. The number of lines results from multiplying 80 by the peak video frequency of the videocassette recorder. For instance, if a VCR can record and play back three million cycles per second (3 MHz), then that machine is listed at 240 lines. Black-and-white and color resolution are usually listed separately in specifications. With a video camera, resolution refers to the maximum number of clear lines discernible on a resolution or video test chart. Resolution depends on several factors. Industrial or commercial equipment, for instance, generally produces higher resolution. Different consumer formats yield disparate ratings. The ED-Beta format can generate more than 500 lines of horizontal resolution; laserdisc players, 425 lines; S-VHS format (both camcorders and VCRs), more than 400 lines; broadcast television, between 300–330 lines; Super-Beta, as much as 300 lines; conventional Beta, VHS and 8mm units average about 240 lines. See HIGH RESOLUTION, HORIZONTAL RESOLUTION, VERTICAL RESOLUTION.

Resolution Chart. See VIDEO TEST CHART.

Response Control. A function on an IMAGE ENHANCER (or MINI-ENHANCER) designed to cut the high frequency noise of a video signal. The response control usually works in conjunction with an enhance control. First, the enhance (sharpness) button or dial is adjusted to a point where the picture appears sharp, but slightly grainy. Then the response control is used to reduce the noise as much as possible without affecting the sharpness or detail in the picture. The control can often decrease 3 MHz signals, composed chiefly of noise, from zero to almost half their enhancement. See ENHANCE CONTROL, VIDEO SIGNAL.

Retentivity. In videotape, the ability of the particles that form the magnetic medium to keep their magnet charge. A tape with high retentivity characteristics demonstrates high quality. See COATING, COERCIVITY, VIDEOTAPE.

Retrace. See HORIZONTAL RETRACE.

Retrace Blanking. Refers to the blanking of a TV picture tube during vertical retrace intervals to deter the retrace lines from appearing on the screen. See VERTICAL BLANKING INTERVAL.

Reverse Image Switch. A video camera feature which permits changing the image from positive to negative. This switch may be used to reverse photographic negatives to positive pictures on the TV screen.

Reverse Polarity Adapter. An accessory cable which restores the original functions of a video camera that is used with an incompatible portable VCR. For example, although both units may be VHS in format, the multi-pin connectors may not match. A green light in the camera viewfinder may go on when the VCR is off and may go out when the machine is operating. To reverse these functions, some video supply houses sell reverse polarity adapters.

Reverse Polarity Control. See REVERSE IMAGE SWITCH.

Reverse Scan. See VISUAL SCAN.

Review Button. On some cameras, a control which plays back, both in reverse and forward, the last few seconds of a tape through the electronic viewfinder before the recorder reverts to Record/Pause mode. The button is usually located on the handgrip of some cameras.

Rewind. A control button on all VCRs to rewind the videotape in the cassette either partially or fully. On top-of-the-line machines the Rewind solenoid control has an additional function. When pressed half-way while the VCR is in Play mode, the reverse VISUAL SCAN feature is activated and the picture appears on the TV screen in reverse without sound. This function is sometimes listed on machines simply as REW or, shorter still, as an arrow (<—).

Rewinder. An accessory for rewinding videotape. The use of a rewinder keeps the tape away from the video and audio heads in this mode, thereby prolonging their life to some degree. The rewinder is also useful for uninterrupted use of the VCR. Since the accessory is a separate unit, it can rewind one tape while another is playing or recording in the VCR. In addition, the rewinder, which is a relatively inexpensive item, saves on the wear and tear of the rewinding mechanism of a VCR. Repairing or replacing these parts can result in a relatively expensive repair bill. Rewinders may be purchased for VHS or Beta formats. At least one company sells a rewinder that can handle both formats. Some models offer additional features, such as automatic stop, soft eject, forwarding, digital counter and an on/off LED light.

Figure 35. A rewinder designed only for VHS cassettes. (Courtesy Vivitar Corp.)

RF (Radio Frequency). A method of transmission utilizing the radio spectrum to carry audio and video signals. Although videocassette recorders and TV sets can receive and transmit RF signals by way of antenna inputs and outputs, video or line inputs and outputs produce superior results. See RF SIGNAL.

RF Adapter. See RF CONVERTER.

RF Amplifier. A device designed to maintain RF (radio frequency) signal strength, especially when that signal is split. For instance, an amplifier may be necessary when an antenna signal is divided between two or more TV receivers with exceptionally long wires separating them; or when a VCR is feeding a few TV receivers. The RF amplifier is connected between the antenna line and the splitter in the first example above and between the VCR and the splitter in the second example.

An RF amplifier differs from a VIDEO DISTRIBUTION AMPLIFIER in that the former produces RF signals that can be fed directly to TV receivers.

RF Converter. An accessory that converts or changes any video source such as a portable VCR or video camera to an RF signal for direct hook-up to a receiver's input. The alteration is necessary since a video signal is different from and not compatible with an RF signal. RF converters are sold separately or as combinations with IMAGE STABILIZERS. See COMPOSITE VIDEO SIGNAL, RF SIGNAL, VIDEO SIGNAL.

RF Modulator. An accessory designed to combine separate audio and video signals from VCRs, IMAGE ENHANCERS, VIDEO CAMERAS and other IMAGE PROCESSORS into an RF SIGNAL which can then be played through a TV set. In a VCR, the RF modulator places a low-power RF signal over the audio/video signals to trigger the TV into accepting the transmission as a normal TV signal. A home VCR requires an RF modulator so that the machine can be connected to a standard TV receiver which receives only broadcast or RF signals. See COMPOSITE VIDEO SIGNAL, SATELLITE TV MODULATOR, VIDEO SIGNAL.

RF Signal. The radio frequency signal that mixes together audio and carrier signals; also, audio and video signals that can be transmitted. Radio frequency refers to the range of frequencies used to transmit electric waves. RF, audio and video signals are different in nature. For example, IMAGE PROCESSORS accept video signals but do not affect RF signals. On the other hand, most consumer TV receivers accept RF signals, but not audio and video signals. These must first be converted to RF. See RF CONVERTER, SIGNAL, VIDEO SIGNAL.

RF Switcher. See SWITCHER.

RGB Input. A feature on some MONITORS that feeds the individual red, green and blue signals and a sync signal directly to a picture display. By feeding the primary colors directly into this input, over 400 lines of color detail can be seen instead of the usual 40 or so on conventional TV sets or 150 lines when using a typical video input. Presently no equipment (except computers) has RGB outputs to connect to this feature which first appeared on Sony, Panasonic and NEC monitors in 1982.

Ringing. Refers to a series of ripple-like effects that appear at the edge of a high-contrast object. Ringing tends to be a peculiarity more of tube-type than solid-state video cameras. The problem results from random capacitance within the circuitry. A VIDEO TEST CHART can determine whether a camera produces ringing in its images.

Ripple Effect. A condition of videotape that is stored and not played for long periods of time. Because of this inactivity, the tape may develop wrinkles during recording. These wrinkles tend to develop the ripple effect. Tape technicians usually recommend running tapes through VCRs once in a while so they don't "memorize" particular patterns during their settling period.

Ripple Filter

Ripple Filter. See SMOOTHING FILTER.

Ritcheouloff, Boris. Russian-born inventor, visionary, designer of a video camera, receiver and a video recorder. While working in London, the Russian filed patents in 1927 with the British Patent Office. His inventions went unrealized, but his ideas and technology were ahead of their time. He incorporated into his process such items as lenses, light-sensitive material, electromagnetics, a photo cell, transmission of radio waves to a conventional vacuum tube, etc. No working models of his TV transmission and recording systems were ever built, and his patents were never renewed.

Roll. Refers to the TV picture moving up or down because of the loss of vertical sync. Roll may also result from anti-copying devices placed on commercial prerecorded videotapes. See VERTICAL SYNC PULSE.

Rolling Dropouts. Horizontal lines, resembling scratches, that move rapidly through the screen image. The problem is associated chiefly with videodiscs. Occasionally appearing in clusters, these rolling dropouts are often the result of a faulty aluminum coating or a defective videodisc stamper.

Rose, Albert. Developer of the image orthicon tube, which replaced the ICONOSCOPE. The latter was invented by VLADIMIR ZWORYKIIN in 1923 and led to the first working model of a video camera. See CAMERA TUBE, ORTHICON.

Rotary Tuner. See TUNER.

Rotary Wipe. A special effect produced by a MIX/EFFECTS SWITCHER in which a rotating radial line replaces one video image with another. The result is similar to that of a clock hand rapidly moving around the dial. See WIPE.

Rotation. Refers to a waveform monitor function designed to adjust the instrument to the magnetic field at a shooting site. Rotation control is one of several basic features found on such units as portable waveform monitors, or OSCILLOSCOPES.

Routing Switcher. An accessory unit that directs several audio and video connections into one box so that signals can be matched with various destinations. Routing switchers often incorporate a distribution amplifier that splits one input into more than one output for purposes of copying with little signal loss. Some switchers handle six VCRs, an auxiliary source, an external processor and a TV monitor/receiver. These units offer more flexibility, including mixing a VCR picture with the audio of a compact disc and taping the result on a separate VCR. Consumer models range in price from about $100 to $1,000. Professional routers, which retail for several thousand dollars, may feature HDTV compatibility at 30 MHz, remote control and more than one level of switching. See SWITCHER.

R-Y Signal. The red-minus-luminance color-difference signal in color television. The R-Y signal, together with the Y (luminance) signal, creates the red primary signal.

S

SAP (Secondary Audio Program). A separate channel of a television stereo system known as MTS, or multichannel television sound. SAP was designed as an additional soundtrack for bilingual programs or other such purpose without interfering with monaural reception of current TV receivers. The SAP feature usually can be controlled from the remote control of a monitor/receiver or TV set.

SAP Indicator. A feature, located on the front panel of some VCRs, that lights up when a second, or separate, audio program broadcast is received.

Sarnoff, David (1891–1971). Became president of RCA in 1930; built TV station in Empire State Building in 1930; demonstrated TV programs in 1936; displayed and demonstrated RCA television receivers at the 1939 World's Fair in New York. Not a scientist, engineer or inventor, Sarnoff led RCA between 1935 and 1938 into field testing television. To publicize this relatively new communications medium, he chose the opening day of the World's Fair in New York (April 30, 1939) to usher in the age of television. Franklin Delano Roosevelt became the first president to appear on the new medium. TV sets, equipped with screens from 5 to 12 inches, sold at the RCA pavilion for from $200 to $600.

Satellite. In television, a communications instrument placed in orbit 22,300 miles above the earth's equator so that it travels at the same speed as that of the earth's rotation. Satellites require two systems of antennas, one to receive signals transmitted from earth stations (UPLINK) and the second to return converted signals to specially equipped receiving stations (DOWNLINK). Each satellite uses a group of transponders, each one allocated to a specific TV channel. The satellite also amplifies the signals transmitted to it and converts them to the required microband frequency range of 3.7 to 4.2 GHz (downlink). Among those who have placed communications satellites in orbit are AT&T, the Canadian government, G.E., RCA and Western Union. See DOWNLINK, GEOSTATIONARY, SATELLITE SIGNAL, SATELLITE TV, UPLINK.

Satellite Dish. A special antenna used in conjunction with a TV satellite system. See PARABOLIC ANTENNA, SPHERICAL ANTENNA.

Satellite Focus. The area on earth covered or focused upon by a particular satellite's transmission. The satellite signal may vary from narrow to broad in the zone or land area it reaches. For example, one type, the domestic satellite, transmits its signal to a confined area, such as one country. It is also known as a spotbeam satellite. A second kind, the hemispheric satellite, transmits to an individual hemisphere. It, too, has another name: zone satellite. The global or international satellite transmits to large areas of earth. Some satellites may have signals that cover two or all of the areas above. See SATELLITE, SATELLITE TV.

Satellite History.

Oct. 4, 1957: Russia's satellite, *Sputnik*, launched into orbit.
Feb. 1, 1958: First United States satellite, *Explorer*, launched.
Aug. 12, 1960: *Echo* launched; first satellite to contribute to international communications.
July 10, 1962: *Telstar* becomes the first satellite to carry TV signals.
Feb. 14, 1963: *Syncon*, the first GEOSTATIONARY satellite, orbited at 22,300 miles above the equator.
Apr. 6, 1965: *Intelsat I*, or Early Bird, becomes the first international commercial satellite.
Apr. 13, 1974: *Westar*, the first domestic (U.S.) communications satellite.
Sept. 30, 1975: Home Box Office transmits first satellite TV signals to a cable TV operator.
Feb. 26, 1990: A new consortium, Sky Cable, announces a 200-watt transponder to transmit 108 channels to a 24-inch square home antenna.

Satellite Location. The position of a communications satellite in orbit for the purpose of receiving and transmitting various signals. There are various sources of information for owners of satellite TV systems that reveal where satellites are and which programs appear on each. Home satellite handbooks list the locations of satellites and how they operate. Wall charts position the many satellites and their FOOTPRINTS. Monthly magazines provide a program guide to all the U.S. satellites. In addition, local newspapers list some of the programs of satellite transmissions. See GEOSTATIONARY, SATELLITE.

Satellite Master Antenna Television (SMATV). An alternative over-the-air system of delivering satellite-transmitted channels. Rivaling cable TV and one-channel satellite TV systems, SMATV utilizes a rooftop parabolic antenna. The system selects various sports and film programs and offers clear reception via wires to the TV receivers of apartment dwellers.

Satellite TV

See CABLE TV, PAY TV, SUBSCRIPTION TV.

Satellite Modulator. See SATELLITE TV MODULATOR.

Satellite Receiver. The component of a SATELLITE TV SYSTEM designed to convert the microwave signal to VHF and modulate it for acceptance by the antenna input of the TV receiver. A satellite receiver may be built into the horn assembly of the dish just as the low noise amplifier is, or attached to the base of the dish or kept indoors. Those built into the antenna allegedly minimize signal loss by the absence of cables and additional circuitry. The more sophisticated table-model receivers, also known as integrated receiver/descramblers (IRDs), have a host of special features, including integrated descramblers, on-screen menus for various controls, programmable tuners, digital stereo sound and video noise reduction for clearer images. See INTEGRATED RECEIVER/DESCRAMBLER, LOW NOISE AMPLIFIER, MICROWAVE, SATELLITE TV, SIGNAL LOSS.

Satellite Signal. An FM signal carrying both audio and video information transmitted from an earth station (UPLINK) to a satellite which converts the signal before returning it (DOWNLINK) to a ground station antenna. The signals transmitted from earth range from 5.9 to 6.4 GHz. The satellite both amplifies and converts the signals which now range from 3.7 to 4.2 GHz, the downlink frequency range. Satellite channels need a bandwidth of 40 MHz each. Since the difference in the signal range amounts to 500 MHz, there is space for 12 satellite TV channels. See DOWNLINK, SATELLITE TV, UPLINK.

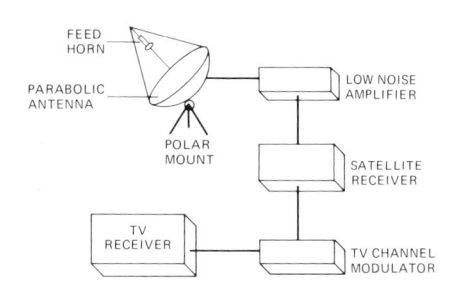

Figure 36. Satellite TV system.

Satellite TV. Refers to any of the satellites in orbit (22,300 miles above the equator), each of which can collect up to 24 different signals from broadcasters and amplify them to five watts before sending them back to earth via the MICROWAVE band. Three basic units of a satellite TV system include a special 10-to-15-foot antenna, a LOW NOISE AMPLIFIER (LNA) and a SATELLITE RECEIVER. The antenna may be either a PARABOLIC or SPHERICAL type. The parabolic is usually mobile while the spherical is permanently mounted. The LNA amplifies the signal from the antenna while the receiver converts the signal to VHF and modulates it for the TV set. Two other devices of a satellite TV system are a control console for changing channels indoors and a remote control accessory to align each satellite. By 1990, almost three million television viewers in the U.S. had purchased their own satellite TV sys-

225

Satellite TV Modulator. In a SATELLITE TV system, that part that permits the output to go directly to the antenna input of a TV receiver. A modulator converts the audio and video signals to an RF signal. Some satellite receivers have a modulator built in. VCRs have a modulator so that they can be connected to TV sets. In a satellite TV system, the modulator is connected between the receiver and the TV set. See RF MODULATOR, RF SIGNAL, SATELLITE RECEIVER, SATELLITE TV.

Saticon Camera Tube. A video camera tube that costs more than other image pickup tubes, such as the VIDICON, but provides higher resolution. Although the saticon has the ability to produce less noise under high illumination, it is more susceptible to IMAGE RETENTION or lag. To compensate for this, different techniques are employed in the design of the tube. See LIGHT BIASING, TRINICON, VIDICON.

Saturation. See COLOR SATURATION.

Saw Filter. An acoustic wave filter built into television sets. Saw filters, or Surface Acoustic Wave filters, are designed to remove ghosts or cross-channel interference, audio/video crosstalk or electrical interference from household appliances.

Sawazaki, Norizaki. Inventor of the HELICAL SCAN principle in the 1960s. Dr. Sawazaki had been employed at Toshiba's Tokyo laboratory when he developed the method of laying down diagonal tracks on videotape, making possible slower speeds on 1/2-inch tape, both of which were necessary elements in the development and success of home video machines.

Scanner. The center portion of a three-part video head drum. The scanner, which contains the video heads, rotates against the videotape while the upper and lower parts of the drum remain stationary. See VIDEO HEAD, VIDEO HEAD DRUM, VIDEO SCANNER.

Scanner/Reader. See BAR CODE PROGRAMMING.

Scanning. The horizontal and vertical motion of the electron beam in the video camera or CATHODE RAY TUBE. In film-to-video transfer, scanning refers to the horizontal movement of the video camera across the film image projected on a screen. To capture wide-screen films on tape, a blank space would have to appear across the top and bottom of the TV screen. Therefore, to fill the screen, some of the film image is deleted by scanning. The best possible smaller image is selected from the wider film image based on esthetics, information (which actor is speaking at the moment), the avoidance of head chopping and other considerations. A professionally scanned TV broadcast appears unobtrusive and gives the viewer the feeling that nothing has been removed. However, no matter how carefully the film has been scanned, there have been voices of criticism from those involved in the creative process of filmmaking who believe that the scanning process compromises the artistic integrity of the original wide-screen work. DIGITAL TRANSFER, an alternative to scanning, offers a more sophisticated technique of converting theatrical films to video.

Scatterwind. The uneven winding of tape, usually resulting in damage to its edges which hold the audio and control tracks. In a videocassette, reel spindles control the tape so that it is packed evenly. Scatterwind is usually caused by defective spindles or other faulty parts of a cassette.

Scene Transition Stabilizer. See AUTOMATIC TRANSITION EDITING.

Schmidt Optical System. The first successful method of producing large-screen television pictures utilizing a CATHODE RAY TUBE, mirrors and a set of reflectors. Developed in the 1940s, this system was employed at first to increase the screen size of small CRT television sets and was later used in theaters for special sports events.

Scophony. An abortive attempt to eliminate the CATHODE RAY TUBE as the display method of large-screen television. Developed in England in the 1940s, scophony employed a composite of electronics and mechanics in its process.

Scrambled Channel. A cable system channel that is unviewable unless a decoder is used. Scrambled channels, of course, are PREMIUM CHANNELS that require an additional monthly fee from the cable subscriber who receives a special decoder for unscrambling the channel's programs.

Screen. In projection TV, the surface upon which the projected image is cast. A special aluminum screen gives optimum performance. The beaded-type screen is barely acceptable since it scatters the light and the flat matte screen gives a washed-out picture under normal lighting conditions. The surfaces of these screens need special care. To decrease bright spots in the picture area and to maximize the reflected light, projection TV screens are often curved. See PROJECTION TV.

Search Mode. A function employed by VCRs, video cameras or videodisc players to locate a desired point in a videotape or disc, in the case of a VDP, for viewing or editing. These three types of units provide a variety of search methods and procedures. The search mode is sometimes referred to as cue/review or cue and review. See ADDRESS SEARCH, AUTOMATIC CHAPTER SEARCH, CAMERA SEARCH, CHANNEL SEARCH, FRAME/TIME SEARCH, HIGH-SPEED SEARCH, INDEX SEARCH, INTRO SEARCH, PICTURE SEARCH/LOCK, SHUTTLE SEARCH, SKIP SEARCH, TIME SEARCH, VARIABLE SPEED SEARCH, VISUAL SCAN, VISUAL SEARCH.

SECAM (Sequential and Memory). A system of TV broadcasting used in France, the U.S.S.R., most Eastern European and a few Arab and African countries. Incompatible with other

See-Through Mode

systems, such as PAL or NTSC, SECAM uses an 819 line-scan picture which provides better resolution than PAL's 626 lines or NTSC's 525 lines. The SECAM and PAL systems had the benefit of time in producing better resolution pictures than those of the NTSC. Europe did not begin widespread TV broadcasting until many years after the United States, which, adopting the NTSC 525-line scan format, began commercial broadcasting in 1948. See HORIZONTAL RESOLUTION, LINE SCAN.

See-Through Mode. A video camera feature that makes visible the background image through the use of superimposition.

Segment Recording. See ONE TOUCH RECORDING.

Self-Timer. A video camera feature that, when activated, delays recording operation for several seconds so that the camera user can enter the scene.

Sensitivity Range. The difference between the least and the greatest amount of light to which a video camera responds while still maintaining good picture resolution. The sensitivity range of a typical camera is 10–10,000 FOOTCANDLES. Some cameras feature a SENSITIVITY SWITCH to increase its response to light, but this often causes the resolution to suffer, resulting in a noisier picture. A NEUTRAL DENSITY FILTER may be attached to the front of the lens to decrease the sensitivity range. See VIDEO CAMERA SENSITIVITY.

Sensitivity Switch. A video camera feature that permits the electronic increase of the camera's sensitivity to light. The switch extends the general amplification of the VIDEO SIGNAL so that the camera can produce a satisfactory picture even in poor light, although the image will be noisier. Each camera has its own sensitivity range. For example, one camera may list its range as 5 to 6,500 footcandles. Most video cameras average about 10–10,000 footcandles. This range can be decreased if desired by adding a NEUTRAL DENSITY FILTER to the front of the lens. On cameras without a sensitivity switch, low light problems are handled by the AUTOMATIC GAIN CONTROL, which otherwise differs in its functions from that of the switch.

Sequence Edit Mode. A digital VCR edit feature that can handle several assemble edits automatically. Sequence edit is usually part of a bevy of edit functions, including edit start and preview and assemble edit.

Sequential and Memory. See SECAM.

Sequential Scanning. The formation of the video image across the face of the TV screen or CATHODE RAY TUBE. An electronic beam, originating in the rear or neck of the picture tube, scans the inside face of the tube or RASTER in a horizontal and vertical motion. This scanning affects the phosphor coating on the face of the CRT, causing degrees of light to appear. The beam scans from left to right (facing the viewer), blanks out until it returns to the left side, ready to repeat its scanning process. When it reaches the bottom of the picture, it returns to the top and starts over, "painting" the 525 line scans of the NTSC standard. See CATHODE RAY TUBE, FIELD BLANKING, LINE BLANKING, RASTER.

Serial Digital Interface. An input/output capability built into advanced professional/industrial VTR machines such as Sony's D-2 videotape recording system. Serial digital interface, a marked improvement over parallel inputs and outputs, can handle signals with up to ten bits resolution for the video signal, with 20 bits resolution four-channel audio. In addition, integrating the model D-1 component digital VTR with the D-2 composite equipment becomes much simpler, since only one cable is required.

Serial Timer. See PROGRAMMABLE TIMER.

Setup Menu. A remote control feature, found only on certain TV receivers, that displays a set of color diagrams of installation connections. The menu, designed chiefly for new owners of the unit, includes color-coded jacks, located on the rear of the TV system, that match the different connections appearing on screen. Setup menus also provide automatic channel search, allow setting of time displays and permit popular channels to be arranged in tuning sequence.

Setup Tape. See MONITOR/RECEIVER REFERENCE TAPE.

SFX. Refers to the special effects of a videocassette recorder, videodisc player, video camera, camcorder, etc.

The two main types are conventional and digital special effects. See DIGITAL EFFECTS.

Shader. A special feature, found chiefly on professional/industrial workstations, that produces various textures, such as marble, wood, glass, etc. These workstations usually incorporate the use of a computer and special software.

Shadow Mask. A perforated metal screen containing hundreds of thousands of small apertures and positioned between the neck or yoke of a TV picture tube and just before the face or RASTER of the CATHODE RAY TUBE. Three electron guns, one for each of the primary colors, are stationed at the rear of the tube. When the electrons leave each of the three color guns, they pass through the shadow mask and affect only the color of their origin. For example, if a red barn is to reproduced, only the electrons from the red electron gun pass through the mask to activate the red phosphor dots on the face of the picture screen. The function of the mask, therefore, is to focus the proper electron beam to the correct phosphor color. Sony's fine-pitch picture tube, known as Fine-Pitch Aperture Grille, is a variation of the shadow mask. It utilizes a series of unbroken vertical slits or stripes to produce more and tinier pixels (picture elements). With the addition of a dark, tinted screen, these TV monitor/receivers offer improved picture definition (as much as 600 lines of horizontal resolution with direct video input) and a higher contrast ratio. See FINE-PITCH PICTURE TUBE.

Sharpness

Sharpness. See IMAGE SHARPNESS.

Sharpness Control. A feature, found on many TV receivers, VCRs, MONITORS, component TV systems, IMAGE ENHANCERS and other image processors, that allows the viewer to vary the bandwidth, thereby affecting the horizontal resolution. The sharpness control can soften as well as bring more detail to a screen image. With some units such as component TVs (which are capable of covering a wider bandwidth and thereby producing an image of high resolution), a sharper picture may also mean more detailed VIDEO NOISE. To minimize this interference, the sharpness control can be turned down, which slightly softens the image detail, but also decreases the unwanted video noise.

Shift. A function of a VCR clock display that is designed to select a programmed instruction. When a particular button is pressed, another program item will appear on-screen ready to be set. For example, when time is set on some machines, the shift button is pressed to prepare the clock display for the next entry, which may be day of week, date of month, month and year. Also, the term "shift" applies to a digital television or VCR feature that allows the viewer to exchange the main image and the inset image of the picture-in-picture function. See IMAGE SHIFT.

Shotgun Microphone. A mic with long, acoustical barrels used to narrow its response into a beam pointed at the subject. A shotgun mic (named for its appearance) is effective in picking up sound from a moderate distance or in noise-filled situations. It is a supersensitive, UNIDIRECTIONAL MICROPHONE, rejecting most sound emanating from the sides. See MICROPHONE.

Shrink-wrap. The tight cellophane wrapping around a blank or prerecorded commercial videocassette. However, some unscrupulous dealers have been known to recycle their less popular rental tapes by rewrapping them in shrink-wrap to make them appear as new merchandise. A careful examination of the corners of these rewraps reveals an unevenness.

Shuttle Search. The standard method of speeding the videotape back and forth through a VCR to find a specific point. More advanced VCRs provide multiple speed search in forward and reverse. With conventional machines, Fast Forward and Rewind do not display any images on the TV screen whereas the advanced models, with their high-speed search feature, show a picture, albeit a fast-moving one, on the screen.

Shuttle Speed. The rapidity at which a video deck can get from one end of a videotape to the other. The speed factor, in relation to access time, is considered a limiting component, especially in the editing process.

Signal. Information converted into electrical impulses. There are RF, audio and video signals, all different in nature. For instance, a video signal is not recorded directly on tape as is an audio signal. Audio and video signals can be directly fed into audio/video inputs for better signal reproduction. But on units without these inputs, such as TV sets, these signals must be converted to RF signals which are accepted into these conventional receivers. See COMPOSITE VIDEO SIGNAL, RF SIGNAL, VIDEO SIGNAL, ANTI-PIRACY SIGNAL.

Signal Amplifier. See AMPLIFIER.

Signal Converter. See CONVERTER.

Signal Generator. A professional/industrial multi-signal-generating instrument designed to produce all the standard video signals used in broadcasting, cable TV and TV set production. These include color bar, crosshatch, flat-field and gray step signals. The generator helps in the testing of TV receivers, monitors, broadcast studios, closed circuit TV studios and similar setups. More specifically, the color bar signal, consisting of the standard six color bars, is used by TV manufacturers to check the color portions of their sets; the cross-hatch signal aids in aligning convergence and horizontal and vertical linearity; the flat-field signal is used to check color balance; gray step helps to check color tracking. See COLOR BAR SIGNAL, CROSS-HATCH SIGNAL, FLAT-FIELD SIGNAL, MUNSEL COLOR CHIP CHART.

Signal Loss. See INSERTION LOSS.

Signal Processor. A device designed to stabilize, enhance or control the degree of fading of a video signal. The three major video signal processors are FADE CONTROL, a STABILIZER and an ENHANCER. These may be purchased separately, although some manufacturers have combined these three functions into one accessory. However, some of these controls may already be built into certain components. Many video cameras have the fade control feature while virtually all late model VCRs have advanced circuitry which overrides ANTI-PIRACY SIGNALS which destabilize the picture.

Signal Splitter. An accessory often used to split an input signal from a conventional 75-ohm cable so that it feeds more than one output. For example, a two-way splitter is used to split an incoming signal from an antenna or a VCR to two TV sets. If the wires run more than a few feet, an amplifier is sometimes added to minimize signal loss which causes a snow image. Another type of signal splitter is the VHF/UHF splitter in which the VHF signal goes to one source while the UHF signal travels to another. See BAND SEPARATOR.

Signal-to-Noise Ratio. See VIDEO SIGNAL-TO-NOISE RATIO.

Simulated Stereo. A process employed on some TV receivers and accessories to give the effect of stereo sound. Some frequencies are distributed to the right speaker while others are directed to the left, giving a fuller sound, although not true stereo. Both speakers are producing the identical sound. The technique is similar to that used in audio when mono recordings are electronically rechanneled.

Simulcast Mode

Simulcast Mode. Permits a stereo VCR to record a video signal from its tuner together with an audio signal delivered to its line input from an external FM tuner. Under normal circumstances, a videocassette recorder tapes either the video and audio signals from its own tuner or the video and audio from its line-level inputs. Simulcast mode is useful in recording TV/FM simulcast programs and FM simulcasts from cable TV movie channels seeking to enhance the sound portions of feature films.

Single Conversion Block Converter. A device utilized by some CABLE TV systems to transmit midband channels so that the picture has a higher frequency than the sound, a reversal from the normal broadcasting procedure. This prevents VCRs and TV sets with VARACTOR TUNERS from picking up both of these signals simultaneously. However, authorized subscribers to the cable system are able to receive these stations. See MIDBAND CABLE TV.

Single Frame Advance. See FRAME ADVANCE.

Single Gun Color TV Tube. A specially designed television tube with three separate cathodes, each producing a green, blue or red electron beam, each contained in a single control grid. An electrostatic focusing system interacts with all three beams. Each color beam must pass through a respective channel. The three channels make up an electrostatic convergence technique that functions similar to the three-gun systems of conventional TV sets. The most popular application of the single-gun picture tube is Sony's Trinitron system.

Single Point Stereo Microphone. A small, single-unit mic with two enclosed components, one to pick up sound on the left, the other to record sound on its right. Employing electret-type condensers, these stereo mics provide clear, well-balanced sound although they lack the high specifications of the more costly studio models. Some manufacturers offer single-point stereo mics but use a different approach. One internal component is aimed forward while the second is capable of picking up sound on both its left and right. This is sometimes known as the MS (Mid-Size) principle. See MICROPHONE, MID-SIZE PRINCIPLE.

Sizing. See CHARACTER SIZING.

Skew Error. A term that describes the differences in the angles or lengths of the diagonal tracks placed on videotape by two different machines. These discrepancies may lead to FLAGGING or bending in the upper portion of a TV picture. If the tracks are slightly longer, the image tends to pull to the left; if shorter, it may pull to the right. See TRACKING, TRACKING CONTROL.

Skip Field Recording. A technique used in video recording to reproduce one FIELD twice instead of normally reproducing the two fields which compose one FRAME. A similar technique is employed in playback with the FREEZE FRAME mode in which the heads pick up one field twice instead of a frame (consisting of two fields).

Skip Memory. A TV or VCR feature that permits the viewer to eliminate unused or unwanted channels during the scanning process so that only desired channels appear. The chief advantage of skip memory is that it removes the static and interference displayed on screen by inactive channels during the selection process.

Skip Scan. A feature, first introduced by Sony on its Beta VCRs, that slows down the Fast Forward or Rewind mode so that the viewer can see a portion of the tape on the TV screen. Other VCR formats have since adopted the function. See SKIP SEARCH.

Skip Search. A VCR search feature that slows down the videotape in Fast Forward or Rewind mode when it senses an electronic index previously placed on the tape automatically or manually. In addition, skip search displays the beginning of the scene for several seconds before continuing on its search. The viewer can activate the Play mode whenever a desired scene is reached. At this point, the VCR returns to the beginning of that scene and plays it back in its entirety. See ADDRESS SEARCH, HIGH-SPEED SEARCH, INDEX SEARCH, SHUTTLE SEARCH, TIME SEARCH.

Sky Cable. A direct broadcast satellite service proposed for full commercial operation in 1993. A joint venture of NBC, Cablevisions Systems, News Corp. and Hughes Communications, Sky Cable will be the first high power Ku-band DBS service in the United States. The service plans to offer 108 channels which will be received by 12-inch home satellite antennas. See DBS, SATELLITE HISTORY.

Slave VCR. In video, that VCR which is used as the recording half of the duplicating process. The playback unit is called the MASTER VCR. The use of the wrong slave units may cause problems in prerecorded tapes. Some VCRs are optimized for the slowest speed. When these machines are used as slaves at standard speed (for prerecorded tapes), the tapes they produce may cause video noise or snow on the screen. These machines have video heads with a narrower gap while the other VCRs used for playback may have heads with a wider gap. Therefore, the wider heads are picking up more than the narrow diagonal track laid down by the narrow-gapped heads.

Sleep-Timer. A feature built into several more costly TV monitor/receivers and front projection TV systems which can be programmed to turn off the unit at a certain time.

Slow Motion. One of the SPECIAL EFFECTS of a videocassette recorder and designed to slow down the tape speed to permit a closer look at a scene. Some VCRs provide variable slow motion, a control which adjusts the rate of slow motion. At its slowest speed, this feature is difficult to watch, since the movement loses a sense of continuity. A speed one-half to one-fourth that of normal retains movement and is the most popular range of this special effect. The slow motion mode,

Slow Tracking Control

like other special effects, is usually accompanied by video noise, such as horizontal bars of snow, on many machines. The newer digital VCRs have corrected this defect. If VCRs are kept in the slow motion mode for long periods of time, the video heads or the tape may become damaged.

Slow Tracking Control. A feature on a VCR designed to improve the image quality produced by the SLOW MOTION mode. It is sometimes located only on the remote control panel. The feature, which adjusts tape speed only during slow-motion playback, is also known as slow motion tracking control.

SLP. See EXTENDED PLAY.

Smear. A video image which displays blurred objects, especially at their edges and beyond. Smear is usually the result of insufficient lighting combined with the idiosyncrasies of the VIDICON camera tube. Smear is also known as streaking.

Smoothing Filter. In video, an electronic circuit designed to eliminate fluctuations in the output current of such devices as semiconductor rectifiers or direct-current generators. Smoothing filters are sometimes referred to as ripple filters. See ACTIVE FILTER.

Snow. A type of TV interference that resembles actual snow, such as that seen with clogged video heads, unrecorded tape, etc. Snow differs from VIDEO NOISE in that the former makes the screen image totally unintelligible. These terms are often used interchangeably. See VIDEO HEAD CLOGGING, VIDEO NOISE.

Society for Private and Commercial Earth Stations (SPACE). An organization devoted to protecting and promoting the rights of home satellite TV owners and manufacturers.

Society of Motion Picture and Television Engineers (SMPTE). A committee that frequently sets standards in its related field. For example, the society issues an official color bar to adjust, test or measure various aspects of a television picture or individual equipment.

Software. The instrument which can hold or already contains information on it for playback. In video it is the cassette or disc. With the computer, it is the floppy disk or the cassette. With video games, it is the game cartridge. VCRs, videodisc players and other pieces of equipment are known as HARDWARE.

Solarization. See POSTERIZATION.

Solenoid Control. An electromagnetic circuit system which uses relays in conjunction with the tape motion buttons to operate various functions on a VCR. Solenoid controls offer distinct advantages over older, conventional keys, among them (1) the convenience of a light touch to change functions, (2) the ability to switch functions without going through the Stop mode and (3) quieter operation in contrast to the clunking piano-type keys previously provided. A machine equipped with solenoid controls, for example, can go from Reverse to Fast Forward

234

mode without first going into the Stop mode.

Solid State Image Sensor. See IMAGE SENSOR.

Solid State Memory. Electronic digital memory that contains no moving parts, such as tape or disc. In editing, sequences marked for placement elsewhere used to be stored on a separate videotape. When all the necessary copying was completed, the editor would assemble the completed, revised tape from the different segments. Solid state memory permits the editor to store the coded segments in memory which would then be called upon to automatically reassemble the final version more rapidly and accurately.

Sony. The Japanese firm that started home video. The first home video machine from the company was the Betamax SL-7200. It used a 1/2-inch Beta cassette that played in only one speed, Beta I. The maximum playing time of the VCR was one hour. It offered no timer features or remote control. The second VCR, the SL-8200, offered two speeds, Beta I and II, for one- or two-hour playing time. These two machines were discontinued with the introduction of the SL-8600, a single-speed Beta II, with wired remote pause control, a built-in electronic 24-hour timer and 3-hour record and play capability with the L-750 tape. Later VCRs by Sony featured the slower Beta III speed, electronic tuner, etc. Sony also introduced the first portable VCR (SL-3000) in 1978.

Sound on Sound. An audio dubbing technique found on some VCRs that permits recording a second soundtrack over an existing one without erasing the original. Conventional audio dubbing automatically erases the previous audio track. Sound on sound can be particularly useful in adding narration to a scene while keeping the natural sounds of the background or the music. See AUDIO DUB.

Sound Retrieval System. Refers to a television receiver that utilizes electronics to manipulate the audio signal for the purpose of widening the stereo effect. Introduced by Sony, SRS provides a stereo effect throughout a room without the viewer having to sit between the two basic speakers. This is accomplished by special circuits that first blends left and right signals and then creates "difference" signals. Each set produces distinct spatial and sound characteristics before they are once again combined. Ordinarily, TV sets and monitor/receivers with stereo and built-in speakers are limited in producing stereo separation since the speakers are restricted to the width of the TV unit.

Source Mode. See PIP SOURCE MODE.

SP Speed. The fastest speed on three-speed VHS-format VCRs. With a basic T-120 videocassette, standard play speed can record and play back for two hours. SP has several advantages over the other two speeds of the VHS format. It provides less VIDEO NOISE than the LP (four-hour) and SLP (six-hour) modes. Copies made from tapes recorded in SP offer better picture res-

SPACE

olution. All VCRs are equipped with this speed, whereas some machines may not have the LP or SLP mode. Finally, when a tape recorded in SP is played on machines by other manufacturers, fewer tracking and other related problems arise. SP is standard for industrial and professional VHS equipment. This speed is recognized as the mode that provides the best audio and video reproduction in the VHS format and is used almost universally by the prerecorded tape industry.

SPACE. See SOCIETY FOR PRIVATE AND COMMERCIAL EARTH STATIONS.

Space Invaders. A popular science fiction arcade and home video game which reached its peak in the early 1980s. First marketed in Japan in 1978 by Taito, Ltd., it was basically a game for one player who, with a rapid-fire laser cannon at his or her disposal, attempts to prevent alien creatures from invading earth. It has been featured in many magazines and sung about by various rock groups. Atari, who released the home version of the game in 1980, provided many variations with the cartridge, including two-player formats and fast-moving bombs. The company sponsored a national competition in the fall of 1980, the result of five regional contests involving more than 10,000 fans.

Special Effects. Refers to the various features on virtually all VCRs, with the top-of-the-line models providing the most sophisticated. By changing the speed of the tape, a variety of special effects are possible. VISUAL SCAN permits skipping over unwanted portions of recorded material. SLOW MOTION provides the viewer with the opportunity to study a particular segment of a tape with greater scrutiny. FREEZE FRAME locks in a single picture so that the image appears as though it were a slide. DOUBLE-SPEED PLAY allows viewing material at twice the normal speed but with intelligible sound. FRAME ADVANCE moves the recorded information along at one frame at a time. Not all effects work in all speed modes. Some VCRs offer several features in slow speed while other models provide the same effects in more than one speed mode. The more special effects on a VCR, the higher the cost of the machine. Video cameras also provide an array of special effects, including, among others, AUDIO/VIDEO DUB, CLOCK/CALENDAR, CHARACTER GENERATOR, DIGITAL SPECIAL EFFECTS, INTERVAL TIMER, SELF TIMER, TRIGGER ALARM and VARIABLE-SPEED ZOOM. Like the VCR, the more special effects a camera offers, the more costly the unit.

Special Effects Editing. A general term applied to a group of video camera features designed to enhance the editing process. For example, wipe, fade, a character generator for titling and date/time insert are some of special effects editing functions that can be utilized with the camera in the field.

Special Effects Generator. An accessory that permits switching between several video cameras and video mixing for various effects. This device permits "keying" or superimposing a high contrast graphic over another video camera image, while recording both simultaneously. The SEG can also produce dissolves, horizontal and vertical wipes, corner inserts, etc. The generator is usually combined with a SWITCHER/FADER, which permits connecting two video cameras to one sync source. See CHARACTER GENERATOR, CORNER INSERT, DISSOLVE, KEY, SUPERIMPOSITION, WIPE.

Spectral Space. The measurement between frequencies of a signal. For instance, a VHS videocassette recorder produces a brightness signal on tape, occupying 1 MHz of spectral space between 3.4 and 4.4 MHz. On the other hand, S-VHS machines provide more picture detail as a result of increased spectral space. The brightness signal recorded on tape with the S-VHS format extends to 1.6 MHz—the difference between 5.4 and 7 MHz.

Spectrum Analyzer Mainframe. A professional/industrial unit designed to measure signal drift, draw low-level signals from noise, check signal changes, close in on the frequency span for closer investigation, read amplitude, etc. Featuring a digital display, this highly technical component, like many other professional pieces of equipment, costs thousands of dollars.

Speech Recognition. A voice-controlled method designed to turn the power on and off, change channels, switch modes and perform other similar functions on VCRs, TV sets and related units. Several manufacturers are presently experimenting with such voice-controlled devices.

Speed Selector Switch. A feature on most VCRs to permit the choice of a tape speed for recording. The Play mode usually operates automatically through electronic circuitry that selects the proper playing speed. On most Beta machines the switch includes Beta II and III for Record and Beta I for Play only. VHS machines usually feature three speeds, SP, LP and SLP (or EP). However, on some VHS models the LP mode functions only as a playback feature, since these machines optimize the video head gaps for either SP or SLP speeds. See TAPE SPEED.

Spherical Antenna. An almost flat, rigidly mounted antenna dish in a SATELLITE TV system. Because of its shape, it can pick up more than one satellite without being moved. Instead, one or more FEED HORNS designed to collect signals are placed in front of the antenna and moved accordingly. See PARABOLIC ANTENNA, SATELLITE, SATELLITE TV.

Spike Suppresser. A device designed to filter electrical lines to prevent fluctuations or "spikes" in the current. These variations may result in motor speed discrepancies as well as picture breakup. See INTERFERENCE, INTERFERENCE SUPPRESSER.

Splice. See TAPE SPLICE.

Split Screen. A technique in video permitting the viewer to watch two images simultaneously. Used prima-

Split Screen Viewing

rily in sports (one portion of the picture shows the pitcher, the other the batter or base runner) and news programs (seeing both the anchor man and the news scenes over his shoulder), the split screen is presently controlled by the studio or station, not the viewer. However, this situation is slowly changing with recent technological developments in digital VCRs. See MULTIVISION, PIP.

Split Screen Viewing. A feature on some image enhancer and processor accessories that allows the user to see the differences in image detail and sharpness between the original image and that modified by the enhancer. The video signal image can be adjusted by turning the Detail and Sharpness knobs or dials. The Split Screen switch is then pressed for a comparison of the two images. See IMAGE ENHANCER.

Splitter. See SIGNAL SPLITTER.

Spotbeam Satellite. See SATELLITE FOCUS.

Stabilizer. See IMAGE STABILIZER.

Staircase Signal. In video, a test signal incorporating several steps at increasing luminance levels. The subcarrier frequency normally amplitude-modulates the staircase waveform which helps to check the amplitude and phase linearities in video systems. The staircase signal is one of the many test patterns produced by such components as the video test signal generator.

Stairstep Linearity. A test designed to measure the capacity of a videocassette recorder to "remember" and reproduce the gradations of gray ranging from black to white. In addition, stairstepping applies to lines drawn at angles other than 45 degrees that appear on raster (picture-tube face) displays.

Stand-Alone Electronic Game. A game console that resembles a portable black-and-white TV set and is capable of displaying three-dimensional effects in its electronic games. The games use a process called vector graphics which provides a visual quality not unlike that of arcade games. The Vectrex Arcade System by General Consumer Electronics was the first stand-alone game to utilize vector graphics. Although the system was limited to monochrome, multicolored overlays were available with each cartridge. See VECTROGRAPHIC DISPLAY SCREEN.

Standard-Grade Videotape. The basic tape that many owners of the 60 million VCRs already sold use for time-shifting, copying or other uses. Manufacturers usually recommend their higher-grade tapes for preserving more important material. The term "standard grade" has a specific meaning whereas other grades of tape, such as high grade and super high grade, tend to be more nebulous. However, many professionals in the video field have narrowed down all these terms to four basic types of videotape: standard grade, high grade, Hi-Fi or stereo, and professional grade.

Standardized Time Code. See TIME CODE.

Standards Converter. A professional/industrial unit designed to adapt one video or broadcast standard to another. The basic converter can usually handle NTSC, PAL, PAL-M, SECAM and several other standards. Some models offer other features as well, such as a built-in PROC AMP, noise reduction, automatic input selection and image enhancement.

Step Picture. A videodisc or VCR feature that permits the viewer to display one video frame at a time. "Stepping," which works only with CAV discs, can operate in both forward and reverse. Some VCRs permit bi-directional stepping at various speeds, including sound.

Stereo. Refers simply to the use of two separate audio channels—one left and one right. A stereo system built into a VCR or camera does not necessarily mean that the unit is capable of producing high fidelity, which implies other parameters and standards, such as offering frequency response over the entire audible range and producing a quiet background free from tape hiss.

Stereo Audio Output. The jacks on some VCRs and TV monitor/receivers that are used to connect cables to the left- and right-channel audio inputs of an external stereo amplifier. This connection produces stereo sound through an external stereo hi-fi system.

Stereo Decoder. An accessory unit designed to bring MTS (Multichannel Television Sound) stereo to monaural TV sets or VCRs. A relatively inexpensive way of capturing MTS stereo broadcasts with mono units, stereo decoders may be used with self-powered speakers or a home stereo system. These units usually have dual stereo audio line outputs for playing back a program through a stereo system while recording it on a stereo VCR.

Stereo-Ready. Refers to a VCR that can record in stereo and one that has a jack to connect to an external stereo-TV decoder. The term "stereo-ready" may be misleading since it does ensure that the VCR is prepared to receive stereo TV broadcasts. For a VCR to receive and decode stereo TV sound, the unit often displays either the abbreviation MTS (Multichannel Television Sound) or BTSC (Broadcast Television Sound Committee).

Stereo-Ready Camera. A video camera capable of producing stereophonic videotapes by means of two mic inputs. This does away with the need to connect two microphones to the stereo VCR. Different connections are required to obtain stereo with conventional video cameras. An AC adapter is necessary when using a one-mic video camera with a table-model stereo VCR. The camera is connected to the adapter which is then hooked up to the recorder's video input. By connecting two mics to the mic inputs of the stereo machine, all is ready for stereo recording. In some cases, the stereo VCR has a built-in camera jack, thereby eliminating the need for an

Stereo Separation

AC adapter. See AC ADAPTER, STEREO VCR.

Stereo Separation. The difference between the right and left channels that produces the stereo sound. The greater this difference, which is measured in dB (decibels), the better the stereo effect.

Stereo Simulator. See SIMULATED STEREO.

Stereo Synthesis. The presentation of a stereo effect by artificial methods. There are various means of producing stereo. One of these processes, for example, transmits many of the low sounds to one channel while much of the high frequency is directed to second. A truer stereo effect may be gotten by utilizing a "phasing" process which directs related sounds to a particular speaker. The most advanced technique, however, separates the audio signal into multiple frequency bands, sending different segments of each band into each of the speakers. The last two of the above processes are more sophisticated than the first but tend to exhibit problems with certain instruments "drifting" from one side to the other. Several accessories, under names such as stereo simulator and stereo synthesizer, are available for use with VCRs and TV sets. Some of these devices have controls which compensate for instrument drift or movement. See SIMULATED STEREO.

Stereo TV. A television receiver capable of reproducing two audio channels, usually through the use of its special tuner or tuners, and playing the discrete channels through dual speakers. Since manufacturers of TV sets as well as TV broadcasting companies emphasized the picture half of TV over the sound portion, the potential of audio remained a low priority for decades. Motorola in 1958 developed a stereo TV broadcast system but with very limited frequency response. Its narrow-band FM frequency proved that the main obstructions to stereo TV in the United States were the limitations of audio and telephone transmission. The 1960s witnessed stereo simulcasts of concerts using one channel via TV and the other over an FM monaural station. In the 1970s two events moved stereo TV closer to reality. First, the Japanese dominated TV set production and studio equipment with their sophisticated audio technology. Second, the technique of DUPLEXING, or carrying two audio channels on video circuits, made quality stereo transmission feasible. The Federal Communications Commission finally approved the standards for stereo TV. Some cable systems offer their own stereo services. Other countries like Japan, Germany and England have had stereo TV for many years.

Stereo VCR. A videocassette recorder capable of recording and playing back two discrete audio channels. Stereo was first introduced into VCRs by AKAI in 1981 in its portable model 7350. JVC followed its VHS competitor with its own stereo VCR in 1982. To obtain stereo, the longitudinal audio track on a videotape is divided into two parts, one for each channel. Because regular mono sound of VCRs is only adequate because of their slow

Still Video

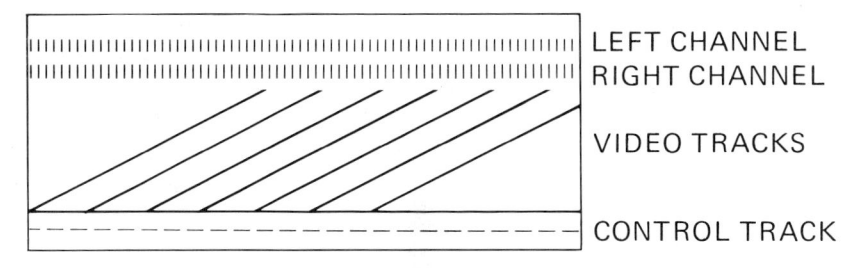

Figure 37. Dual channel stereo audio tracks.

tape speeds inherent in the video process, an even poorer signal-to-noise ratio results when prerecorded tapes are played back on the smaller audio tracks of stereo VCRs. VHS machines, therefore, incorporate a noise reduction system such as Dolby B to help compensate for this. Marantz was first with a Beta stereo recorder. Sony soon followed with its own unique stereo process called Beta Hi-Fi. Instead of using conventional stereo tape heads which produce longitudinal tracks, Sony first FM-modulates the audio signal, combines it with the video and finally the video heads place the audio signal on the tape. The FM modulation retains the audio quality while the use of the video heads sharply increases the dynamic range. The first prerecorded stereo videocassette was MICHAEL NESMITH'S "Elephant Parts." In videodisc players the laser format models sold by Magnavox and Pioneer were the first to offer stereo, with RCA's CED format following in 1982. See BETA HI-FI, DOLBY NOISE REDUCTION SYSTEM, LV VIDEODISC SYSTEM.

Still Adjustment Control. A control knob found on some VCRs designed to stabilize an image that has been locked in by pressing the Pause mode. This control usually is used only if the on-screen picture wiggles.

Still Frame. See FREEZE FRAME.

Still Mode. See FREEZE FRAME, PAUSE.

Still Store. A device that converts an image from analog format to digital and holds the image until it is needed for editing, etc. Used in professional/industrial processing, still store images are often used as backgrounds while foreground images are then placed over the former to produce composites. The results, after going through a variety of changes, including color-correcting, appear as one image.

Still Video. A camera that takes individual images or pictures on an electronic disc rather than on film. Images taken with still cameras can be viewed immediately or, if preferred, can be printed commercially or on a special printer that the user can purchase. In addition, if the user is not satisfied with the result, he or she can shoot over using the same disc. Some discs hold as many as 50 images. Introduced by Sony in 1980, still video cameras have been used by the Japa-

Still Video Floppy Disk

nese to cover the 1984 Olympics and made their formal debut in 1986. Although still video features several conveniences, it does not offer the picture quality of film. In a move to counter this drawback, the ELECTRONIC STILL CAMERA STANDARDIZATION COMMITTEE (ESCC) has introduced the HiVF still video camera, or Hi-Band video standard. Hi-Band is designed to increase resolution to 500 horizontal lines by separating image signals into their brightness (luminance) and color (chrominance) components. The signals are then independently transformed into FM signals that can be recorded on tape. In contrast, a 35mm frame of color negative film, according to Kodak, can produce a comparable print of more than 2,000 horizontal lines. The still video camera market is faced with the same problems that other electronic products had to overcome at their inception—compatibility and competing formats. For example, HiVF and VF are not fully interchangeable. In addition, some later-model cameras are experimenting with digital as opposed to analog, while still others prefer a card medium to that of the disk.

Still Video Floppy Disk. The medium used in still video cameras. The video floppy, or VF, as it is sometimes referred to, contains a coating of metal particles, is slightly less than two inches in diameter and enclosed in a sealed hard plastic cartridge that measures 2.4" x 2.1" x 0.14". The VF, which retails for about $10 each, can record 25 video frames of full resolution or 50 fields (half-resolution).

Still Video Recorder/Player. An electronic unit that can record, play back and store in memory video signals from several video sources. These include still video images, video camera frames, VCR frames, images from video scanners and computer graphics. The unit, which uses a 2-inch floppy disk with rapid random access and is similar to a VCR in appearance, is designed for professional use in broadcasting, video productions, trade shows, etc. See STILL VIDEO, STILL VIDEO FLOPPY DISK.

Stop Action. See FREEZE FRAME.

Streaking. A problem associated with tube-type video cameras and identified by smears that appear at the edges of large, solid objects. Cameras with solid-state pickups rarely display signs of streaking.

Strobe Display. A digital VCR feature that unfolds a quick sequence of still images in electronic slow motion rather than continuous motion. The audio track continues in real time while strobe display is on. Some VCRs can adjust the strobe display rate to present from 1.5 to 8 frames per second. Because of the digital process built into these units, the images contain no noise bars or other video interferences. When in strobe mode, the VCR records one frame at a time and stores the frames in digital memory for predetermined lengths of time prior to capturing more frames.

Studio II. A now defunct video game system produced by RCA. It was taken off the market in the spring of 1978 because of diminishing sales and its obsolete black-and-white format. The play on screen was slow paced for a video game.

Subcarrier. A carrier used for the purpose of generating a modulated wave. This wave is then utilized to modulate another carrier.

Subcarrier Frequency. In video, the frequency that adjusts or modulates the color information. For example, the color information of a video signal, or chroma as it is often called, is grouped in such a way that when it is placed onto a subcarrier the groups reside in spaces between the luminance groups. See CHROMINANCE SIGNAL, DEMODULATED, LUMINANCE SIGNAL, MODULATION.

Subcarrier Pass Filter. In electronics, a particular bandpass filter whose function it is to restrict the luminance information in color TV signals. Subcarrier pass filters are used when it is necessary to separate color information. Another application of these filters is in measuring differential gain distortion. the filter can be connected directly to the signal paths. See BANDPASS FILTER, FILTER.

Subcarrier Rejection Filter. In electronics, a band stop filter which minimizes the level of subcarrier color signals, either NTSC or PAL (3.58 or 4.43 MHz). To reduce the distortion of the luminance signal, these filters utilize phase equalization. The rejection filters are used to prevent flashes of color on the TV screen when black-and-white information is broadcast on a color system. The filter is simply inserted into the video line. It may also be used in conjunction with black-and-white monitors to prevent interference from the color signal. See BANDPASS FILTER, DEFEAT FILTER, FILTER.

Subcarrier Tuning. Refers to a TV satellite receiver that is capable of accepting radio signals from satellite video broadcasts. Since radio transmissions often double up on satellite video broadcasts, the satellite receiver owner can tune in on these transmissions.

Subscriber. Any home or establishment connected to a CABLE TV or PAY TV service. A minimal monthly fee is usually charged for basic cable TV while special pay TV services such as those offering recent films require additional fees.

Subscription TV. A single-channel video broadcasting service which utilizes conventional over-the-air VHF-UHF channels to transmit TV programming in a metropolitan area. With this system a decoder box is required to unscramble a signal. The service usually carries a monthly fee plus installation charges. STV differs from MULTIPOINT DISTRIBUTION SERVICE (MDS) which uses a MICROWAVE antenna on apartment buildings with individual decoder boxes to bring in channels like HBO and SHOWTIME. Both of these systems differ from CABLE TV.

Subsonic Filter. Special VCR electronic circuitry that operates below 20 Hz to eliminate undesirable subsonic signals by accurately tracking the noise reduction system. These filters

Substrate

that work within the VCR's audio system correct any mistracking resulting from inappropriate audio circuitry that is incapable of tracking a wideband signal.

Substrate. A material whose surface contains an adhesive chemical for bonding or coating purposes. Videotape, for example, has a substrate designed to hold oxide or other magnetic particles necessary for recording information.

Sun Outage. A natural phenomenon that occurs when the orbital positions are such that a satellite and the sun are in one line. As a result, the earth station receives signals from both, with the more powerful sun suppressing the preferred signal. This results in a sun outage.

Superband Cable TV. Television channels that occupy frequencies not used for TV broadcasting. Superband channels, which start just above channel 13, usually range from channel J, operating at 216 to 222 MHz, to wherever the cable operators want to take them. See MIDBAND CABLE TV.

Super-Beta. A VCR and camcorder enhancement introduced by Sony and designed to improve the quality of the video image. This is accomplished by compressing more image detail from the video signal within the narrow band of frequencies on the tape. In addition, the video noise reduction system cuts down the flickering and the amount of snow or video noise that often occurs on screen. These improvements add up to the format's ability to produce 280 to 300 lines of horizontal resolution, resulting in clearer and sharper pictures.

Super Capacitor. In video, the memory backup method employed by many VCRs in the event of a power failure. The super capacitor stores an electrical charge which is activated whenever an external electrical outage occurs. This smaller element, which retains VCR program instructions and keeps the clock functioning from about five seconds to about 30 minutes, has generally replaced the more costly nickel cadmium battery cells, which were known to last for several hours. A handful of VCR manufacturers utilize a large super capacitor to provide extended protection against power failure. See also MEMORY BACKUP.

Superhigh Frequency. A frequency range in the ELECTROMAGNETIC FREQUENCY SPECTRUM measured in gigahertz (GHz) rather than in the conventional megahertz (MHz) of the lower spectrum. The superhigh frequency, when opened by the Federal Communications Commission, will accommodate HIGH DEFINITION BROADCASTING (which requires a wider band of the spectrum) somewhere in the 12 GHz band. Before this special transmission is put into operation, however, a set of national technical standards will have to be decided upon. Superhigh frequency extends beyond radar frequencies, such as 2.45 GHz, that are used in microwave ovens.

Superimposition. The overlapping of one image onto another. In video, a CHARACTER or SPECIAL EFFECTS

GENERATOR is used to create this effect. Some video cameras permit electronic superimposition as well as titling. One of the shortcomings of early generators was that one of the cameras (the one supplying the special effect) had to be a black-and-white model. Superimpositions can be tinted any color. See VIDEO DUBBING.

Super-NTSC. See LINE DOUBLING.

Superstation. A local TV station that is picked up by an intermediate "carrier" and transmitted by satellite to cable systems throughout the nation. Basically, it is still a local station with typical local programming of sports, films and other features, tuned in by residents of the city in which the superstation is based. But because of its national exposure, the station can raise its advertising rates. TED TURNER'S Atlanta TV station, WTBS-Channel 17, was the first superstation to appear on CABLE TV. Other superstations include WGN-TV, which began broadcasting in Chicago in 1948 and went on satellite in 1978; WOR-TV in New York; and KUTV in Oakland.

Super-VHS. A VHS format that delivers a sharper picture, produces 400 lines of horizontal resolution and can play back standard videotapes recorded on conventional machines. Although present American broadcast and cable programs produce a maximum of about 330 lines, S-VHS recorders provide better quality reproductions than standard VHS machines that can capture only 230 to 240 lines. Technically, S-VHS accomplishes its improvements by increasing the luminance signal range, reducing the Y (luminance) and C (chrominance) signal crosstalk, using a better grade of tape and narrowing the video head gap. In addition, the format has increased its frequency bandwidth from 3.4 MHz to 5.4 MHz, thereby expanding the amount of information that can be recorded. This results in better detail in the final screen image. To carry the increased data, a high-grade tape capable of holding the finer and more densely distributed particles is required. To fully enjoy the benefits of the S-VHS format, which was introduced in 1987, a high-resolution monitor equipped with separate Y/C video inputs is essential.

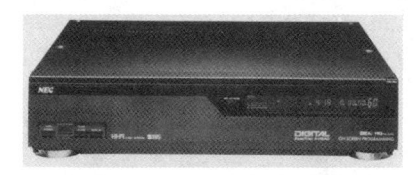

Figure 38. A Super-VHS VCR format featuring digital tracking, HQ circuitry and hi-fi stereo. (Courtesy NEC Technologies, Inc.)

Super-VHS-C. A VHS video recording format that produces 400 or more lines of resolution with a special, higher-priced compact 20-minute videocassette. These mini-cassettes are compatible with full-sized Super-VHS recorders by way of an adapter cassette. S-VHS-C offers certain advantages, including higher resolution, smaller and lighter camcorders, improved copies and editing as a result of reduced signal loss and Y/C video connectors. Some camcorder models in this format are fitted with such ad-

Super-VHS Compatibility

vanced features as high-speed shutters, stop-action capability, built-in character generators, high-fidelity sound and provision for fade effects.

Super-VHS Compatibility. The ability of the S-VHS format to play back standard VHS tapes. S-VHS compatibility has its limitations. Although the format can play back standard VHS tapes (without the additional quality), S-VHS videocassettes will not play back on standard VHS machines.

Super-VHS Tape. A special videotape capable of recording high-frequency signals without the use of a metallic powder formulation that would make standard VHS taping incompatible. Tape designed for S-VHS machines employs a ferric-oxide compound like that found on regular VHS tape. Super-VHS tapes, however, use smaller particles more densely packed and provide higher frequency output.

Super-Video Input. See S-VIDEO INPUT.

Super-Video Output. See S-VIDEO OUTPUT.

Supplementary Lens. An accessory lens placed over the regular video camera lens, changing the viewing angle of the original. With a supplementary lens, a normal lens can be temporarily converted to a wide-angle lens; with a different attachment, a lens can be made to function as a telephoto. However, some of these accessory lenses may not fit a lens of another manufacturer. See DEDICATED DESIGN, EXTENDER LENS, LENS, ANGLE OF VIEW.

Suppresser. See INTERFERENCE SUPPRESSER.

Surface Acoustic Wave Filter. See SAW FILTER.

Surface Integrity. Refers to the face or surface of a parabolic antenna that is part of a satellite TV system. To receive the best possible signal from a satellite, the surface of the large antenna dish should be as close to perfect as possible, thereby concentrating the reflected energy into one point. Deviations in the parabolic curve cause this energy to stray or miss the feed horn (or focal point) which is mounted exactly in the center and front of the antenna. Tolerances in this area are rather small. If the surface integrity is off by more than about 1/16 of an inch, reception may be adversely affected. Tolerances can be checked by using a solid parabolic form called a template, sometimes supplied by the antenna company. See FEED HORN, PARABOLIC ANTENNA, SATELLITE TV.

Surround Decoder. A technique designed to decode the rear-channel audio track of theatrical films that have been encoded with Dolby Surround. See AUDIO DECODER.

Surround Sound. Special electronic circuitry, built into TV monitor/receivers, that is designed to enhance the stereo audio portion of the system to simulate a theatrical or concert-hall effect. This is usually accomplished with only the two built-in speakers of the TV unit. More than 2,000 theatrical films have been made with Dolby stereo surround-sound tracks; many have been transferred to prerecorded tapes with these tracks intact. With the proper components, this surround effect, with its directional and ambient sounds, can be reproduced in the home. Surround sound gives the effect that the audio is coming from the front, sides and rear of a room. One setup sends the principal left and right audio signals (music, speech, for example) to the two speakers at the sides of the TV screen while another channel with ambient sounds (automobile, airplane, etc.) are sent to a pair of rear speakers. Surround sound systems often require a decoder and a separate amplifier. Walt Disney's production of *Fantasia* (1940) was one of the earliest theatrical films with multichannel sound. The technique involved two synchronized films running simultaneously, with the second containing four-track sound. The Dolby Surround system currently used in theaters gained its popularity as a result of the successes of *Star Wars* (1975) and *Close Encounters of the Third Kind* (1977). See AUDIO DECODER, SOUND RETRIEVAL SYSTEM.

S-VHS. See SUPER-VHS.

S-Video Input. A special connection on a TV monitor/receiver designed to separate the Y (brightness) and C (color) signals and to process them without their interfering with one another. The S-Video input helps to bring out the high resolution of ED-Beta, Hi8 and S-VHS videocassette recorders.

S-Video Output. A connection on a TV monitor/receiver that helps to deliver high-resolution Y (brightness) and C (color) signals to other units such as VCRs. See Y/C CONNECTOR.

Sweep Generator. An electronic circuit that employs voltages to the deflection components of a cathode ray tube as a means of examining, comparing and measuring electron beam deflection against other factors. The sweep generator, for example, can help in checking and aligning the color bandpass amplifier response of a TV receiver. See COLOR BANDPASS AMPLIFIER.

Sweetening. A technique employed to enhance the audio portion of a TV show. The process can be applied during or after the recording of the program and involves audio control, recording, mixing and post-production work. To sweeten or enrich the audio, highly specialized and complex components are usually required, such as echo-producing units, audio tape recorders, equalizers and audio-mixing consoles as well as high-quality microphones.

Switcher. An accessory that permits routing any of several RF input signals to any of several outputs. For example, inputs usually include antenna, VCR, video game, pay TV decoder box, etc. Outputs may include main TV, a second TV set, VCR (for duplicating tapes from the input VCR), etc. A simple switcher permits viewing standard TV while recording encoded pay TV or vice versa—at the same time. It also allows monitoring the output of either VCR during the process of duplicating a tape from one machine to the other. A switcher simplifies the complexity of connecting various components and

Switcher Contact

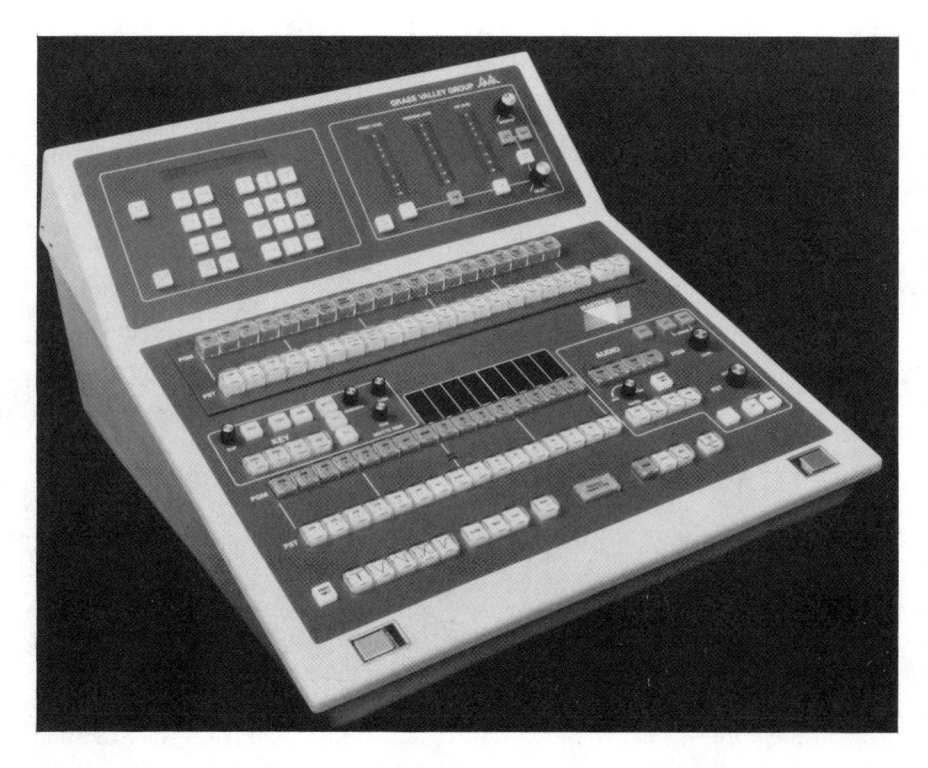

Figure 39. A switcher designed for professional/industrial studios. (Courtesy The Grass Valley Group, Inc.)

eliminates the entanglement of cables and wires around and behind the TV set. A passive switcher is one which has no amplifier or other device to change or modify the signals. Switchers are also listed by their number of inputs and outputs. One with four inputs and three outputs, for instance, is known as a 4 × 3 switcher. Switchers, sometimes called video switchers, are generally rated by their ISOLATION, listed in decibels (dB). The higher the number, the better the isolation of signals. Production switchers are professional/industrial units that provide a vast array of sophisticated features. See DIGITAL SWITCHER, MID-RANGE SWITCHER, PRODUCTION SWITCHER.

Switcher Contact. That part of a SWITCHER that activates the changing of signals from one input/output to another. Contacts can be mechanical or electronic. Mechanical contacts may eventually develop dirt build-up, thereby diminishing the resistance-free electric circuit necessary to avoid noise. Switchers with electronic contacts are considered preferable since this type retains its effective signal-to-noise ratio (measured in decibels) indefinitely. See SWITCHER.

Switcher/Fader. A device that allows the connection of more than one video camera into a video system. The switcher/fader permits two cameras or more to operate from the same sync source, thereby eliminating a discontinuous signal which causes picture roll and other problems. The fader portion of the accessory permits the slow transition from one camera to the other.

Switching Matrix. Refers to the number of inputs and outputs on a SWITCHER. Switchers have various inputs and outputs. A typical model may feature four inputs (any combination of antenna, videodisc player, video game or VCR) and three outputs (two TV receivers and a VCR, for example). This unit would be designated as a 4 × 3 switching matrix. See SWITCHER.

Sync. A shortened form used for referring to synchronous, synchronizing, etc.

Sync Generator. A device employed in all home video cameras to synchronize the pulses required to regulate or control a video system. A sync generator also provides pulses to synchronize a television system.

Sync Processor. A professional/industrial unit designed to reproduce the correct sync, blanking and burst information, including their proper levels. The sync processor can help to correct such problems as those resulting from noisy off-air signals as well as from other video sources. Some models offer additional features, including switchable line bypassing and a locking system that allows the processor to replace any missing sync.

Sync Pulse. A signal that is part of the composite video signal and responsible for the precise timing of video signals inside a VCR or video camera. Sync pulses, which can be horizontal or vertical, help the picture scanning process to re-form and control the image on the TV screen. These pulses, along with other such signals as horizontal and vertical blanking, and drive pulses, are located on the lower portion of the video signal. The longitudinal speed of a videotape should be in sync with the playback video head. See COMPOSITE VIDEO SIGNAL, HORIZONTAL SYNC, HORIZONTAL SYNC PULSE, VIDEO SIGNAL.

Sync/Test Generator. A professional instrument designed to synchronize with all standard composite video signals so that various aspects of these pulses can be tested and measured. The S/T generator usually contains a color bar, "staircase," white-on-black window, convergence, alignment and color raster displays. Some versions feature multiburst frequencies while others provide graduated video sweeps. Other functions include variable control of luminance and chroma, interlace and progressive scanning, outputs of composite video, subcarrier, black burst, etc. In addition, many models measure sync amplitude, equalizer width, vertical pulse width, horizontal and vertical blanking widths, burst amplitude and number of cycles in burst. See COMPOSITE VIDEO SIGNAL.

Synchro Edit. A VCR player or recorder control that permits the start/pause buttons on the unit to operate

Synchro-Edit Input

the start/pause controls of another unit. This feature solves the problem of releasing Pause on two units simultaneously. The syncro edit function may be found on many EDIT CONTROLLERS or external EDITING CONSOLES.

Synchro-Edit Input. A special VCR jack that permits the videocassette recorder to be connected to a camcorder or another VCR of the same format for editing or dubbing purposes. See SYNCHRO EDIT.

Synchronized Editing. See SYNCHRO EDIT.

System. An array of units or components joined together by some type of organization that results in interacting as a whole. Projection TV and stereo are just two examples of systems that tie together several different components.

T

Take-Up Reel. The reel upon which the tape is wound as it leaves the supply reel during recording or playback on a tape recorder.

Tape Deck. The basic component of a tape recorder, comprising the tape transport and a head assembly. Some tape decks provide playback-only amplifiers and are usually listed as tape players. See VIDEOCASSETTE PLAYER.

Tape-Free Editing. See NON-LINEAR ELECTRONIC EDITING.

Tape Guide. In video, a metal and plastic free-spinning spindle designed to keep the tape aligned with the rotating video heads. Tape guides are sometimes mistakenly identified as TWIST PINS which serve a different function. A tape guide has a wide base and head, similar to a spool of thread, whereas a twist pin has a tapered top.

Tape Length. See VIDEOTAPE LENGTH.

Tape Remaining Indicator. A VCR or video camera feature that registers how much recording time is left on a particular videotape. Usually displayed in minutes, the indicator works regardless of the tape speed. This function was first introduced in 1980 by Sharp on its VHS video recorder. The VCR contained a switch which had to be set according to the cassette tape length. Sony further refined the function in 1982 on its SL-2500 model. The feature, known also as tape time remaining indicator, appears on about half of all video cameras produced today.

Tape Repair. See TAPE SPLICE.

Tape Speed. The speed at which videotape travels through a VCR. The tape speeds are different for the two major home formats, Beta and VHS, as well as for speed modes within each system. Sony's original one-hour playing time Betamax had a tape speed of four centimeters per second (Beta I). Its Beta II speed on its next generation of VCRs cut the tape speed in half to two cm/sec with a two-hour playing time. Finally, to compete with VHS, Sony

Tape Splice

once again reduced the tape speed to 1.35 cm/sec with its Beta III mode, offering three hours maximum with a standard L-500 cassette. VHS provides two hours of playing time in its SP mode with a tape speed of 3.34 cm/sec, four hours in its LP mode at 1.67 cm/sec and six hours in EP or SLP with a tape speed of 1.14 cm/sec, all with a standard T-120 cassette.

In addition, tape speed may vary within a specific format, depending upon the special effects built into the VCR and activated by the user. These include SLOW MOTION, DOUBLE SPEED PLAY, several visual SEARCH MODE speeds and full-speed search without picture.

Tape Splice. Physically connecting two ends of videotape. Splicing is usually not recommended because of potential damage to video heads, but kits and special splicing tape are available. Tape cannot be spliced like movie film which, when held up to a light, reveals specific frames. It is almost impossible to locate one frame on tape since images are placed down electronically as a diagonal magnetic signal. See FIELD, FRAME, EDITING.

Tape Tension Guide. The first tape guide near the supply reel of the videocassette. It is specially designed and adjusted to maintain the appropriate skew of the tape. A faulty tape tension guide can cause video as well as audio distortion during record and playback.

Tape Transport Control. A multi-function feature that transfers the forward and reverse VISUAL SCAN and SINGLE FRAME ADVANCE function from the VCR to the video camera. The built-in control is standard on some cameras and optional on others.

Tape Transport System. A method of moving and aligning videotape within a VCR. The individual parts include tape guides, capstan, pins, tension guides, etc. A faulty transport system results in poor picture reproduction, damaged tape and other related problems. See TAPE GUIDE, CAPSTAN.

Target Area. In reference to a video camera, the face of a CAMERA TUBE. Also, with a CATHODE RAY TUBE, the area on which the image is formed and transformed into an electronic signal. Home video cameras that use a camera tube usually have a tube with a target area 2/3-inch in diameter. The RASTER differs from the target area in the sense that it (the raster) refers to the pattern formed on the target area (the flat surface area coated with a light-sensitive element).

TBC. See TIME BASE CORRECTOR.

TBS (Turner Broadcasting System). See WTBS.

Tele Macro. A video camera feature that allows the user to fill the entire frame of the picture with an object at distances of between 2 and 3 1/2 feet. Tele macro virtually closes the gap between the minimum normal focus distance and the maximum macro setting. With a conventional camera in the normal mode, the operator would have to move in to about 4 feet of the subject or object.

Telecast. A television program; the broadcasting of a TV program. The word "telecast" is formed from "television broadcasting." See SIMULCAST.

Telecine Adapter. A device usually containing a mirror, a lens and a small screen used for making FILM-TO-TAPE TRANSFERS. A film or slide projector is aimed at a lens and/or an angled mirror which transfers the image onto a small screen, which in turn is recorded by a video camera. These adapters vary in complexity and price.

Figure 40. A telecine adapter for converting slides and movie film to videotape. (Courtesy Vivitar Corp.)

Telecorder. A combination television monitor and videocassette recorder. One drawback of this otherwise compact, convenient unit is that the TV half may not have its own tuner. This means that if the unit is recording one channel from the VCR tuner, the viewer cannot watch a program on another channel.

Teleconferencing. A method of communicating from one central studio through the use of a communications satellite to many sources, usually located in different cities. For instance, a manufacturer may want to unveil its newest product to its thousands of distributors and retailers. By using a studio and earth station in its base city, the company beams its program via satellite to dozens or hundreds of locations. Teleconferencing may consist of one-way video/two-way audio or two-way video/audio and may employ various techniques such as full motion, compressed and freeze frame video. Teleconferencing user conferences have been held annually since 1981, with experts gathering to exchange ideas and inspect the latest technology. Teleconferencing is also known as video teleconferencing and videoconferencing.

Telephone Cable Network. An experimental project designed to provide various video services to homes and businesses over normal telephone lines. The telephone network has the capacity to carry dozens of video channels, banking and shopping cable TV services and several pay-per-view movie channels. The first experiments with telephone cable service were conducted in 1990 in California. Legislation written in 1984 protected cable TV owners by prohibiting telephone companies from transmitting video signals to the consumer. But the introduction of fiber optics, which has the potential of providing interactive education and advanced telecommunications, especially to rural areas, has boosted the prospects of telephone companies entering the video

Telephone Programming

field. By 1990, more than a dozen phone companies will have begun experimenting with fiber optics cable in the home. See FIBER OPTICS.

Telephone Programming. See REMOTE TELEPHONE PROGRAMMING.

Teletext. A system that provides the broadcasting of text and graphic material in the form of a television signal. Teletext, the least expensive method of obtaining VIDEO DATA, can be carried as part of a TV signal, over a telephone line or transmitted as a subcarrier on an FM signal. Teletext is a flow of digital information concealed within the vertical interval, or the black bar usually seen with a defective vertical hold. Special accessories "read" the otherwise invisible information as it rushes through the signal, translates it onto the screen either in a superimposed format, a full screen image or relegated to the bottom portion of the picture. The information appears as words or graphics and in color. The system, available for about a decade, has been used in England and France but has been slow in gaining acceptance in the United States. The viewer must purchase hardware to decode the signal transmitted to the TV set but usually pays nothing for the service. Because of its signal limitations, teletext must send its information in order, page by page, rather than simultaneously, such as the instant access transmission of VIDEOTEX. Although both systems are similar, teletext is not as complex or sophisticated. See VERTICAL HOLD, VERTICAL BLANKING INTERVAL, VIDEO DATE, VIDEOTEX.

Teletext Decoder. A feature, built into some TV sets, that displays different kinds of information from cable services. In addition, several other broadcasting systems provide teletext information, all of which appears on screen in the form of letters.

Teletext Printer. An accessory used in conjunction with TV sets capable of receiving Teletext information. The printer reproduces text that simultaneously appears on screen, thus saving the information for future reference. Teletext printers are usually activated by the remote control that is supplied with the TV receiver. See TELETEXT.

Television. A method or system of telecommunication for transmitting images of stationary or moving subjects.

Television Broadcast. A monthly magazine reporting on "television equipment, news, applications and technology." Using a large format that resembles a tabloid newspaper rather than a magazine, the periodical, using slick paper and an abundance of color and graphics, covers developments in such areas as teleproduction, postproduction, graphics and transmission technology. Departments focus on international news, new products and people in the industry.

Television History.
1923: Vladimir Zworykin patents electronic TV camera tube.
1927: Philo Farnsworth patents electronic TV system.
1928: RCA granted first TV station permit.

1934: Federal Communications Commission comes into being.
1938: First U.S. 14" electronic TV introduced by Dumont.
1939: Public demonstration at N.Y. World's Fair (April 30).
1940: First use of TV for a college course.
1941: First commercial license issued to a TV station.
1947: Invention of the first transistor.
1948: Network and cable television begin.
1950: Pay TV broadcasts introduced in the U.S.
1951: First videotape recording demonstrated.
1952: UHF-TV introduced in the U.S.
1954: Mexico introduces first three-dimensional TV show.
1955: TV introduced in Vancouver, Washington, high school.
1956: First successful demonstration of videotape recorder.
1957: U.S.S.R. launches *Sputnik*.
1959: Walt Disney show offers first stereo-TV simulcast.
1960: U.S. launches first communications satellite.
1962: Satellite carries TV signals for the first time.
1965: First 1/2-inch consumer video recorder by Sony.
1967: First portable video camera introduced by Sony.
1968: First appearance of Trinitron picture tube.
1972: Introduction of home video game, home video projector.
1974: High definition TV demonstrated in the U.S.
1975: Sony markets first Betamax.
1976: VHS VCR system introduced.
1978: Laser videodisc player introduced.
1980: First consumer camcorder demonstrated.
1982: First Beta Hi-Fi brought out by Sony.
1984: 8mm video and VHS-C introduced in the U.S.
1989: First compatible three-dimensional TV broadcast.

Television/Radio Age. A biweekly trade magazine for those involved in TV and radio programming, program production and sales. Main articles zoom in on network events, innovations and personalities, success and failure of new shows and new ideas in programming and advertising, with the preponderance of coverage on news stories. Some of the many departments of the magazine, which has been around for several years, include related international and national news items, Wall Street and Washington reports, radio and TV news, TV and radio business "barometers" and plenty of reporting on TV and radio personnel, including station managers and executives.

Terminator. A plug connected to unused outputs of distribution amplifiers and other open terminals to prevent power loss. The 75-ohm type is generally used.

Terrestrial Interference. In SATELLITE TV, interference from local MICROWAVE systems, often from telephone companies that transmit in bands similar those of satellites. It can affect one TV channel or more. Sometimes relocating the PARABOLIC ANTENNA minimizes the problem of terrestrial interference. A special TI (terrestrial interference) filter or circuit helps to eliminate microwave and other unwanted signals from satellite transmissions.

Test Monitoring Switcher. A professional accessory used in conjunction with video cameras and a video recorder and a WAVEFORM MONITOR and a COLOR VECTORSCOPE. The special switcher permits checking a recorded signal and the individual camera signals. The test monitoring switcher differs from, but can be used along with, a production switcher. The cameras are hooked up to the production switcher, which is connected to the test switcher, which in turn is plugged into the vectorscope. Finally, the waveform monitor is connected to the vectorscope.

Test Pattern. A geometric design incorporating a set of lines and circles that is used for judging the performance of a TV receiver or similar unit. Test patterns usually test the characteristics of such signals as horizontal linearity, vertical linearity, aspect ratio, contrast, interlace, horizontal resolution and vertical resolution.

Test Signal Generator. An industrial video unit that provides dozens of test patterns in composite, S-VHS, RGB and several other output formats, with RF channel coverage of all broadcast and cable channels. Some models have a menu-driven multipurpose LCD readout control panel and storage capability for 100 events. Standard patterns include multiburst, video sweep, SMPTE color bars, modulated/unmodulated staircase, raster convergence and crosshatch. The unit is known also as a video test signal generator.

THD (Total Harmonic Distortion). Refers to the distortion of the audio part of video equipment such as VCRs and videodisc players. THD is measured in percent figures, with each play/record speed having its own designated number. For instance, a typical VHS machine may be rated as follows: THD of 2.9 percent SP (Standard Play) and 3.1 percent at LP (Long Play) and EP (Extended Play). The smaller the number, the better the performance. In comparison, some LV videodisc players have a THD of less than 1 percent.

Thermomagnetic Recording. An experimental process designed to store information more densely on videotape using DIGITAL VIDEO. To produce this, a video head is combined with a laser which heats the tape so that is can accept a greater number of recorded tracks. See DIGITAL VIDEO.

Three-Dimensional Digital Video Effects System. A sophisticated professional/industrial device designed to simplify the creation of cubes and other solid shapes on screen. The process involves the use of three separate inputs that will map a signal in three dimensions onto a form during one pass on a single channel unit. See DIGITAL VIDEO EFFECTS SYSTEM.

Three-Dimensional Television. After the popular success of 3-D films in theaters during the 1950s, several companies made attempts to develop a 3-D system for television. The major setback has been its incompatibility with normal, two-dimensional TV. Mexico experimented with 3-D in 1954, but by the following year the process was dropped. Japan and Australia also tried to introduce 3-D, but these attempts also failed. All of the above systems used the special two-color glasses familiar to moviegoers in the 1950s. In 1975 a company called Mortek demonstrated a system without the use of glasses. In 1979 an optometrist, Dr. Robert McElveen, presented his glasses-free 3-D system which was compatible with 2-D TV. The only problem was the picture flickered. James Butterfield, chief scientist of 3-D Systems, Inc., presented his "3-D Video" on subscription TV in Los Angeles in December 1980. But once again viewers needed glasses. New York City had its day when a local channel in the summer of 1982 presented the 1954 movie *Gorilla at Large* in 3-D, requiring viewers to purchase glasses at local chain stores. Meanwhile, other systems continued to emerge, including PARALLAX STEREOGRAM; AUTOSTEREOGRAM; Digital Optical Technology Systems, which used only one lens and a special iris; and a Japanese entry employing CHROMOSTEREOSCOPY and the PULFRICH ILLUSION.

Three-Gun Projection TV. A large-screen TV system using three separate lenses or tubes to project each of the three PRIMARY COLORS (red, green, blue) onto a screen. Since single-gun or single-lens projection systems limit the degree of picture sharpness and the amount of light reaching the screen, the more costly three-gun systems are preferred. This process separates the TV signal into basic color elements which are projected through quality lenses, retaining a sharp and bright image, so essential in projection TV. Each color image is enlarged by a refractive lens system utilizing an F opening of f/1.0 or f/1.3 with four-element, five-inch optics. The Kloss Novabeam One projection system employed f/0.7 lenses, one of the "fastest" lens openings available. HENRY KLOSS developed the three-gun technique. See HENRY KLOSS, PRIMARY COLORS, PROJECTION TV, REAR PROJECTION TV.

Through-the-Lens Optical Viewfinder. A video camera viewfinder with the advantages of displaying exactly what the lens sees and offering adjustable focus. It is more expensive than the simple optical finder but less costly than the electronic type, which has become the dominant system used on home video cameras. See OPTICAL VIEWFINDER, ELECTRONIC VIEWFINDER.

TI Filter. See TERRESTRIAL INTERFERENCE.

Tier. A PAY TV channel carried by a CABLE TV service and offered to its subscribers for an additional monthly charge. For example, HOME BOX OFFICE is one tier commonly carried by many cable TV systems which can offer one or more tiers.

Time Base. A voltage induced by the sweep circuit of a cathode ray tube indicator. The waveshape trace may be linear (with respect to time) or nonlinear (in reference to a known time).

Time Base Corrector

Time Base Corrector. An electronic unit designed to transfer information between various videotape formats by synchronizing both machines. More specifically, the TBC can dissect an electronic tape signal and reconstruct it to adapt to broadcast standards—a requirement set decades ago by the Federal Communication Commission. It does this by using digital techniques. For the video professional and hobbyist, the unit enhances picture quality and is an economic boon to the video editor. In addition, the time base corrector has become a valuable tool in the hands of a VIDEO ARTIST. Built-in time base correctors were introduced into top-of-the-line home VCRs in 1989, but because of the higher standards demanded of professional equipment TBCs have remained separate devices. Time base correctors can be units that perform only this single function. These models generally divide into two types—those with the minimal basic functions and TBCs with many sophisticated features. The latter units begin at about $3,000. In addition, several multi-purpose units have found their way into the marketplace, such as the TBC/freeze, TBC/synchronizer, TBC/noise reduction system, TBC/framestore and TBC/digitzer.

Time Base Error. A video signal defect in which picture lines begin their scan too soon or too late. Time base error results in vertical screen images that appear broken as horizontal lines shift to the right or left. One of the functions of a TIME BASE CORRECTOR is to make certain that all picture lines, each having a duration of 63.5 microseconds, begin and end at the same time. The term "time base error" is often used interchangeably with that of "time base instability," the latter term reserved by video professionals and technicians for mechanical malfunctions. See TIME BASE INSTABILITY.

Time Base Instability. A general term for technical problems such as faulty tape guides, tape tension discrepancies between VCRs, worn tape heads and defective tape—all leading to an unstable TV picture. Time base instability, or time base error as it is sometimes incorrectly called, also refers to mechanical speed variations or the differences, however minor, in the actual timing of two VCRs. Since no two machines are exactly the same, a dubbed tape may often produce a picture with jitter or vertical roll. See JITTER, TIME BASE ERROR.

Time Base Stability. The control and maintenance of the scanning process to extremely close tolerances. See SEQUENTIAL SCANNING.

Time Code. An electronic or digital address that individually identifies each frame and permits random access to each one. Time code is particularly significant in the editing process. Often, professional videographers use portable units hooked up to their camcorders or VCRs to record the time code on a separate address track, thus freeing both audio channels. Amateurs, however, face difficulties in recording the SMPTE time code because the simultaneous recording of the code and microphone signal by way of the Audio Dub mode often causes "bleeding." The stan-

Time-Lapse VCR

dardized time code made off-line editing efficient and functional for professionals. See OFF-LINE EDITING.

Time Code Analyzer. An external test instrument used for locating time code errors during editing. In addition, the unit can match color frames, help set tape speed, realign video playback heads and check for "wow" on an audio synchronizer. The time code analyzer presents a readout of each time code. The unit, known also as a time code reader/generator, often comes equipped with video key and LED displays.

Time Compression. A process which permits intelligible viewing of and listening to a tape at a faster than normal speed (usually double speed). The technique, using special integrated circuitry, eliminates distorted sound, known as the DONALD DUCK EFFECT, by dropping the pitch to normal. Time compression is useful in recording odd-length films on standard cassettes. For example, a 128-minute movie can be placed on a two-hour cassette without editing using time compression. In advertising, the process can pack more information than usual into a 30-second commercial. Some VCRs offer a feature called DOUBLE SPEED which basically performs this function.

Time Constant. The amount of variation in the rate of sync pulses that automatic circuits of a TV set will accept. The longer the time constant, the slower the TV set reacts to a different sync situation, and, conversely, the shorter the time constant, the quicker the set's response.

Time/Date Superimposition. A video camera/VCR feature that automatically records the time and date over an image. With a VCR, the time, date and channel are written at the beginning of any recording. See SUPERIMPOSITION.

Time Lapse. The shrinking of large periods of time into shorter ones. Similar to time lapse photography, time lapse video entails recording action intermittently over an extended range of time (such as sunrise to sunset) so that during playback the action appears speeded up. Used mostly for special effects, the technique is more difficult in video than in photography because keeping the VCR component in pause for longer than five minutes may damage the tape or the video heads. A special timing device called an INTERVALOMETER may be employed in this process to turn the video camera on and off automatically. The first home video camera with a built-in time lapse feature was distributed by Akai. It had the capability of automatically shooting a subject for up to eleven days at 90-second intervals with a conventional VHS cassette.

Time-Lapse VCR. A videocassette recorder designed to record the video signal from a surveillance camera either in two hours of real time or up to hundreds of hours in time-lapse recording mode. Some of these professional/industrial time-lapse VCRs utilize computer software that permits adding on-screen text to the recording.

Time Phase Circuit

Time Phase Circuit. See AUTOMATIC TRANSITION EDITING.

Time Search. A VCR feature that permits the locating of a scene or portion of a videotape by selecting the exact time (hour, minute and second). Time search is especially helpful when trying to find a scene or program on a commercial prerecorded tape or a tape not made on your machine—two types that may not respond to ordinary index searches. See SEARCH.

Time Shift. In video, the ability to watch a TV-broadcast program at the viewer's prerogative, not at the officially scheduled time. This is done by setting the VCR to record the program for playback at the viewer's discretion. Because of the VCR's recording capabilities, the machine differs from the videodisc player (which does not record) in that the VCR can play back prerecorded tapes and time shift programs whereas the VDP can handle only prerecorded discs. VCRs are often advertised as allowing the viewer to set up his own "prime time." Time shift, therefore, is one of the VCR's strongest features.

Timer. See PROGRAMMABLE TIMER.

Tint. See COLOR TINT CONTROL, HUE, HUE CONTROL.

Title 17. The Federal Bureau of Investigation warning that precedes or follows commercial prerecorded material on videotapes and videodiscs. Title 17, United States Code, Sections 501 and 506, reads as follows: "Federal law provides severe civil and criminal penalties for the unauthorized reproduction, distribution or exhibition of copyrighted motion pictures and videotapes." The F.B.I. has been known to investigate allegations of criminal copyright infringement.

Titler. See CHARACTER GENERATOR, SUPERIMPOSITION.

TNN (The Nashville Network). A CABLE TV advertiser-supported network specializing in country music. The station features shows, entertainers and occasional western films. In 1990 TNN was listed as the eighth most popular cable network, with more than 48 million subscribers.

TNT. An advertiser-supported CABLE TV network specializing in old films, television reruns, some sports and occasional original productions. Owned by Turner Broadcasting Services, TNT periodically produces its own miniseries and films. Recently, the National Basketball Association games were shifted from WTBS, the Turner superstation, to TNT to lure cable operators to the latter cable network.

Tokyo Video Festival. An international videotape contest designed to promote VIDEO ART productions. Founded in 1978 and sponsored by JVC (Victor Company of Japan), the annual contest is open to amateurs and professionals. Besides the grand prize, the festival presents awards in different categories. A recognized panel of video artists and art experts acts as judge. See NEW YORK EXPO OF SHORT FILM AND VIDEO, VIDEO EXPO NEW YORK.

Total Harmonic Distortion. See THD.

Touch Sensor Button. A feature found on TV sets and VCRs and utilized in conjunction with electronic tuners and remote control accessories. A slight touch of one of these buttons changes a programmed channel or another function on the VCR. See RANDOM ACCESS.

Track Intro Scan. A feature, found on some videodisc players, that permits the viewer to retrace the contents of a disc by moving sequentially through each track and chapter. In the case of combination LD/CD players, the track intro scan reviews the contents of CD discs. The feature is similar to FRAME/CHAPTER SEARCH.

Tracking. The method employed by the video playback head to follow or track exactly the helical signal or path encoded by the recording head. Poor tracking results in VIDEO NOISE while complete tracking loss causes picture BREAKUP. Tracking in a VCR is similar to framing on a movie projector. See VIDEO HEAD ALIGNMENT.

Tracking Control. A knob or control usually on the front of a VCR used to correct VIDEO NOISE, picture instability, FLAGGING and other video anomalies caused by tracking problems. For example, sometimes a tape recorded on one machine will not play correctly on another of the same format because of the way one VCR records and plays back the diagonal tracks. In other cases, the problem may lie within the tape tensions of the machine, especially at slower speeds, or in a defective cassette, both leading to a tracking error so that an adjustment becomes necessary. The tracking control lifts or lowers the video heads so that they track or "read" the signal recorded on the tape. See VIDEO HEAD ALIGNMENT.

Tracking Level Meter. A VCR feature that displays the strength of the recorded control track signal on videotape. Usually, a poor control track affects future editing. For example, a weak tracking level produced by a video camera may result in greater instability and more glitches. Super-VHS camcorders tend to lay down a stronger timing signal than the more compact S-VHS-C models. A tracking level meter may be found on virtually all industrial VCRs designed for editing and some top-of-the-line home models.

Tracking Weight. In the now defunct CED VIDEODISC SYSTEM, the weight of the stylus assembly as it tracks the grooves of the disc. The CED player operated much like a phonograph—the stylus made physical contact with the disc. This differed dramatically from the LV videodisc system in which no contact is made with the disc; instead, a laser beam of light "reads" the information implanted into the disc. Although the tracking weight of the CED system was only 65 milligrams, the stylus would eventually have to be replaced while repeated plays would take their toll on both the picture and sound.

Trailer. A strip of extra-strong non-magnetic tape attached to the end or recording tape. Professional/indus-

Transceiver

trial trailers, which come is several colors, often have one surface available for writing.

Transceiver. A component used in electronic still video and designed to send images over telephone lines. See STILL VIDEO.

Transfer. See FILM-TO-TAPE TRANSFER.

Transformer. An audio accessory, composed of several coils of wire, that modifies the impedance of an audio signal. It is often used to match the impedance of a microphone to the input of a VCR, mixer, etc. In other words, a transformer converts the output impedance of a mic from LOW-Z to HI-Z or vice versa. The accessory is also known as a matching transformer or line matching transformer. In video, a BALUN performs a similar function. See IMPEDANCE.

Transient. See MICROPHONE.

Transient Response. Refers to the outlining of objects in a TV image. Poor transient response causes brightness or a "halo" effect around objects. It can also create a smearing effect around edges. Transient response results from changes in the voltage of the video signal. TV sets that cannot accommodate such changes produce a wavering effect on the edges of objects. Transient response is sometime measured with an oscilloscope or waveform monitor, which produces an electronic image of the voltage. The voltage image, represented as a horizontal waveform, is then compared to a pulse-and-bar pattern for accuracy and deviation.

Transition Status Display. A professional/industrial switcher feature that permits keying into any TV monitor and presents on screen the operating status of the switcher. Different windows of the video graphic device report various parameters, including the on-air source, source name, pre-roll name, position of the automation feature and transition type and duration. Windows can be concealed by the operator. See SWITCHER.

Transitional Digital Video Effects. Refers to a series of special visual creations that can be generated by certain sophisticated devices such as a DIGITAL VIDEO EFFECTS SYSTEM. Transitional effects like DISSOLVES are used in POST-PRODUCTION editing to shift smoothly from one scene to another. Warp, prism, curvilinear, montage, mirror, mosaic, sparkle, trailing, decay, drop shadow, multifreeze and rotation are some of the more popular effects a DVE system can produce. These differ from NON-TRANSITIONAL DIGITAL VIDEO EFFECTS.

Transmission. The conveying of a signal, message or other means of intelligence from one point to another by electrical processes.

Transmission Ability. The amount of light that a filter will allow to pass through it, usually expressed as a percentage. A filter with an 80 percent transmission ability, therefore, will admit 80 percent of the light while rejecting 20 percent.

Transmit Button. A VCR remote control feature designed to send programming information from the remote control to a VCR. Ordinarily, to program newer model VCRs equipped with on-screen displays to record a certain channel at a given time and day, the TV receiver must be on. However, remote controls with LCD displays can be programmed directly, even while in another room, and sent to the VCR by way of the transmit button without going through the TV set.

Transmitter/Receiver System. See VCR TRANSMITTER.

Transponder. Part of a communications satellite. Broadcast signals are transmitted from earth to the satellite where they are then passed through a transponder and sent back to earth. Each satellite may have 12 or more of these transponders. One can transmit thousands of typhoon messages back to earth— simultaneously. But the same transponder can hold only one or two TV signals because of the wider television band. See SATELLITE, SATELLITE TV.

Trap. Refers to special circuitry used by a cable company to lock out certain channels that customers have not subscribed to.

Traveling Shot. Moving the video camera about while it is recording the scene. Professionals use a dolly or tracks to produce smooth traveling shots. Amateurs can get satisfactory results by simply holding the camera firmly while moving forward or by placing the camera on a small cart with wheels.

Triaxial Cable. A technique of expanding the transmitting possibilities of cable systems. Still in the experimental stages, triaxial cable will make possible the transmission of dozens of CABLE TV channels as well as INTERACTIVE TELEVISION.

Trichlorotrifluoroethane (TF). A video head cleaning solvent. It evaporates more quickly than alcohol, leaving no residue to attract dirt. TF is the most popular solvent used in video cleaning kits. It is available in liquid form and spray cans. If the spray is used, it is applied to the swab, not directly to the heads. TF is also called Freon or Freon TF, a trademark of E.I. DuPont.

Tri-Electrode Tube. A video CAMERA TUBE that utilizes three internal lenses, instead of the conventional one lens, to isolate the three primary colors. Sony's Trinicon image pickup tube uses this system. Other types of camera tubes include the most widely used VIDICON and the SATICON. See CAMERA TUBE.

Trigger Alarm. A video camera feature that alerts the user by means of a beeping sound when a recording begins or ends. This function, sometimes listed as beeper feedback, appears on relatively few cameras. See AUDIO ALARM.

Trinicon Camera Tube. A video CAMERA TUBE that allegedly has higher RESOLUTION than other tubes. The Trinicon tube is a single tube system comprised of three color filters (primary colors) which separate the incoming light. Used by Sony, it is

Tripod

a modified version of the SATICON. The average user, however, would probably not be able to detect any significant advantages in any one tube over the others. See CAMERA TUBE.

Tripod. A support for a video camera. Video tripods differ from those designed for film cameras. A good tripod consists of sturdy legs that can contract to a size small enough to be considered portable, a center column that can be raised or lowered, supports that protrude from the base to the legs and a camera mount with pan (left/right) and tilt (up/down) controls. More costly and professional tripods feature fluid heads, instead of the conventional countersprings, for controlling the pan and tilt controls. A tripod, which can vary in price from $40 to hundreds of dollars, has a screw on top of its base to which the camera connects. Some models provide a wheel base, converting the tripod into a dolly. Other models feature a combination camera support and carrying cart.

Tsuno, Kaiko. VIDEO ARTIST, co-producer of documentaries, co-winner with her husband JON ALPERT of the 1981 TOKYO VIDEO FESTIVAL grand prize. The couple created the Downtown Community TV Center in New York City in 1972. Their segment of a PBS TV documentary, "Third Avenue: Only the Strong Survive," won the highest award at the Tokyo competition devoted to short video films. See VIDEO ART, VIDEO ARTIST.

TTL Optical Finder. See THROUGH-THE-LENS OPTICAL VIEWFINDER.

Tuner. The electronic component that accepts and permits the selection of VHF and UHF frequencies. The tuner is usually incorporated into a TV set but may also be a separate unit of a COMPONENT TV system. The channel selection of a tuner is either by the almost extinct two rotary knobs, one for VHF and the other for UHF, that work mechanically, or by touch sensor buttons which operate electronically. There are two types of electronic tuners: the analog which are adjustable and the FREQUENCY-SYNTHESIS tuners which are pre-set for all cable-ready channels. TV monitors do not have tuners. See ANALOG TUNING, DIGITAL TUNING, FREQUENCY-SYNTHESIS TUNER, MONITOR.

Tuner/Controller. A unit that permits owners of projection TVs to receive cable or broadcast signals without the use of a VCR tuner. Selling for several hundred dollars, the tuner/controller often comes with remote control; on-screen audio and video control; and several inputs and outputs, including those for S-VIDEO and RGB, the latter for driving video projectors.

Tuner/Line Switch. See INPUT SELECTOR SWITCH.

Tuner/Timer. The second half of a two-part portable VCR system. The tuner/timer permits recording off-the-air programs at predetermined times by pre-setting the timer. This unit usually remains behind while the VCR deck, which contains the battery, is taken into the field along with a video camera. Some manufacturers have sold each unit separately while others offered only the complete system. Tuner/timers have virtually faded from the home video market with the introduction of the one-piece camera-recorder, or camcorder. See PORTABLE VCR.

Turbografx-16. A video game that uses 16-bit computer chips for better graphics and sound than its competitors, such as Nintendo's 8-bit system. The new video game, introduced in 1989 by NEC, sells for almost twice as much as its competition. Sega's Genesis game system also employs the 16-bit computer chip. See NINTENDO ENTERTAINMENT SYSTEM.

Turner, Ted. Businessman, promoter, owner of the Atlanta Braves, winner of the America's Cup for the United States in 1977, TV station owner. A controversial personality to TV broadcasting, he built his Atlanta station, WTBS-Channel 17, into a superstation in 1976 by placing its signal on RCA's satellite, Satcom, enabling WTBS to be picked up by virtually any cable system in the country. He also introduced the first 24-hour all-news cable station, Cable News Network (CNN). His insistence on colorizing old black-and-white movies, some of them recognized classics, has unleashed a storm of angry protest from film directors, performers and others. His critics claim that he is tampering with the original creative processes of those who turned out the original works. See SATELLITE, SUPERSTATION.

TV Antenna. See ANTENNA.

TV Broadcast Signal. The composite radio frequency (RF) signal that is transmitted from television stations. These signals are picked up by a TV antenna, pass through the antenna cable to the TV tuner which is used to select the channel or frequency. The signals are amplified and converted to a lower frequency signal by the tuner so that they can be fed to the TV picture tube. The broadcast signals are then divided into their audio and video parts by a VIDEO DETECTOR and again amplified. See COMPOSITE VIDEO SIGNAL, TV RECEIVER, TUNER.

TV Camera. See VIDEO CAMERA.

TVCR. A television set/videocassette combination, sometimes described as a TV/VCR. Some models are quite portable, offering a 3.3-inch screen and weighing only five pounds. Other TVCRs have a 19-inch or 27-inch screen with hi-fi stereo sound. See TELECORDER.

TV Interference Tuner. See FILTER.

TV Memory. See MEMORY.

TV Monitor. See MONITOR, MONITOR/RECEIVER.

TV Receiver. A unit that collects radio frequency broadcasts and reproduces them in their original audio and video forms. A TV receiver has a TUNER, either mechanical or electronic, which a MONITOR lacks. Radio waves transmitted from a TV station enter the TV set as an RF (radio frequency) signal through the antenna input. The COMPOSITE SIGNAL car-

TVRO

ries video, audio and sync information and is separated by various components. Each channel frequency enters an IF AMPLIFIER which increases the signal that then travels to a VIDEO DETECTOR where it is separated into a video and audio signal. The video portion is carried to a video amplifier for more processing until it ends up displayed on the CATHODE RAY TUBE or screen. The audio is sent to its own amplifiers and finally to the speaker.

TVRO. A television-receive-only satellite system consisting of a large PARABOLIC ANTENNA (usually 10–15 feet in diameter), a LOW NOISE AMPLIFIER, a receiver and a tuner. The receiver converts the audio and video microwave signals and feeds them to a TV set or monitor. See SATELLITE TV.

TV Signal Generator. See SIGNAL GENERATOR.

TV Still. See MEMORY.

TV/VCR Selector. A control on a videocassette recorder that is used to select either the antenna or VCR source of programming. The selector permits (1) watching regular TV with the VCR off, (2) taping a TV program while watching another, (3) taping a TV program while watching that show and (4) playing a recorded videotape. With a recorded tape set in the VCR, the machine will automatically switch to VCR mode when Play is engaged.

Tweaking. A term used for the technical modification of recording devices in which the AUTOMATIC GAIN CONTROL is deliberately bypassed. Since the AGC attenuates and boosts all sounds, it may present problems in certain situations. For example, if a subject being interviewed is prone to long pauses, the AGC will boost these intervals, causing extraneous sounds to be recorded. In addition, when the subject continues the interview, his remarks will begin at an excessively high level. Tweaking restores control of the audio level to the operator of the VCR or other recording unit.

Twin Digital Tracking. See DIGITAL TRACKING.

Twin Lead Cable. An antenna and video cable that needs no connectors and is less costly than the 75-ohm COAXIAL. The 300-ohm twin lead is more susceptible to radio interference and has a propensity for more signal loss than its counterpart. It is also available in a shielded version which better maintains signal quality.

Twist Pin. A thin metal guidepost that works in conjunction with a TAPE GUIDE at different stress points of the tape path, positioning the tape at the same angle as the head drum. Twist pins are usually located next to tape guides.

Two-Channel Sound. On a VCR, the ability of the machine to record and play back audio on two separate tracks. Of course, this is the basis of stereo sound in VCRs. Two-channel sound also permits audio dubbing in stereo. Mono sound is produced by a single sound track (in home video, the audio track is located near the top edge of the videotape). To obtain two-channel sound, that portion of the tape allotted to audio is divided into two, one part for the left channel and the other half for the right channel, with a small separation band between the two tracks. Because each audio track is so small, the sound quality would suffer without the aid of some noise reduction system. See AUDIO DUB, NOISE REDUCTION SYSTEM, STEREO VCR.

Two-Field Metering System. Refers to an electronic process employed by some camcorders as a means of providing a balanced exposure. The system involves one separate reading of the entire field and another reading of the central zone. Both are then automatically calculated by the camcorder to produce an accurate exposure. See FIELD METERING SYSTEM.

U

UCM. A 1/4-inch home video camera (Ultra Compact Machine) developed chiefly by Japanese manufacturers in the early 1980s. Small and light, UCM cameras were one-piece units, combining camera and recorder, and had a one-hour recording time. A small videocassette fit directly into the camera which produced a picture of satisfactory resolution. This mini-camcorder differed from Technicolor's CVC (Compact Video Cassette), which came out at about the same time but was a two-piece unit. These two systems were incompatible with each other and the VHS-C format, another mini-system which used a small VHS cassette that fitted into a conventional VHS holder for playback on any VHS machine. See CVC FORMAT, VHS-C FORMAT.

UHF. Refers to the Ultra High Frequency channels 14 through 83. Found on most television sets, UHF has its own channel selection knob on mechanical rotary tuners or touch sensor buttons on some electronic tuners. UHF transmission costs are higher than those of its counterpart, VHF.

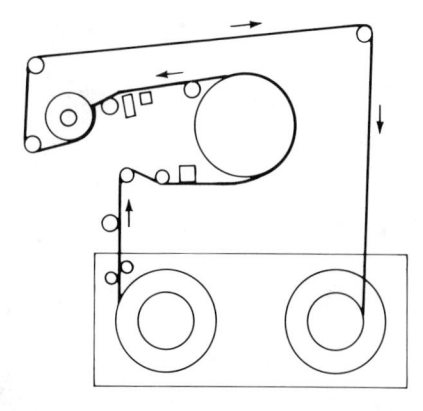

Figure 41. U-load of Beta machines.

U-Load. The Beta format loading system. The videotape is threaded around the video head drum in a U shape. A typical home Beta VCR withdraws about 24 inches of videotape from the cassette, wraps the tape around the head drum and directs it in a U-turn around various tape guides before it returns to the take-up reel in the cassette. The VHS format uses the M-Load system. See FULL LOAD, HALF LOAD, M-LOAD.

U-Matic. A 3/4-inch videocassette recorder designed by Sony. These machines are used primarily for commer-

UNDERSCAN   NORMAL SCAN   OVERSCAN

Figure 42. Underscan and overscan.

cial and industrial purposes. They are sometimes found in schools.

Unattended Recording. A feature, found on virtually all VCRs, that permits the machine to automatically record a program at a pre-set time. This is performed by setting the DIGITAL CLOCK and PROGRAMMABLE TIMER. Unattended recording can refer to programming any home VCR or portable model with a built-in automatic timer capable of taping one event within a 24-hour period to recording eight events from as many channels over a twelve-month period. See TIME SHIFT.

Underscan. A condition that occurs when an image projected by an electron gun does not fill the entire screen. An underscanned picture has a black border at the edges of the screen where the image normally extends to. This is sometimes done deliberately on MONITORS to make certain that no part of the picture is lost. See OVERSCAN.

Unidirectional Microphone. A cardoid-type microphone designed to minimize unwanted sounds from behind and the sides of the mic. See CARDOID MICROPHONE, MICROPHONE.

Unified Remote Control. A single control unit designed to operate several different components of the same brand, including a TV set, a VCR, an LP turntable, a stereo amplifier, an AM/FM tuner, an audio cassette deck and a compact disc player. One drawback of the unified control is that all the components must be from the same manufacturer since the coded commands are of the same pattern. A more versatile accessory is the universal remote control. See MULTI-PURPOSE REMOTE CONTROL, REMOTE CONTROL, UNIVERSAL REMOTE CONTROL.

Uninterrupted Power Supply (UPS). An alternative method of assuring that a VCR will operate during unattended recording even after a power failure. The UPS unit is an external accessory designed to sense power loss, in which case it automatically becomes active and provides the necessary power to keep a VCR running for a limited time. Although most VCRs come equipped with either a built-in nickel cadmium battery or a super capacitor, two very dependable memory backup methods, they usually offer less storage time than the more costly UPS device, which is available in several power sizes and retails for several hundred dollars.

Universal Lens Mount

See MEMORY BACKUP, UNATTENDED RECORDING.

Universal Lens Mount. In video, a uniform bayonet-mount standard designed for interchanging lenses regardless of the camcorder format. Japanese manufacturers of VHS, VHS-C and 8mm camcorders early in 1990 decided to turn out interchangeable lenses and camcorders containing microprocessors which would permit the camera to control the autofocus, power zoom and auto-iris of any lens.

Universal Remote Control. A single control unit that can operate a variety of components, including those of different brands. The universal control has the capability of "learning" different pulse patterns which it then transmits to various equipment. This is accomplished by placing the universal control next to each remote control from other units. The universal control then "listens" and memorizes all the necessary commands of each remote. A built-in sensor in the TV set, VCR or other unit then translates these pulse patterns into electrical signals so that individual instructions can be executed. Some URCs provide several built-in weekly timers and can store up to 38 sequences of up to 14 commands each. See MULTI-PURPOSE REMOTE CONTROL, REMOTE CONTROL, UNIFIED REMOTE CONTROL.

Universal VHS VCR. A videocassette recorder, in the VHS format, designed to record and play back different broadcast standards of tape. In addition, the universal unit can convert one standard to another so that the signal can be duplicated on another machine. One could, for example, copy a tape encoded with the PAL standard by playing it on the universal unit which would deliver an NTSC (National Television Standards Committee) signal to another VCR with the NTSC standard. Although there have been other multiformat VCRs, none have been able to offer all the functions of the universal VCR. See MULTIFORMAT VCR.

Up Converter. An electronic device which changes or "converts" VHF, midband and superband cable TV signals to conventional UHF channels. The converter permits a programmable VCR to record all channels including those on CATV. For example, midband cable channels A through I on a cable box now become channels 47 to 56 on UHF while superband letters J through R are translated into UHF channels 63 to 71. The up converter is also known as a block converter or cable converter.

Uplink. The component of a SATELLITE TV system on earth which transmits the original signal to the SATELLITE. The uplink signals that are transmitted usually range from 5.9l to 6.4 GHz, with each channel requiring a bandwidth of 40 MHz. Since the difference in the range amounts to 500 Mhz, there is room for 12 satellite TV channels. The spherical dish which receives the return signal from the satellite is called the DOWNLINK.

UPS. See UNINTERRUPTED POWER SUPPLY.

URC. See UNIVERSAL REMOTE CONTROL.

USA Network. A diversified CABLE TV network offering family movies, TV series, professional and college sports, health and fitness programs and shows for children and teen-agers. Owned by Madison Square Garden, UA/Columbia Cablevision, the advertiser-supported service started in 1977 and programs its events almost around the clock. By 1990 USA became the fourth most popular cable TV network, with more than 50 million subscribers.

V

Vapor Deposition. A technique of placing magnetic oxides on tape without the use of a BINDER, resulting in fewer DROPOUTS and other tape-related problems. Evaporated metal is placed on the backing while in a vacuum. This process combines metal foil with the backing, eliminating the binder on which oxide or metal particles are normally paced. Doing away with the binder results in a stronger, much thinner tape, providing longer lengths for extended playing time. In addition, vapor deposition offers the potential for a higher density of magnetic material, leading to better performance. See OXIDE, VIDEOTAPE.

Varactor Tuner. On some VCRs, a feature that, among other things, permits tuning in MIDBAND CABLE TV channels D through I on the highband range. In some cases, the VCR can tune in as low as channel B and as high as Channel I of the midband channels. A varactor tuner is sometimes blocked from receiving mid- and superband channels by a few CABLE TV systems that employ SINGLE CONVERSION BLOCK CONVERTERS.

Variable Audio Line Output. A VCR remote control feature that permits the user to control the volume of several TV receivers along with the TV set connected to the VCR. This feature is usually restricted to UNIFIED REMOTE CONTROL units.

Variable Audio Output. A TV monitor/receiver feature designed to control the volume of the unit even when it is connected to an external stereo amplifier and speaker system. Variable audio output can be operated by the TV's remote control pad.

Variable Focal Length Lens. See ZOOM LENS.

Variable Power Zoom Lens. See ZOOM LENS.

Variable Slow Motion. A feature of many VCRs that simply permits changing the rate of speed of the slow motion control—usually starting from FREEZE FRAME. The result on the screen is often accompanied by various noise bars, except in some VCR models that offer variable noiseless slow motion and other machines with

272

digital effects. Slow-motion playback speeds often range from 1/30 to 1/6 of normal speed. Variable speed slow motion, as it is sometimes called, was first introduced by JVC. See SLOW MOTION.

Variable Speed. See TAPE SPEED.

Variable Speed Control. See TIME COMPRESSION.

Variable Speed Playback. A videodisc player feature that allows the viewer to control slow motion to a fraction of a second or view fast play at several times the normal speed. Variable speed playback, also known as variable speed display, performs other special-effects tasks such as advancing one video frame at a time.

Variable Speed Search. A VCR feature that permits steady images during the searching process at different speeds. The results of special electronic circuitry and uniquely designed video heads, variable speed search gives the viewer the choice of moving from about 3 to 21 times the normal speed—in either direction—while retaining a stable picture.

Variable Speed Shutter. A video camera feature that allows the camera user to select a setting commensurate with the action of the subject in different lighting conditions. Some variable speed shutters range from 1/60 to 1/4000 of a second. The larger the number (faster the shutter), the sharper or clearer the moving subject will appear on screen.

Variable Window Filter. A noise reduction device designed to eliminate HISS. The filter permits loud sound (which covers hiss) to pass through, but when sound is low and hiss is detectable, part of the treble range is cut.

VASS (VHS Address Search System. An electronic indexing system that marks videotape during recording or playback so that the viewer may later quickly locate specific points on the tape. See ADDRESS SEARCH, INDEX SEARCH.

V-Cord. A now-defunct videocassette recorder system introduced by Sanyo in the late 1970s. Because it had it own videocassette format which was not compatible with either Beta or VHS, it failed to win consumer approval and was dropped by the company to be replaced by a Beta-format VCR.

VCP. See VIDEOCASSETTE PLAYER.

VCR. A general term for videocassette recorder; also, a unit that (1) can record programs off the air or from cable TV and store them, (2) can play back prerecorded tapes and (3) with the aid of a video camera, can make video movies. Unlike audio, video recording requires a higher tape speed to attain the higher frequency range necessary for video signals. To do this, the VCR employs moving video heads attached to a cylinder or drum which rotates at 1,800 rpm. This HELICAL SCAN technique permits 1/2-inch tape, which itself moves slowly, to hold the high frequency signals. The two most popular systems, Beta and VHS, work similarly, although

VCR Deck

they are incompatible. Tape is drawn from a cassette, wrapped partially around the drum by tape guides, is controlled by a pressure roller and capstan, and is eventually fed onto a take-up reel in the cassette. In the recording process, the tape first passes an erase head that clears previous information from the tape, then moves past a composite audio/control head that writes audio information and a control signal and, finally, passes the video heads which lay down video signals. During playback, the audio/control and video heads reproduce the information from the tape so that the signals appear in the form of sound and pictures through the TV receiver. See AUDIO HEAD, CAPSTAN, VCR FEATURES, VIDEO HEAD, VIDEO HEAD DRUM, VCR HISTORY.

VCR Deck. The unit of a two-piece portable VCR that is taken into the field and to which a VIDEO CAMERA is usually connected. It contains the videocassette housing; operating keys such as Play, Stop, Eject; tape counter; tracking control; and other essential elements that affect the tape movement, recording and playback. The second unit, which usually remains behind, is known as the TUNER/TIMER. With home video systems, the one-piece camcorder has virtually replaced the two-piece video camera.

VCR Features. The basic functions or controls on virtually all VCRs and various options provided by higher-priced models. Basic features include the following: EJECT, REWIND, STOP, PLAY, FAST FORWARD, RECORD, AUDIO DUB and PAUSE. On the tuner section there is either a mechanical rotary knob or sensor buttons for VHF/UHF channel selection, AUTOMATIC FINE TUNING (AFT), an automatic timer and a clock. Other basics include an On/Off switch (for power), a stand-by light, a TV/VCR SELECTION SWITCH or button, a tape speed selector, a memory/index counter/search control, an input signal selector or camera/auxiliary line switch, TRACKING CONTROL, microphone input, a remote control jack and a channel 3/4 switch. Some VCRs provide additional features, such as VISUAL SCAN, SLOW MOTION, FREEZE FRAME, SINGLE FRAME ADVANCE, DOUBLE SPEED, WIRELESS REMOTE CONTROL, four video heads, STEREO audio recording and playback and CABLE READY capability. Many of these more advanced features have eventually filtered down to lower-priced models.

VCR History. The VCR as we know it today was invented and introduced by Sony. Ampex introduced the first reel-to-reel video tape recorder (VTR) in 1956. Sony, in the early 1970s, introduced the U-Matic system, a 3/4-inch VCR, used mostly by industry. But this machine was bulky and expensive. Sony followed this in 1975 with the first successful home VCR, the Betamax, which had a maximum recording time of one hour. It wasn't long before JVC challenged Sony by coming out with its own system, the VHS (Video Home System) format, incompatible with that of Sony's. Although both systems use 1/2-inch tape, the size of the cassette and the recording speeds are different. There are portable VCRs, professional models and mini-VCRs. By the late 1980s

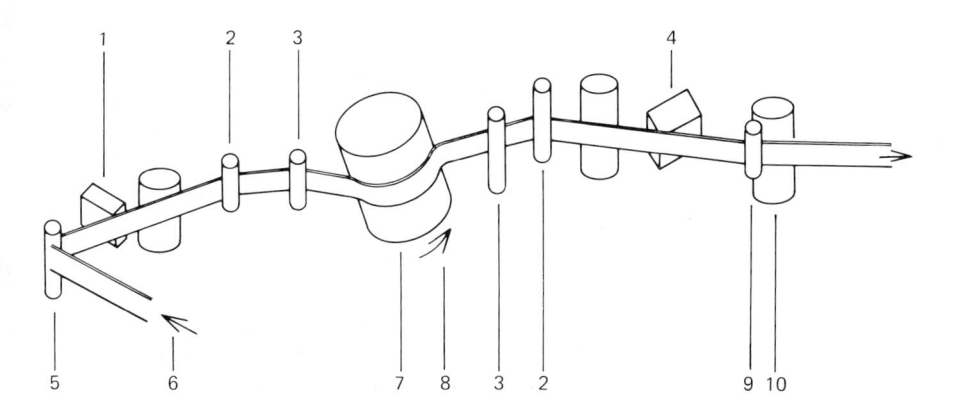

1. Erase head
2. Tape loading roller
3. Slanted tape guide
4. Audio record/playback/erase head assembly
5. Tape guide
6. Tape direction
7. Head drum
8. Head drum rotation
9. Pressure roller
10. Capstan

Figure 43. VCR videotape path with helical scanning.

consumers had a choice of about 40 different brands and over 100 models of VCRs. Reports state that almost two out of every three homes in the U.S. have at least one VCR. Some 1990 estimates place the number of VCRs in use at more than 60 million. See BETA, VCR FEATURES, VHS.

VCR Manufacturers. A handful of Japanese manufacturers provide the VCRs sold under various brand names in the United States. The largest supplier, Matsushita, not only owns Panasonic and Quasar, but makes sets for several companies, including Canon, General Electric, Harmon Kardon, Instant Replay, J.C. Penney, Magnavox, Philco, Philips, Quasar, Scott and Sylvania. JVC supplies Kenwood, Marantz, Teac and Zenith as well as its own U.S. distributors. Sharp, Tatung, Teknika and Mitsubishi make models only under their own name.

Hitachi supplies Minolta, Pentax and RCA. NEC provides Pioneer, dbx and Yamaha with their VCRs. Orion makes machines for Emerson; Samsung for RCA; and Toshiba for Montgomery Ward, Sears and Vector Research. In the Beta format, Sony used to make VCRs for Zenith and Sanyo, both of whom switched to the VHS format.

VCR Modification. See MODIFICATION KIT.

VCR Sales in the U.S. The videocassette recorder has become one the most popular electronic units purchased by American consumers during the 1980s. The table below lists the total number of American households with one or more VCRs from 1984 through 1989.

| 1984 | 10 million |
| 1985 | 20 |

275

VCR Transmitter

1986	30
1987	45
1988	55
1989	60+

VCR Transmitter. A video accessory that sends an audio/video signal from one VCR to several TV sets. For example, this unit permits a viewer to watch a regularly scheduled TV broadcast on one TV receiver while another person views a videotape on another TV set not accompanied by a VCR. In other words, the device allows every TV set in a home to utilize one VCR. The average range of a VCR transmitter is usually between 100 and 200 feet.

VCR/TV Selector. See TV/VCR SELECTOR.

VCS. See VIDEO COMPUTER SYSTEM.

Vectorscope. See COLOR VECTORSCOPE, OSCILLOSCOPE.

Vectrographic Display Screen. A home video game system that uses a screen similar to that of an arcade game. Vectrographic screens are capable of producing multidirectional lines, including diagonal ones, adding more complexity to the game. One such system, Vectrex, employed this process and did not have to be connected to a TV receiver since it contained its own black-and-white screen.

Velocity-Modulated Scanning. An electronic technique designed to increase the sharpness and contrast of a TV monitor/receiver image. This is accomplished by means of a series of electromagnetic coils encircling the neck of a CATHODE RAY TUBE so that the speed of the electron beam is altered. By changing this speed, the scanning process allows the beam signal, which must pass from black to white to produce image contrast, to remain longer on white. This bolsters the picture sharpness. Variations of velocity-modulated scanning, or velocity scan modulation as it is sometimes called, include DELAY LINE APERTURE CONTROL and HORIZONTAL IMAGE DELINEATION.

Vertical Blanking Interval. A fraction of time during broadcasting and playback of videotape in which the screen goes blank. This blanking occurs during the time the VCR switches from one video head to the other. This interval is imperceptible to the viewer but important in electronics. For example, during FREEZE FRAME special circuitry moves the tape so that any noise bars are hidden during the vertical blanking interval, which is also known as vertical interval.

Vertical Centering Control. A feature found on waveform monitors designed to place blanking at zero (0) IRE so that accurate waveform interpretation can be attained. Vertical centering control is a basic function of portable waveform monitors made for field use. See OSCILLOSCOPE, WAVEFORM/COLOR PICTURE MONITOR.

Vertical Hold. A control that regulates and stabilizes the TV image and keeps it from rolling. Many TV manufacturers have eliminated this external feature, depending on internal circuitry to halt vertical image problems. However, many anti-piracy signals work on the principle of a weak vertical hold so that VCRs cannot duplicate tapes with this track. But some TV sets without the external control cannot lock into these prerecorded tapes. IMAGE STABILIZERS, sold for this purpose, control the vertical signal both in TV sets and in VCRs, thereby defeating the purpose of most anti-copying signals.

Vertical Interval. See VERTICAL BLANKING INTERVAL.

Vertical Interval Reference Signal. A broadcast signal that, when received by a VIR-equipped receiver, helps to control automatically both tint and color. The VIR signal, developed in 1977 by G.E., is broadcast on one of the vertical blanking interval lines. It contains a color reference bar, color sync, a black level reference and the horizontal sync pulse. The appropriately equipped receiver matches its red and blue color and black level with those of the incoming signal and automatically eliminates any color distortion. Matsushita produces a VIR II circuit chip for all manufacturers. The VIR is considered a significant development in color television. See BLACK LEVEL, COLOR SYNC, HORIZONTAL SYNC PULSE.

Vertical Interval Time Code. In video, a special signal recorded between video fields. One of its uses is in relation to some edit controllers which depend on the VITC for accurate editing. These controllers utilize the vertical interval time code to find and return to specific edit points. See EDIT CONTROLLER.

Vertical Resolution. Like HORIZONTAL RESOLUTION, a method of measuring detail in an image. The more HORIZONTAL LINES, the better the vertical resolution or detail and quality of the picture. With many home video components a vertical resolution of 400 horizontal lines is considered excellent. Although the U.S. standard produces 525 lines, very few pieces of equipment come close to that number. Home video cameras, for example, have a general range of from 250 to 280 lines. The average VCR produces only about 240 lines. See HORIZONTAL LINES, HORIZONTAL RESOLUTION.

Vertical Sweep. The downward motion of the scanning beam from the top to bottom of a televised image.

Vertical Sync Pulse. A small pulse at the end of each video signal line. These pulses, of predetermined size and length of time, activate internal circuits of various video components, providing stable and viewable images. If the vertical sync pulses are modified (for example, by some anti-piracy signal device), then the picture rolls. See COMPOSITE VIDEO SIGNAL, VIDEO SIGNAL.

VF (Still Video Floppy Disk). The medium used in still video cameras. See STILL VIDEO FLOPPY DISK.

VHD Videodisc System. A now-defunct format that used a rotating disc to produce a picture and sound by means of a TV receiver or monitor.

VHF

The Video High Density player used a stylus similar to that of the RCA system and a grooveless disc like that of the LV format. The VHD player, which had a playing capacity of one hour per side, offered fast and slow motion, FREEZE FRAME, rapid picture scan and several other features. It played in stereo, similar to other formats.

VHF. Refers to Very High Frequency channels 2–13. VHF uses less energy during transmission than does UHF (Ultra High Frequency) channels.

VHF/UHF Splitter. See SIGNAL SPLITTER.

VHS Address Search System. See ADDRESS SEARCH.

VHS Format. A home VCR system introduced in 1976 by JVC to compete with Sony's one-hour Betamax machine brought out in 1975. The early Video Home System utilized a larger cassette to hold about one-third more tape and decreased Sony's Beta I tape speed (four centimeters per second) to 3.34 cm/sec., thereby offering two hours of maximum playing time in standard play. When Sony countered with Beta II, which provided a two-hour maximum, VHS returned with its own Long Play mode (LP), raising the playing time to four hours. And, to make certain of its victory over its competition, VHS in 1979 followed with its six-hour mode. This new speed was labeled EP or SLP (Extended Play or Super Long Play), slowing down the tape speed to 1.14 cm/sec. Beta retaliated with a five-hour maximum by using a thinner tape base and a slower speed (1.35 cm/sec.), calling the mode Beta III. But by this time—with Sony offering too little, too late—the battle was over. VHS began to dominate the home video market. The format differs from Sony's in yet another area—its loading system, called M-LOAD. Other advancements in the VHS format include the introduction of hi-fi in 1983, HQ in 1985, CTL coding in 1986 and Super-VHS in 1987. Some experts estimate that more than 200 million consumer VHS machines have been sold since its introduction. Matsushita, the largest manufacturer of VHS recorders, produces machines for several other companies.

VHS Hi-Fi. The use of two additional recording heads built into the video head drum to produce high-quality audio. The added recording heads, which make the VHS method of producing hi-fi different from that of Beta Hi-Fi, record the audio signals on the same area as the video signals. Because the azimuth positioning of the additional heads differs from that of the video heads, crosstalk between these tracks is negligible.

VHS HQ Circuitry. See HQ CIRCUITRY.

VHS Index Search System (VISS). See INDEX SEARCH.

VHS Speed Mode. A speed at which videotape plays or records in a VHS-type VCR. The three basic speeds are SP (Standard Speed or normal play), LP (Long Play) and EP or SLP (Extend or Super Long Play). SP plays for two hours on a conventional T-120 cassette and is the mode used for best resolution and prerecorded tapes. LP, which has a four-hour range, has been phased out as a recording mode by some VHS manufacturers who consider the speed superfluous. These companies often provide two additional video heads designed especially for the EP mode. With a thinner-type tape, the VHS maximum play/record time has been extended to eight hours.

VHS/VHS-C VCR. A videocassette recorder that can accommodate both standard VHS cassettes and VHS-C compact cassettes without the required additional adapter. Normally, the smaller VHS-C cassettes, to play in conventional VHS machines, must first be placed into a cassette adapter which is then inserted into the VCR. The new decks have a dual-loading system consisting of a built-in VHS-C adapter in the main transport compartment and special transport devices to handle the thinner tape.

VHS-C Format. A unique method of using VHS tape and equipment, yet providing an opportunity for compactness and light weight in camcorders. The VHS-C (for Compact) uses a mini-cassette about the size of a pack of cigarettes that is inserted into the camcorder. The mini-cassette can later be installed into a special VHS housing or adapter to be played on conventional VHS machines. Besides offering a compact and light-weight camcorder, the format allows editing, with the addition of an optional editor, onto a regular VHS cassette. The format was originally restricted to SP mode for quality recording. Some more recent VHS machines have a permanent built-in adapter for VHS-C cassettes. With the manufacture of new and thinner videotapes, some camcorders have expanded the running times of VHS-C and Super-VHS-C cassettes from 20 to 30 minutes in fast-speed mode and 60 to 90 minutes in slow-speed mode.

Video. Refers to picture information or a medium which uses television to transit and receive video. Video has emerged to a position in which it is no longer synonymous with television. For instance, video does not have to be sent from long distance as TV does. Participating in a video game or playing back a videodisc or videotape is not watching TV. Television is, or has become, only one form of video, albeit a major one.

Video. The first magazine devoted entirely to the home video user. Published at its inception as a quarterly in 1978, *Video* changed to a bimonthly and then switched to a monthly as consumer interest in VCRs and video cameras grew. Highly respected for its technical articles, it is strong in product testing, timely features and entertaining columns. Its covers have often been devoted to VIDEO GRAPHICS and VIDEO ART. The magazine initially printed a helpful glossary of technical and video terms on its last page, but this has since been dropped. The writing level appears to be designed for a readership with an above-average education. Its weakest point is its coverage of films and television shows, a minor factor considering *Video's* superiority and accuracy in its technical information.

Video Accessory

Video Accessory. See ACCESSORY.

Video Action. A breezily written periodical that lasted for a relatively short time as a monthly, changing to an in-depth quarterly in 1981 before folding several months later. As a monthly, it featured articles on a variety of video-related subjects, columns and some good product test reports. But it specialized in neither technological matters nor video hardware. If anything, the early issues were diversified; articles appeared on computer art, Henry Kloss, the American Hero exposed, repairing video equipment, medical video, TV churches and how to be a producer. One unusual item was its column on vintage TV. The quarterly seemed to be a different magazine before its demise. See VIDEO PERIODICALS.

Video Amplifier. A special VCR circuit that increases the strength of a VIDEO SIGNAL before it is sent to a TV receiver for display. A video amplifier differs from an RF AMPLIFIER which transmits RF SIGNALS only.

Video Animation. See PIXILATION.

Video Arcade Game. A coin-operated electronic game console containing its own controls and screen. An arcade game may be located in places as diverse as grocery stores, bars, movie theater lobbies and diners as well as arcades where many machines line the walls and aisles. These machines represent big business. By 1981, $1 billion was spent on the purchase of these machines which generated revenues of more than $7 billion a year. This figure breaks down to over 76 million quarters a day fed into these game consoles. A game that is successful in coin format is often adapted to home video game use. Notable examples include Space Invaders and Pac-Man. Sega Enterprises introduced the first three-dimensional video arcade game called SubRoc-3D. Other innovations in arcade games include the use of videodiscs for realism and the process of holography to produce 3-D images. Much controversy has emerged from the growth of the arcade game industry. Parents and community leaders view this aspect of the electronic amusement business as having a negative effect on school-age children. Others believe that the games provide clean, healthy stimulation, mental challenge and a test of one's reflexes.

Video Art. A creative use of video technology including lasers, videotape, computers, TV furniture and sound expressed, often abstractly, in stills, moving pictures, sculpture and sometimes architectural forms. Artists first experimented with video as a means of artistic expression in the late 1960s. The most notable proponent of this art form is NAM JUNE PAIK. Video art, which has a limited market, has been presented in galleries, museums and on public television. To some VIDEO ARTISTS, the term is considered derogatory; they prefer to be known as "artists who use video." As for a definition, still others prefer a more classic one such as "a meaningful work done with consummate skill." Museums that have exhibited video as an art form include the Museum of Modern Art, the Whitney Museum, Anthology Film Archives, the Donnell Library Center and the Kitchen Center for Video, Music and Dance, all located in New York City. Both the MOMA and the Whitney have a video department.

Video Artist. Any artist who works full- or part-time with any number of components related to video and/or its technology. With the accessibility of less costly and relatively portable video units during the 1960s, more and more creative people started to experiment with video as an art form. In 1969 an exhibition center for the works of video artists was made available at the HOWARD WISE Gallery in New York City with the first major video show, "TV As a Creative Medium." Artists who have worked or are working in and with video include NAM JUNE PAIK, Charlotte Moorman, Shuya Abe, Charles Atlas, Shigeko Kubota, RON HAYS, Douglas Davis, David M., JOHN REILLY, Laurence Gartel, Stephen Beck, JULIE GUSTAFSON, Richard Bloes, Ed Emshwiller, Bill Viola, John Sanborn, Kit Fitzgerald, Rhona Ronan, Shi Sun, Norman Pollack, Chip Lord, Phil Garner, Margia Kramer, Cary Hill, Hermine Graad, Ruth Rotko, John Keller and Isamu Tesuka.

Video/Audio Amplifier. See AUDIO/VIDEO AMPLIFIER.

Video Cable Tester. A device designed to check cables for broken conductors, continuity and shorts and whether the problem stems from the shield or the center conductor. The tester usually accepts BNC and UHF cables as well as a combination such as BNC to UHF. There are separate AUDIO CABLE TESTERS available.

Video Camera. A video recording system component that collects images (through a lens onto an image pickup tube or similar device) and sound (through a mic) and converts them into electrical signals that are then changed to magnetic signals by a VCR. There are several types of video cameras produced by manufacturers: professional, industrial, surveillance and home models. Those that are used for professional broadcasting can weigh up to a few hundred pounds and can cost tens of thousands of dollars. Industrial cameras can sell for up to $50,000 and provide for electronic editing, have a color electronic viewfinder and also provide quality pictures. Some camera models serve as surveillance cameras in banks, stores and other places requiring high security. Home video cameras are smaller, more portable, have fewer features and sell for less than $2,000. Those at the high end of the price range usually have many special features.

The first home cameras, introduced in 1969, for use with 1/4-inch reel-to-reel videotape recorders, were simple instruments that recorded in black-and-white only. These were followed in 1977 by home cameras that worked with VCRs. The first home color cameras, introduced by JVC in 1977, ranged in price from $1,500 to $3,000. The following year JVC brought out a portable battery-operated home video recorder. Sony in 1980 displayed its VideoMovie, a one-piece camera/VCR combination.

Early home video cameras required a two-piece system made up chiefly of

Video Camera Components

the lens, image pickup tube and viewfinder in one unit and a portable tape deck in the second unit. The one-piece camcorder, which houses both components in one compact unit, has virtually replaced the two-piece home models.

Video Camera Components. The basic elements that make up all video cameras, professional or home models. The components include a LENS (either fixed focus or zoom); a VIEWFINDER (either optical, through-the-lens or electronic); a CAMERA TUBE or metal-oxide semi-conductor chip; amplifiers; control circuits; limiters and a built-in microphone. Other features vary according to the price and model of the camera. See VIDEO CAMERA, VIDEO CAMERA FEATURES.

Video Camera Features. Any functions, controls and/or switches excluding the basic components, such as lens, of a video camera. The number and sophistication of features vary, of course, with each camera. Some offer the most popular features, such as AUDIO/VIDEO DUB, which replaces part of the old audio and video; BACKLIGHT CONTROL, which slightly boosts exposure; CLOCK/CALENDAR to superimpose time and/or date; SELF TIMER, which allows the camera user to step into a scene; and TAPE TIME REMAINING INDICATOR, which shows how much shooting time is left on the cassette. Other camera models provide additional features, including, among others, AUTO IMAGE STABILIZER, which moves the lens assembly to help steady the video picture; CHAR-ACTER GENERATOR for making titles; COLOR VIEWFINDER; DIGITAL ENHANCER to boost contrast for better images in low light; DUAL CAMERA RECORDING, which mixes pictures from two different cameras; FLYING ERASE HEAD, which helps to prevent video noise and glitches between scenes; MICROPHONE MIXING; MONITOR SPEAKER, which replaced headphones; and TRIGGER ALARM, which warns user at the beginning and end of recording.

Video Camera Sensitivity. The ability of a camera to reproduce a usable image with a minimal amount of light. Sensitivity is usually measured in FOOTCANDLES or LUX. For instance, the sensitivity of a particular camera may be rated at 50 lux (five footcandles). Many video cameras feature a sensitivity switch, which increases sensitivity. To decrease sensitivity, a NEUTRAL DENSITY FILTER is attached to the front of the lens. All video cameras have a sensitivity range, the average being approximately 10–10,000 footcandles.

Video Carrier. The television signal that carries the picture, sync and blanking signals within its modulation sidebands.

Video Compositing. A system, similar to chromakeying, in which one or more images is combined with another image to form a final and different picture. For instance, a shot of the Grand Canyon can be used as background. Then a family is recorded against a blue background. When the two shots are integrated, it appears as though the family were at the site. See CHROMAKEYING.

Video Computer System by Atari. A video game system made up of a console, two remote controllers, one game cartridge and an AC power supply. Options such as wired paddle, keyboard, driving controllers and joysticks could be connected to the game console to accommodate up to four players. At one time the most popular of the game systems, VCS had more cartridges available than any of its competitors. In addition, each cartridge contained several variations. The game system also provided sound effects.

Video Controller. A professional/industrial device that provides frame-accurate control of most videotape and videodisc units. Similar to an edit controller or editing console, the video controller offers several additional and unique features. A composite video switching function, for example, allows frame-capturing or frame-recording to and from the same unit for rotoscoping purposes. In the field, the video controller can pilot two machines. See EDITING CONSOLE.

Video Conferencing. See TELECONFERENCING.

Video Converter. A professional/industrial instrument designed to convert video graphics to various broadcast standards signal specifications, such as NTSC and PAL. Video converters may operate in genlock mode or independently with composite, Betacam, S-VHS and MII formats.

Video Copy Processor. A unit designed to produce full-color hard copies (print-outs) from video images. Often selling for several thousand dollars, these professional/industrial machines incorporate a sublimation-type printing process, can store images and data and produce 640 pixels X 480 lines NTSC dot resolution. See VIDEO PRINTER.

Video Crosstalk. See CROSSTALK.

Video Data. Information transmitted electronically and displayed on a TV screen. There are three systems of providing video data. Cable TV offers a one-way passive system in limited areas to paying subscribers only. Computer networks, available to subscribers nationally, offer the most sophisticated video data services. This data can be stored or processed by a machine into hard (printed) copy. This system also permits data input. The third method consists of signals carried by telephone lines or broadcasts available presently only in a few areas. The above three sources of video data vary in programming, from simple news printout, financial information and transportation schedules to ordering merchandise electronically and reserving a seat on a commercial airline.

Video Design. The introduction of new shapes, styles, modes and functions to television sets, videocassette recorders and other video equipment. Recent innovations and prototypes promise a new look in conventional video units. They include TV screens without the standard frames, VCRs simpler to program, television sets

Video Digital Memory

and monitor/receivers with no function controls on the front to distract from the picture, motorized stands to change the position of the unit and operated by remote control, and equipment controlled by voice commands. Some designers predict that units will be offered in various colors and innovative shapes.

Video Digital Memory. See DIGITAL MEMORY.

Video Digitizer. A professional/industrial instrument designed to display images of 30 frames per second in real time on a computer monitor. Video digitizers can usually accept multiple inputs, such as RGB, Super-VHS and composite video signals from NTSC or PAL, with the capability of displaying them at the same time.

Video Distribution Amplifier. See DISTRIBUTION AMPLIFIER.

Video Documentary. A video essay or story; a genre of video art which borrows from the film documentary but more often explores and applies new techniques peculiar to its own medium. Like its film counterpart, the video documentary deals with real people, actual events and natural settings to tell its story. But unlike the well-prepared script and finished, slick look of film, the video documentary is more the result of a freewheeling, unscripted presentation dependent more on the interaction of its subjects. Inherently more subjective than the film documentary, the American video documentary has received greater recognition in Europe than at home. Some VIDEO ARTISTS working in this genre have received well-deserved recognition, awards and grants for their works, most notably JOHN REILLY and JULIE GUSTAFSON. See VIDEO ART.

Video Dubbing. A special feature found on some videocassette recorders that permits the insertion of a scene or title onto an existing scene. Video dubbing with three VCRs provides superimposed effects without picture breakup at the beginning or end of the newly added material. With conventional machines, recording over previous information automatically erases whatever audio and video signals are on the tape. Video dubbing also applies to the technique of replacing a portion of recorded video information while keeping the original sound track. See SUPERIMPOSITION.

Video Effects Titler. A stand-alone unit designed to superimpose computer-generated color images over another video image coming from an external source such as a VCR, video camera or videodisc player. The VET permits the user to superimpose color titles over a camera image as well as add titles to prerecorded videotape while editing. Most industrial models include Genlock (for locking in to other video sources without time base correction), a keyboard, expansion port and computer interfacing. The keyboard may have a color key to change the color of characters, backgrounds or objects; an object key to display, change or delete an object; and a page key to shift to any page in the unit's memory. The VET usually retails for several hundred dollars and includes several fonts, the capability of storing many pages of titles in memory and special function keys such as color key, object key, page key and clear key. Although a video effects titler is similar in many respects to the more costly character generator, the latter has more sophisticated features, such as direct and sequential page access, edit capability with full cursor control, screen blank function and the ability to generate its own sync without the presence of video. In addition, an industrial-model character generator may also feature preview output, video fade control, key output, loop through and BNC connectors.

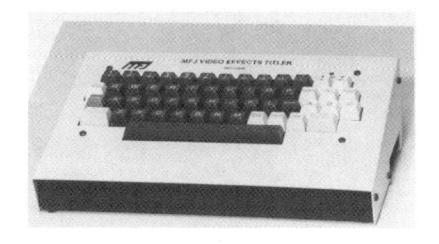

Figure 44. A video effects titler for superimposing computer-generated color images over another video image coming from an external source such as a video camera, VCR or videodisc. (Courtesy MFJ Enterprises, Inc.)

Video Encryption System. A method of encoding video material for security reasons. There are several systems employed for video encryption. Some are designed to prevent access to unauthorized persons; others have the capabilities to destroy video and audio data; still others are accessible through the use of certain decoder devices. Both transmitted signals and videotapes can be encrypted, depending on the system used. Some systems, such as Macrovision, which configures the vertical blanking lines, are chiefly used with consumer videocassettes. Other, more sophisticated systems, are used exclusively in professional situations. Current professional video encryption systems, which because of the high cost can usually be rented, utilize either analog or digital encryptors. See ENCRYPTION.

Video Enhancer/Stereo Audio Mixer. An accessory unit that combines the capabilities of an image enhancer with selected audio features. The stereo audio portion of the unit allows the user to add narration and/or background music to a home video tape. These accessories usually offer several stereo audio inputs and separate volume control knobs along with video signal control. See IMAGE ENHANCER.

Video Equalizer. See EQUALIZER.

Video Expo New York. An annual video exposition at which manufacturers, dealers and other related companies display their video equipment and supplies. The Video Expo, which held its first show in 1969, is aimed chiefly at the professional video user. Each show usually features more than one hundred exhibition booths. See NEW YORK EXPO OF SHORT FILM AND VIDEO, TOKYO VIDEO FESTIVAL.

Video Feedback. A simple special video effect created by aiming a video camera at a TV set used as a monitor

Video Frequency Converter

during taping. By zooming in on the image, the camera records an infinite number of images of itself videotaping itself. By changing positions, one can create an endless variety of images.

Video Frequency Converter. An electronic accessory designed to convert the color frequency of a composite video signal to a different frequency. For example, in some foreign systems in which a 4.43 MHz frequency is used, the video frequency converter can change this to be compatible with the NTSC domestic frequency of 3.58 MHz. A professional/industrial component, the converter permits the connection of multi-standard video recorders to NTSC-standard monitors and recorders for both playback and recording. See COMPOSITE VIDEO SIGNAL, PAL, SECAM.

Video Frequency Response. See FREQUENCY RESPONSE.

Video Gain. See GAIN.

Video Game Cartridge. Software for a video game system. There are various types of games. Some require fast thinking and manual manipulation, such as sports games. Others consist of board-type games like checkers and chess, requiring strategy. There are "slice-of-life" games, such as Lost Luggage. In addition, there are games of chance, like Blackjack, and action-humor games like Frogger. Some games, like those made for the Nintendo, Sega and Socrates systems, are more interactive than others, permitting the players to affect what happens on the screen. All video games require external buttons, joy sticks or other forms of control. The games require either one or more players and often contain variations of different levels of difficulty. One cartridge produced for one particular system will not fit another game system. Some games were successful first in arcades before becoming home video cartridges. Games, like most videodiscs and prerecorded tapes, attain a degree of popularity and quickly fade into oblivion.

Video Game Hardware. The system upon which video games are inserted and played. Hardware may be dedicated game systems (such as those manufactured by Atari, Nintendo and Sega), personal computers, combination computer/game arrangements or VCRs.

Video Game History. The first electronic video games designed to operate with TV sets were introduced in 1971 by Magnavox. Developed by Ralph Baer, the company brought out an unsuccessful series of simple games that required plastic overlays to be attached to the TV screen. In 1972 NOLAN BUSHNEL brought out his game, PONG, through his newly founded company, Atari. Other companies joined in the success of Atari but, like Pong, all these systems were non-programmable until Fairchild's Channel F came along. Fairchild's entry failed commercially, but Atari and Magnavox were quick to introduce programmable systems. By the late 1970s game consoles not only accepted cartridges which offered a greater variety, but color was introduced. Atari presented its VCS (Video Computer System), Magnavox returned with Odyssey[2] and Mattel marketed Intellivision in 1980. Programmability ended obsolescence and added flexibility to the video game market, which generated revenues of over $3 billion by the early 1980s. But by 1985 the game industry screeched to a virtual halt, generating only $100 million. By the end of the decade the market rebounded, thanks in part to advanced microprocessing and Nintendo, the successful game system that imposed its quality control upon companies licensed to produce games for its system. The new games boasted superb graphics, sound and sophistication that challenged arcade-type games. Atari, Nintendo and Sega were the leading companies in the expanding field, with Nintendo chalking up the most sales. Its system has sold in the millions, as has its cartridges. Although home video games originated in the United States, the more complex arcade video games are Japanese products and licensed to American firms.

Video Game Recording. The output of a video game placed on tape by way of a VCR. The RF output of the game may be connected to the VHF input of the recording machine. The VCR is set at an open channel, either 3 or 4, and the Record button is pressed.

Video Game Sales. The following figures represent approximate sales of video games in the United States.

1980:	$500 million
1981:	$1 billion
1982:	$3 billion
1983:	$2 billion
1984:	under $1 billion
1985:	$100 million
1986:	$500 million
1987:	$1 billion
1988:	more than $2 billion
1989:	more than $3 billion
1990:	$5 billion (projected)

Video Game Software. The games, or game cartridges, disks and videocassettes, that fit the various video game consoles or systems. The system manufacturers not only supply games for their own consoles but produce games for other systems. Many independent software companies also turn out games for one or more systems. For example, about 54 companies are licensed to market games for the Nintendo Entertainment System.

Video Grand Prix Awards. An annual presentation sponsored by *AudioVideo International* magazine. Dealers nominate various equipment models, and a panel of prominent journalists who test equipment judges the models. The following criteria are given to both dealers and panelists: fidelity of video reproduction, excellence of design engineering, reliability, product integrity and craftsmanship and high value-for-price ratio. The categories include color TV to 21-inch screens, color TVs 25- to 27-inch, color TVs with over 30-inch screens, projection TVs, LCD TVs, videodisc players, non-MTS VCRs, MTS VCRs, full-size camcorders, compact camcorders, blank tape, accessories, engineering awards, special citations.

Video Graphics. See GRAPHICS.

Video Graphics Generator. An electronic accessory designed for commercial displays, closed circuit TV

Video Head

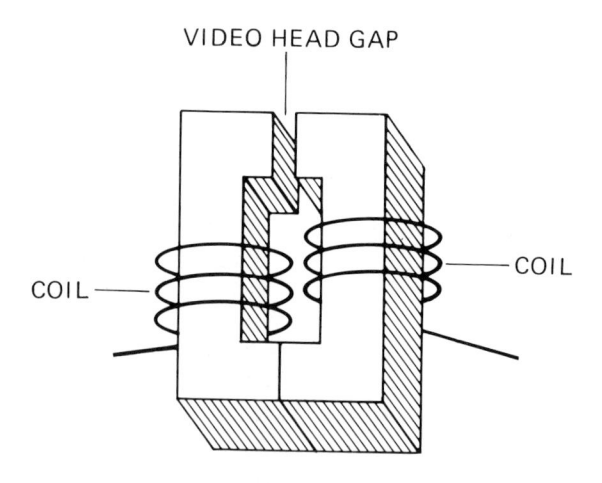

Figure 45. Video head.

applications and, through the use of color graphics and characters, the creation of pictures on the TV screen. Some graphics generators permit storing up to 15 pictures in its memory, provided the power is not turned off, while other models allow superimposing images on present recordings. Generally, the graphics produced by these generators lack the detail of computer graphics. The units resemble the detachable keyboards of computer terminals. Some low-cost video graphics generators do not reproduce the same quality image on videotape as they do on the television screen. See GRAPHICS.

Video Head. A magnetic unit consisting of a small metal housing, a coil through which a signal is passed and a narrow gap from which a magnetic force places video information on the videotape as it passes by. The basic VCR has two heads, each of which places down a diagonal field on the tape. Some VHS machines have four heads, two for the fastest speed and two with a special gap for the slowest speed. Still other VHS recorders have four heads, but with different functions. One pair is for recording and playing back while the other two heads are for special effects such as slow motion and freeze frame. In some Beta VCRs, two heads record and play back while one head is reserved for producing clear special effects. Video heads are very delicate and require special care and cleaning.

Video Head Alignment. See ALIGNMENT.

Video Head Cleaner. An accessory, such as individual swabs, complete kits and special cleaning cassettes, that dissolves or removes dirt, oxide particles, dust and grease from heads and other critical parts of a VCR. For those owners who are willing to open the top of their machines, special foam or chamois swabs and cleaning solvents are available with instructions on how to clean critical elements. There are kits for sale with swabs, solvent, cloth and other items. For others who prefer not to tamper with the VCR, cassettes can do the cleaning work for them. There are two types. The dry method cassette is simply inserted for a few seconds, just like an ordinary cassette, and a cleaning cloth wipes clean the important areas. The wet system requires dampening the special cloth in the cassette with drops of solvent before the cassette is inserted and played. Some technicians claim that the dry method is more abrasive and can eventually affect the heads. The frequency of cleaning has remained a controversial topic for several years. Some advocate cleaning at regular intervals; e.g., every month or after 40 or 50 hours of play. Many more claim that cleaning should only occur when a problem such as "snow" or video noise appears on screen.

Video Head Clogging. A condition caused by an accumulation of oxide particles peeling from tape, dirt, dust and other foreign matter on the head of a VCR. Clogging results in snow appearing in the TV image and other signs of picture interference. See VIDEO HEAD CLEANER.

Video Head Drum. A cylindrical component that holds the video heads. Located inside the VCR, the drum rotates at 1,800 rpm and is set at an angle. The diameter of the drum varies according to the format of the machine. Beta VCRs usually have a 74.5mm drum while the VHS format is slightly smaller. The Beta drum has three sections, the top and bottom, which remain stationary, and the center portion containing the heads. This latter part is called the SCANNER and is the section that rotates. In the VHS system the entire drum rotates.

Video Head Gap. See GAP.

Video Head Separation. The position of the video heads as they appear on the head drum of a VCR. A typical two-head machine has each of the heads mounted 180° apart around the drum. Most VCRs with four heads have each head equally located 90° apart. Some machines with four heads, on the other hand, place two two heads adjacent to each other while their two counterparts are on the opposite side of the head drum. This radical arrangement minimizes NOISE BARS, thereby improving such special effects as FREEZE FRAME. See VIDEO HEAD, VIDEO HEAD DRUM.

Video Head-Switching Noise. See HEAD-SWITCHING NOISE.

Video High Density. See VHD VIDEODISC SYSTEM.

Video Home System. See VHS.

Video Image. See IMAGE.

Video Image Compositing. A professional/industrial system that employs special screen correction circuitry to produce improved multi-layered compositing in post-production work and composites with imperfect blue screens. In addition, the system allows for realistic composites with natural shadows in a multi-layered image without the usual darkening in the corners.

Video Input

Video Input. A jack or receptacle on a VCR or other component that accepts video signals. It is often used in place of an RF input to connect the machine directly to the video signal, thereby producing a picture of higher DEFINITION. If a video signal is modulated to RF (as when using a TV receiver), then demodulated to its original form, some loss of definition occurs. The video input is also utilized for recording from another VCR's VIDEO OUTPUT. See COPYING, DIRECT AUDIO AND VIDEO INPUTS, VIDEO SIGNAL.

Video Insert. A video camera feature that permits recording a new image over the previous one without affecting the audio portion of the tape. On some cameras the insert is performed in the following way. The camera in placed in VCR mode and the tape advanced to the part where the insert is to start. Next, the Pause control is pressed to stop the tape. Then Insert is activated. When the camera trigger is pressed, the new video information will be recorded while retaining the original audio material. The video insert feature is especially useful in adding titles to recorded videotape.

Video Inserter. A stand-alone accessory designed to superimpose messages and graphics onto video images. Chiefly a professional/industrial device, the video inserter conforms to several standards, including NTSC, PAL and SECAM. Some models have the capability to display information from integral memory or from external sources.

Video Jukebox. A commercial console containing a 25-inch TV screen and a video recorder with tape for 40 to 48 selections. The jukebox, no longer available, offered the customer who deposits coins the opportunity to see the entertainers as well as hear them. The video jukeboxes were expensive and were targeted for bars, record stores, airports, restaurants and movie theater lobbies. The concept was developed by a Los Angeles firm, Video Music International.

Video Keyer. A professional/industrial post-production unit that can combine several key sources over a single background picture. Other features include individual clip adjustment for each input, advanced edging effects and full linear keying through a a wide-range gain control. See KEY.

Video Level. Refers to the degree of brightness or contrast in the screen picture. If the picture is too bright, the detail in white portions of the image are lost; if the picture is too dark, detail in the shadows or dark portions of the image suffers.

Video Line In. See VIDEO INPUT.

Video Moire. See MOIRE.

Video Music. A technique employed to display music graphically on a TV screen. The intensity of music being played on an average stereo system is translated into visible patterns, shapes and colors. The TV image constantly changes as it responds to the music, allowing the viewer to "visualize" the sound. In 1978 Atari sold a six-pound unit for approximately

$100 and called it Video Music. It contained five controls and 12 pushbuttons and worked in conjunction with any TV set as well as with any stereo system. Video music also refers to a creative process using computers, the laser and other techniques to accompany films, video performances and live shows. Videocassettes such as Blondie's "Eat to the Beat" featured this form of music. Video or visual music can be said to date back to silent films when experimenters worked with animation set to music. The next important development was Walt Disney's film *Fantasia* (1940). Today many independent filmmakers and video artists are using new techniques to bring video music to the attention of large audiences. Hundreds of videotapes have already appeared on standard and CABLE TV worldwide. See VIDEO ART, VIDEO ARTIST.

Video Noise. Interference or an unwanted signal in the television picture, usually in the form of dark or light horizontal lines, bars, etc. Video noise, which causes "snow" in black-and-white pictures and dark blotches in color picture, is measured in decibels (dB). For instance, on a TV MONITOR the difference between a picture received through an RF antenna terminal as opposed to one picked up through the preferred direct audio/video inputs could represent a drop of five or six dB in picture noise. Video noise affects color clarity and image sharpness. The higher the number of the signal-to-noise ratio, the better the image. Thus, 45 dB and up is considered better than average; 43 dB, average; and below 40 dB can be considered as below average. NEC brought out its digital video noise reduction system in 1986. The electronic technique constantly compares video information on a field-by-field basis in an attempt to improve luminance signal-to-noise ratio. The term "digital" refers to the translation of a signal into numbers. And since numbers are constant and remain intact when transmitted or recorded, they protect the information from external video noise and other interferences—at least, theoretically. No present video noise reduction system can eliminate all video noise; what it can do is improve the viewing of recorded images and those tranmitted from fringe areas.

Video Noise Reduction System. See CORING NOISE REDUCTION, DIGITAL VIDEO NOISE REDUCTION SYSTEM, FIELD CORRELATION.

Video-on-Sound (VOS). A VCR editing feature that permits the user to record a video signal over an earlier recorded Hi-Fi audio track. This provides more versatility in mixing audio and video. Hitherto, picture and sound tracks were recorded simultaneously since they are both written in the same area on the videotape.

Video Output. A jack or receptacle on a VCR or other piece of equipment to transmit the video signal to another unit. For example, when COPYING a videotape onto another machine, a cable is connected from the video output of the player to the VIDEO INPUT of the recording VCR. (The audio input and output are also connected.) Video jacks usually accept RCA phono plugs. See VIDEO SIGNAL.

Video Palette

Video Palette. A special electronic console introduced in 1980 and developed by Ampex consisting of a steel pen, a white tray palette and a TV monitor. The device, called the Ampex Video Art, allows the artist to create color graphics electronically. See DIGITAL EFFECTS VIDEO PALETTE.

Video Periodicals. Magazines that deal essentially with video software and hardware and related areas. VIDEO, which began operations in 1978, was the first consumer magazine on the market. It went from a quarterly to a bi-monthly, then to a monthly. VIDEO REVIEW soon followed as a monthly, then came HOME VIDEO, followed by another monthly, VIDEO ACTION, which eventually switched to a quarterly in 1982. During this time other magazines emerged: VIDEOPLAY and VIDEOGRAPHY, VIDEO SWAPPER, which changed its name to VIDEO ENTERTAINMENT in 1982, VIDEOMAKER, a bimonthly devoted entirely to the video camera user, and a host of quarterlies, buying guides and annuals. In a related area, video game magazines began to appear in 1982, the first going under the title Electronic Games. Soon these periodicals found their own niche on the racks.

Video Periodicals: Industrial/Professional. The increasing use of video in business and education has given rise to various related literature, including monthlies, quarterlies and annuals—all aimed at the users, dealers, distributors and manufacturers of industrial/professional video hardware and software. Some magazines like AUDIO-VISUAL COMMUNICATIONS and AUDIO VISUAL DIRECTIONS have focused more on photographic production and equipment and only peripherally on video, while others such as VIDEO USER and VIDEOGRAPHY have been totally committed to video. Video Store calls itself "the journal of video retailing." Other magazines, such as AV VIDEO, CORPORATE VIDEO DECISIONS, INVIEW, AUDIOVIDEO INTERNATIONAL, BROADCAST ENGINEERING, BROADCASTING, MILLIMETER, PRESENTATION PRODUCTS MAGAZINE and TELEVISION/RADIO AGE deal with trade news, satellite communications and broadcasting in general. Television Factbook and The Video Register are annuals, the latter including a complete list of video publications.

Video Piracy. The illegal duplication and sale of copyrighted material or receiving pay TV programs for free via an illegal decoding device. Video piracy began almost as soon as VCRs proliferated in the consumer market. The demand for Hollywood "blockbuster" movies was met with illegal tape copies made from master prints or from copies distributed to theaters. The Motion Picture Association of America estimated that hundreds of millions of dollars were lost to the film industry each year because of the illegal market. To combat these thefts, the MPAA organized its film security office. The pirates only responded to public demand. When the owners of Star Wars announced originally that the film was not going to be released on tape, the film became an underground best seller by way of pirated tapes, each going for hundreds of dollars. Finally, in the early 1980s, the authorized tape version was released. In the early years of video piracy, when the topic received little notice, the law was relatively lenient on the usurpers. The piracy problem has abated since major film companies have agreed to release their films for rental and sale at relatively low prices. The ease with which copies can be made, the availability of duplicating equipment and the readiness of the public to pay a premium price taught the studios a harsh, economic lesson. Illegal decoders, on the other hand, have given rise to another type of video pirate. This viewer is usually one of the several million owners of a satellite TV system. Enterprising outfits have sprung up advertising an array of decoders that can unscramble such cable services as Home Box Office and Bravo. However, the pay services are constantly retaliating by altering their signal codes so that these devices eventually become useless. See ANTI-PIRACY SIGNAL.

Video Printer. A device capable of producing printed pictures in color from a VIDEO CAMERA, a TV set or a VCR. The invention of the video printer changed the entire concept of video, previously restricted to an image on a screen. Printers operate in various ways. Some work on the principle of a thermal printing head receiving video signals. Depending upon signal intensity, the head generates different degrees of heat to dye sheets. As the dye evaporates, it is transferred to paper. Color video printers operate in much the same way, dividing each of the three primary colors into 64 gradations that offer more than 260,000 colors. These units have controls for color, tint, contrast and brightness and can print a picture in about 90 seconds. Personal color printers, either thermal or inkjet, allow the user to bring up an image on a TV screen, fine tune the color and contrast and print out a hard copy. See VIDEO COPY PROCESSOR.

Video Processing. Refers to the electronic alteration of a video signal. This may take the form of adjusting the color or brightness of the signal during copying. See IMAGE PROCESSOR.

***Video Review*.** The second oldest video consumer periodical on the market. Making its debut in April of 1980, the magazine usually features excellent graphic design and was known for its emphasis on movies before it began concentrating on its testing of VCRs, camcorders and other video equipment. Its film reviews have been written by such major critics and historians as Rex Reed and William K. Everson. *Video Review*, with its technical articles, product testing and occasional how-to articles, is considered the most successful video magazine.

Video Scanner. A computerized CCD unit designed to copy flat artwork, text, maps, photographs and other pictures. These graphic images can be altered with new colors and superimposed on video recordings.

Video Signal. A range of voltage made up of specific line scans, many of which carry specific information.

Video Signal-to-Noise Ratio

Some lines carry picture information, others are for synchronizing video components, several are assigned to the image-making beam and still others may carry TELETEXT data. The signal also has vertical lines (some of which are sync lines) and end-of-line pulses that are pre-set in size and duration. The bottom level of the video signal, for example, carries the SYNC PULSE, blanking and other control signals. Approximately 25 percent of the signal contains other than picture information. A video signal is not recorded directly on tape as an audio signal is. It is first modulated and then recorded. Also, the video signal differs from an RF SIGNAL. Video signals are affected in various ways by devices such as VIDEO DISTRIBUTION AMPLIFIERS, VIDEO SYNTHESIZERS, IMAGE ENHANCERS, IMAGE STABILIZERS and ANTI-PIRACY SIGNALS. See COMPOSITE VIDEO SIGNAL, SYNC PULSE.

Video Signal-to-Noise Ratio. A method of measuring the amount of distortion or VIDEO NOISE in a picture. Signal-to-noise ratio is expressed in decibels (dB). The higher the dB ratio, the less noise or distortion in the picture. Most home video components tend to fall into the 35 to 45 dB range.

Video Single. A prerecorded videocassette introduced by Sony in Japan in 1982 and in the United States in 1983. Video singles, which are available in both Beta and VHS formats, contain two to four songs each and have an average viewing time of about 15 or 20 minutes. They are similar to 45 rpm records in concept.

Video Software Dealers Association. A group founded early in 1982 and catering to the thousands of retail stores that sell and rent prerecorded tapes. The VSDA holds annual conventions which draws its members of the home video industry from the U.S., Canada, the Far East and Europe. Each convention consists of seminars and workshops and exhibitors representing prerecorded video companies, accessory manufacturers, distributors, media publications and computer software firms.

Video Stabilizer. See IMAGE STABILIZER.

Video Still Camera. An electronic camera designed to use a small magnetic videodisc instead of photographic film. After recording, the disc is inserted into a viewer that displays still pictures on a TV set. Prints can be made on an accessory printer. Many companies have experimented with these cameras, including Sony, who demonstrated the first video still camera in 1982. Similar in looks and feel to a 35mm model, the video still camera uses a single chip CCD (charge-coupled device) instead of a picture tube to take video pictures. Power stems from rechargeable nickel-cadmium batteries. The camera offers high picture resolution, re-usable discs, instant replay on TV and individual prints from a special optional copier.

Video Switcher. See PRODUCTION SWITCHER, SWITCHER.

Video Synthesizer. An electronic console designed to transform video signals from various ordinary sources into an elaborate, boundless variety of patterns. Developed by VIDEO ARTIST NAM JUNE PAIK and Shuya Abe, the synthesizer, of which there are several versions, is intended to turn TV viewing from its passive absorption of images into an active medium. Another video artist working with this technique is STEPHEN BECK.

Video Systems. A monthly trade magazine mailed free to professional and industry personnel. Specializing in professional/industrial techniques and equipment, the publication offers articles on such topics as video projectors, character generators and choosing a video encryption system. Departments cover video news and business, audio topics, computer techniques, people in the video world and new products. The informative periodical uses plenty of color photos and graphics to support its feature articles.

Video Teleconferencing. See TELECONFERENCING.

Video Telephone. An experimental project that allows telephone callers, using a specially equipped booth, to see one another as they speak. The booth contains a small screen with a split image, each displaying one of the persons. Video phone booths may offer other services, such as transmitting and receiving fax messages and recording videotapes. The first public use of video telephones occurred in 1989 in Berkeley, California. The cost of a video phone call—ten dollars for three minutes.

Video Test Chart. A method of testing or measuring one aspect, function or feature of a video component. There are various test charts. For example, a color bar chart may be used for a video camera set-up or for recording a color reference on tape. Containing the primary and secondary colors, it checks the color accuracy of lenses, color separation and other related areas. A resolution chart measures the sharpness of camera images; in addition, it checks other aspects of a video camera, including camera streaking, ringing and aspect ratio. A linearity chart measures a camera's scanning linearity and assists in its adjustment. A registration chart is designed to test cameras for scan height, skew, width, rotation, linearity and centering of each channel by using a special grid pattern. A gray-scale chart, progressing in steps from black to white, measures differences in phase and gain. A flesh tone reference chart checks for skin and hair color, avoiding the use of a live subject during camera setups. A multi-burst chart is used in testing and adjusting camera system response. There are also multi-chart systems for professional use. One two-chart system, for instance, contains a registration and color balance chart designed to check alignment, registration and color balance. Video test charts are not inexpensive. They can cost as much as $300 or more. See COLOR BARS, GRAY SCALE, LINEARITY, REGISTRATION, RESOLUTION.

Video-to-Film. A process in which theatrical productions are first produced on videotape and then transferred to film. Using the latest video technology including HIGH DEFINITION TV and ELECTRONIC BEAM SCANNING, film studios can lower production and post-production costs. Early experiments in video-to-film have produced the T.A.M.I. Show (1964) and the Big T.N.T. Show (1966), taped rock shows transferred to film for theatrical release. Film directors George Lucas and FRANCIS FORD COPPOLA have experimented with video-to-film techniques.

Video Waveform. The display of a video signal on an OSCILLOSCOPE. The signal is dissected into its various components which can then be examined for integrity. See VIDEO SIGNAL, WAVEFORM.

Video Windowing Display Controller. A professional/industrial editing controller that integrates real-time video with computer-generated text and graphics on a monitor. NTSC and PAL real-time video appears on screen as a window which can be positioned, scaled, clipped and superimposed with computer graphics. The resulting images can then be digitally stored for future reference. Video windowing controllers, which accommodates several signals connected at the same time, can usually receive inputs from a video camera, recorder, live TV or interactive videodisc. See EDITING CONSOLE, EDITOR.

Videocassette. A two-hub, plastic shell or housing which holds tape. The average videocassette consists of between 35 to 40 parts. There are consumer and industrial cassettes, the latter more firmly constructed both externally and internally. Although many cassette shells and tapes are made by one manufacturer and sold under various brand names, many companies like TDK, 3M, Fuji and Maxell make their own shells and tape. The quality also varies among consumer videocassettes. There are differences in the quality of the plastic, in construction and in internal parts such as clamp locks.

Videocassette Adapter. An accessory that allows VHS-C mini-cassettes to be played in conventional VHS machines. Normally, the compact videocassette is placed into the adapter which is then inserted into VCR housing that accepts conventional-size cassettes. Some machines provide built-in adapters that accept the compact cassettes directly.

Videocassette Player. A videocassette deck that has the appropriate functions to play back prerecorded tapes but lacks those features needed to record. Obviously designed as a second unit or a primary machine for those who just want to view tapes, these players often have additional features, such as automatic rewind, on-screen displays (for the time counter), double speed play and high-speed search. Dubbing, of course, can only be done from a VCP machine to a VCR unit and not vice versa.

Videocassette Recorder. See VCR.

Videocassette Rewinder. See REWINDER.

Videodisc. A record that plays sound and pictures through a conventional TV receiver. Two major types of vid-

Videodisc Player

eodiscs are the LaserVision and the Capacitance Electronic Disc (CED). The grooveless LV disc is decoded by means of a laser; no arm or head makes physical contact with the disc. The CED system, which has virtually disappeared from the marketplace, used a grooved disc that required a stylus to read the information on the surface. The videodisc, popular in the early 1980s, fell into disfavor for most of the remainder of the decade, and made a strong comeback in the late 1980s. Its rebirth has been made possible by several factors, including technical advances in videodisc players, large-screen TV sets and a demand by consumers for a better-quality picture. A videodisc produces 400 horizontal lines for a clearer and sharper screen image than the conventional 240 lines of videotape, especially noticeable on large screens. Laserdiscs have another advantage. Because the audio and video information is protected under an acrylic shield and no stylus or head makes physical contact with the surface, the disc is virtually impervious to age or wear. Prerecorded programs generally sell for about the same price as prerecorded videocassettes. See VIDEODISC PLAYER.

Videodisc Player. An electronic unit, resembling a record player, that plays back pictures and sound from a prerecorded disc to a TV receiver. Although it cannot record, the VDP features some advantages over videotape, such as direct random access, longer-lasting software and better resolution. The two major systems are the LaserVision and the CED (Capacitance Electronic Disc). The la-

Figure 46. A videodisc player, using laser technology, with digital audio and video features. (Courtesy Pioneer Electronics.)

ser optical system uses a grooveless disc and offers stereo sound. The CED process employs a diamond stylus with a grooved disc similar to that of audio.

The first consumer videodisc system appeared in England in 1928. Invented by JOHN LOGIE BAIRD, the primitive disc produced poor picture quality and ran for only two or three minutes. AEG-Telefunken (Germany) introduced a videodisc playback system in 1970, similar to that of RCA's, but had limited success. The product was dropped after a few years of sagging European sales. Meanwhile, in the United States, a practical system was developed by RCA in the 1960s, but was not introduced in the marketplace until the 1980s. The LaserVision (LV) videodisc player was introduced in the U.S. on a limited scale in 1978, then nationwide in 1980. Like Baird's discs that sold in stores for almost ten years before the system faded into history, and Telefunken's, that lasted an even shorter time, the two present systems have had problems with capturing the attention of the consumer. Having to compete with less costly VCRs—which offered recording features as well as playback, the VDP dropped in sales during the mid-1980s and practically disappeared from the market. But by the

Videodisc Recorder

late 1980s, consumer interest had once again awakened in the videodisc player. Newer models play an array of disc sizes including the popular CD disc and offer a host of additional features. It has been estimated that in 1990 there were 350,000 videodisc players in use in the U.S. alone, as opposed to more than 65 million VCRs.

Videodisc Recorder. A professional/industrial machine that can produce 12″ videodiscs. Selling for about $20,000, these units record on both sides, provide standard 30-minute recordings of 54,000 frames per side and offer remote control. Videodisc recording, because of its inherent complexities and relatively high cost, presently remains outside the realm of the average consumer.

Videodisc Speed. The LaserVision has two speeds: CLV (Constant Linear Velocity) discs play for 60 minutes per side while CAV (Constant Angular Velocity) discs play for only 30 per side. Both CLV and CAV discs are compatible with all LV machines, but the 60-minutes-per-side disc cannot be used with freeze frame and other special effects. The CED format has a playing speed of one hour per side. See CED VIDEODISC SYSTEM, LV VIDEODISC SYSTEM.

Videographer. A term sometimes applied to one who uses a video camera or camcorder. The term comes from video photographer. Estimates of the number of fans who make their own home video movies range as high as six million.

Videography. A monthly magazine aimed chiefly at video professionals involved in telecommunications and production and post-production work. Articles and columns concern such topics as technology, production, video facilities and products. Technical articles may go into details about non-linear editing, audio-for-video and animation techniques using computer graphics and video. The publication covers various industry shows such as that of the SMPTE (Society of Motion Picture and Television Engineers).

Videomaker. A bimonthly consumer magazine targeted chiefly at the beginning and advanced camera user. The magazine provides occasional buyer's guides to equipment, feature stories on specific topics and problems such as generation loss, and how-to articles. Other information pertains to selecting camera accessories, directing video productions at home and producing shows for local cable stations. The regular departments and columns tend to concentrate on applying video camera equipment to specific uses and include many helpful tips. Although much of this information is familiar and elementary, it serves as a review for the experienced camera user and is especially helpful to the neophyte. Many of the articles are accompanied by black-and-white illustrations and photos.

Videotape. The software on which VCR video heads record and playback information. The tape consists of several elements: a BACKING that resists stretching and decomposition; a

Videotape Evaluator

COATING of microscopic particles that can be easily magnetized, hold their magnetic charge and resist shedding; and a BINDER that causes the particles to adhere to the base. Beta tapes usually contain chromium dioxide particles while VHS brands mostly utilize ferric-oxide particles that have been treated with cobalt. Early videotape stemmed from the technology of audio tape, with a handful of manufacturers supplying the many tape companies. But the quality improved rapidly, especially with the second generation, HG (high grade) tapes. Present tapes often consist of several layers, such as the magnetic particles and binder, an adhesive, a film base, carbon for added opacity, another film base followed by another adhesive, and a backcoating for improved reliability. With more manufacturers and increased competition, the prices have dropped sharply. See COERCIVITY, HG, HIGH DENSITY VIDEOTAPE, RETENTIVITY, VIDEOTAPE BRANDS.

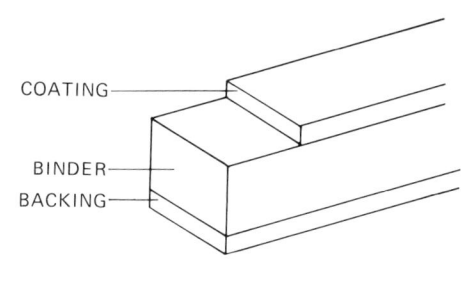

Figure 47. Videotape.

Videotape Brands. Only a handful of manufacturers produce videotape. Some of them sell their tape under their own brand name and also distribute tape to other companies who then package it under another name. Most of the brands provide a variety of lengths; e.g. T-60, T-90, T-120 and T-160 for VHS; L-125 L-250, L-500, L-750 and L-830 for Beta.

Videotape Chronology.

1956: Ampex introduced videotape recorders at the National Association of Radio and Television Broadcasters Show.
VTRs used 2-inch tape.
First TV network use of videotape on the CBS news show "Douglas Edward and the News."

1957: First use of videotape for TV commercial: Dennis James for Kellogg.

1965: First low-noise color videotape introduced by 3M company.

1966: First use of instant replay during the 1966 Rose Bowl Game. Taped TV shows and commercials began winning Emmy Awards.

1972: The 3M Company introduced videocassettes.

1976: Sony introduced 1/2-inch home video recorder format.

1980s: Manufacturers add higher grades of tape.

Videotape Cleaner. A device designed to clean and polish a videotape in several minutes. Some videotape cleaners provide additional features such as inspection and rewinding. Other models have a tape information display that gives tape length and the number and position of physical defects. See REWINDER.

Videotape Coating. See COATING.

Videotape Evaluator. A table-top device that utilizes LED read-outs to count tape defects. A videocassette is

Videotape Format

inserted into the evaluator that fast-forwards the tape at 25 times the normal speed. Various models are designed for U-Matic, Beta or VHS formats. The machine also cleans the tape during the defect-measuring process. At least one company (Research Technology International) provides an optional printer which supplies a permanent record of the conditions of each tape.

Videotape Format. Refers to the size, length and method of enclosure of videotape as well as to the speed at which it progresses past video heads. There are formats for both consumer and professional/industrial use. Generally, consumer formats come in cassettes or cartridges and include Beta II, Beta III, VHS, Super-VHS, Super-VHS-C, 8mm, Hi8; popular professional tape formats include D-1, D-2, 1-inch Type B, 1-inch Type C, 3/4-inch, 1/4-inch SP, Betacam, Betacam SP, MII, industrial Beta, HDTV.

Videotape Grade. A method a manufacturer uses to mark the quality of videotape. Although some video users believe there are no differences among the many grades of tape, manufacturers assert there are distinctions in their tape formulas and processes and that higher-grade tapes provide better audio and video results. Tests conducted by some leading video magazines tend to show some differences, but the technicians are quick to point out these differences generally are slight. Better-grade tapes usually are accompanied by such relative descriptions as "pro," "super," "high" or "extra" on their boxes. Some tests often list a high-quality standard tape as superior or equal to some high-grade tapes.

Videotape History. The 3M Corporation first demonstrated a recording of a two-inch videotape early in 1956. It was developed by three employee-engineers—Joseph Mazzitello, Melvin H. Sater and Andrew Nordloff. The tape was first used commercially on the CBS news program "Douglas Edwards and the News" on November 30 of that year. Then CBS relayed the tape three hours later to Hollywood for the West Coast viewers. Tape became popular with broadcasters because of its lower cost and instant replay ability, whereas film was more expensive and took time to process. See VIDEOTAPE CHRONOLOGY.

Videotape Lengths.

Cassette	Beta I	Beta II	Beta III
L-125	15 minutes	30	45
L-250	30	60	90
L-500	60	120	180
L-750	90	180	270
L-830	110	200	300

	(VHS) SP	LP	EP/SLP
T-30	30 minutes	60	90
T-60	60	120	180
T-90	90	180	270
T-120	120	240	360
T-160	160	320	480 (8 hrs.)

Videotape Quality. The ability of videotape to withstand certain pressures, resist others and retain information recorded on it. Videotape performance can be measured in three basic areas: video, audio and physical prop-

Videowall

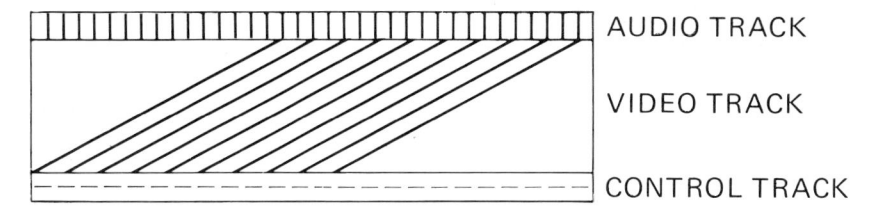

Figure 48. Videotape tracks.

erties. Video characteristics include VIDEO SIGNAL-TO-NOISE RATIO, CHROMA SIGNAL-TO-NOISE, video FREQUENCY RESPONSE, DROP-OUTS and FM LOSS (affecting the number of replays). Audio properties encompass audio signal-to-noise ratio, audio frequency response and several others. The physical aspects of tape that are often judged involve, among others, length, width, strength, stretching and evenness of tape edges. Tapes of extra length tend to be thinner and therefore more delicate. Some VCRs are not too gentle in removing tape from its cassette during the process of positioning the leader around the drum head. The width has to be as exact as possible to avoid the tape's moving up and down the tape guides and other critical parts. If the tape is overly wide, the edges may curl, thereby affecting either the audio track or the control track. Tape strength is equally important, especially if it is to be fast-forwarded and rewound often. If a tape stretches too much, the video and/or audio information may become distorted. Tape edges should be even, within close tolerances, or the top portion containing the audio track as well as the bottom part that holds the control track may be affected. See HG, HIGH DENSITY VIDEOTAPE.

Videotape Recorder. See VTR.

Videotape Thickness. A measurement that determines how much additional tape can be stored in a videocassette for extended playing time. Beta tapes use at least three different thicknesses. The tape in the L-500 and shorter-length cassettes measures about 20 microns, the L-750 tape measures 14 to 15 microns and the L-830 cassettes use 13-micron tape. VHS tape applies two thicknesses. The T-160 cassette uses 15.7-micrometer tape while the standard T-120 and others utilize tape measuring 20 micrometers.

Videotape Tracks. The diagonally placed signals that carry the video information recorded by the video heads of a VCR. The introduction of the AZIMUTH process of electronically printing video tracks diagonally, each with a different pattern, has largely discontinued the need for guard bands. The VCR during playback converts the recorded magnetic tracks on the videotape into a signal for display on a TV screen, usually with the help of an amplifier and speakers. The same video heads used for recording also "read" the information on tape during playback.

Videowall. An assembly of closely placed video screens working in unison and designed to display various

Videowall Controller

special visual effects such as single, "split" images across all monitors or multiple images. Presenting larger-than-life images in stores, shopping malls, exhibitions and other shows with large attendances, videowall presentations are instant attention-getters. They have gained popularity in Japan, Europe and the U.S. and were first seen in the early 1980s. By the mid-1980s videowall shows became professional productions originating in European trade shows. Various suppliers offer an array of videowalls, whose systems chiefly depend on digital video signals which can be stored in semiconductor memory, called a FRAMESTORE. Displays may be simple, using a small number of monitors which present a large TV image, or costly, highly sophisticated setups in which each TV monitor contains its own memory so that the videowall can display a variety of effects. Programs can be live or prerecorded on tape or videodisc. Layouts are described by the number of monitors; e.g., five screens across and three screens high (15 monitors) is commonly known as a 5x3 videowall.

Videowall Controller. A sophisticated electronic instrument or unit designed to program and manipulate the images of a group of closely placed TV screens usually on display at trade shows, special events and other places with large audiences. Videowall controllers usually provide such special features as interactive program selection and simultaneous control of several videodisc players.

Vidicon. A video camera tube that is the basic or standard tube used in most tube-type cameras. Introduced in 1952 by RCA, the vidicon tube in many ways was an improvement over the then-standard image ORTHICON. The vidicon had a longer life, was much less costly, required less light for producing a quality image and was only one-third the size. While both tube designs improved over the years, along with other types, the vidicon became the most widely accepted for non-broadcast television use. Two supposedly improved models were the SATICON and the TRINICON. Many home video camera owners claim that the average user would not be able to detect any significant advantages in any one tube over the others. See CAMERA TUBE.

Viewfinder. That part of the video camera that displays by various methods the scene that will eventually be recorded. There are three types of viewfinders. The OPTICAL, easiest to use and the least costly, is also the most limited since it cannot accommodate different focal-length lenses or zoom lenses and doesn't "see" exactly what the lens sees. The THROUGH-THE-LENS OPTICAL FINDER (TTL) has advantages over the optical type and costs more. It sees exactly what the lens sees, and when the zoom lens changes its angle of view, so does the finder. But the TTL finder does not have the special features of the third kind. The ELECTRONIC VIEWFINDER can be made mobile and can be used as a playback monitor. This finder is the most expensive even though its image is in black-and-white only. Viewfinders provide other types of information, such as whether the recorder is on, the condition of the battery, whether the amount of light is correct, the position of the fade control and other related features.

Viewfinder Diopter Adjustment. A video camera feature that allows the a near- or farsighted user to adjust the finder so that he or she can record without the need to wear glasses.

Viewfinder Inversion Switch. A control on some video cameras to permit the user to change the position of the viewfinder so that it can be used by either the right or left eye. This arrangement also allows for special upside-down effects.

VIR. See VERTICAL INTERVAL REFERENCE SIGNAL.

Virtual Editing. A technique used by an editor to record only decisions concerning editing, not audio or video. This editing system, used chiefly with industrial equipment, usually requires several decks and an editing machine. Selected scenes are recorded on different machines, each scene with extended footage at the beginning and end. These scenes are then shuttled to and from each machine. If a scene needed to be lengthened, the information was available. Finally, all the scenes would then be assembled in their proper sequence and length on one machine for the finished edited version. For an early version of virtual editing, see POSITION IDENTIFICATION.

VISS (VHS Index Search System). See INDEX SEARCH.

Visual Scan. A feature on a videocassette recorder that permits the rapid viewing of a tape. By first pressing the Play mode and then Fast Forward (or Rewind), the viewer can visually scan the material, skip undesirable portions or locate particular segments. When the fast scan mode is operating, the picture is "noisier" than usual and the sound track is muted. Introduced by Sony as Betascan, it is also known as fast scan, fast search, rapid picture search or high-speed picture search, depending on the brand and model.

VITC. See VERTICAL INTERVAL TIME CODE.

Voice Generator. A VCR feature that utilizes a microchip to produce a synthetic voice that takes the viewer through each step of the programming process. Once he or she has finished entering the necessary data, the ersatz voice reviews the information before it is saved. This voice-coaching method of programming the VCR was designed to help users who are often baffled by the complexities of setting up their machines for recording future off-the-air shows. The feature is also known as voice-prompt remote control programming.

Voice/Music Switch. See MUSIC/VOICE SWITCH.

Voice-over. Spoken narrative added to a picture either during recording or after the video camera has completed shooting. Some cameras have a rear microphone attached for adding voice-over while shooting.

Voice-Prompt Remote Control Programming. See VOICE GENERATOR.

Voice Synthesis Module. A video game accessory that adds speech to particular game cartridges. First introduced by Mattel for its INTELLIVISION, it livened up its series of games and proved strong competition for other game systems.

Voltage Spike Protector. An accessory plug that absorbs voltage fluctuations usually attributed to power line surges while permitting normal current flow. The spike protector, used to protect home computers and stereos as well as video equipment, plugs directly into the wall socket.

Voltage-Synthesized Tuner. A tuning system, installed on some VCRs, that requires manual tuning of those channels that are not part of the conventional VHS broadcasting band. These include many cable and UHF channels. These tuners, as opposed to frequency-synthesized tuners, provide a restricted number of presets that can be kept in memory. However, some VCRs offer as many as 100 presets that can be selected. See FREQUENCY-SYNTHESIS TUNER, TUNER.

Volume Control. In video, a variable resistor that can be manually adjusted to modulate the loudness of an amplifying unit or the audio portion of a TV receiver. The volume control may appear on the front of the set as a conventional rotating knob or small flat buttons while on the remote control panel the volume control may be two buttons (one to increase, the other to decrease loudness).

VOS. See VIDEO-ON-SOUND.

VTR. A videotape recorder utilizing an open reel format. The VTR consists of a supply reel and a take-up reel. VTRs cannot accommodate VIDEOCASSETTES or CARTRIDGES. Today, VTRs are mostly used industrially and professionally. Some of these machines utilize direct-drive capstan servo motors that permit a VTR to be synchronized electronically with any other format for editing purposes. In addition, these motors permit remote control, required when connecting the deck to an editor. VTRs may have other features, such as variable voltage power, not found on home VCRs. The introduction of digital technology to VTRs has enhanced videotape recording possibilities in both audio and video areas. Because the quality of digital recording remains constant, the process solves such problems as dropouts and moire, which have plagued conventional analog recording. In addition, digital tape life is much greater than analog.

VTR Type B Format. A professional/industrial reel-to-reel one-inch videotape system still popular in Europe but superseded in the United States, especially with TV networks, by the C format. Relegated primarily to production use, the B format still boasts of many loyal users who emphasize its special virtues. They find it useful in post-production work and, because it was the first format to feature long play, they prefer it for producing film-to-tape transfer masters. Also, the format is more economical, consuming over 200 feet of tape less per hour than its counterpart. Perhaps most important, its users consider it more reliable in the field. But the B format faces a few shortcomings and some major problems. It uses what is called a segmented head system, necessitating the addition of an expensive digital storage unit for certain functions such as FREEZE FRAME. The C units, on the other hand, use a less costly time base corrector to obtain these same effects. The B machines lack the ability to produce variable fast speed functions which the C units perform with relative ease. Finally, many large studios and networks have adopted the C format as their standard, thereby providing a modicum of commercials and programs in B format. See TIME BASE CORRECTOR, VTR TYPE C FORMAT.

VTR Type C Format. A professional/industrial reel-to-reel one-inch videotape system used widely in the U.S., particularly by TV networks, local stations and studios. The type C, which has become the industry standard, has largely replaced the B format. Although the two formats are similar in term of the spec sheets, the C units can produce variable fast speed functions for broadcasting and less costly functions such as freeze frame. In addition, a larger number of commercials and programs are produced in this format, more companies offer a wider range of C models and the units are less expensive than their B counterparts. See VTR TYPE B FORMAT.

VU Meter. A device that registers audio loudness or softness. Some VU (volume unit) meters have gauges divided into segments that represent DECIBEL levels while others are divided simply into a safe zone and a red (distortion) zone.

VX Format. A now defunct VCR system introduced by Quasar. The videocassette recorder was not compatible with any other format. Although the sets have not been marketed for years, Quasar has continued to supply tapes for the VX machines.

W

Wave. A physical action that moves periodically up and down or forward and backward as it journeys through a medium.

Wave Filter. A transducer (component that converts energy from one form to another) that separates waves depending on their frequency.

Wave Resonance. A technique employed chiefly in some compact television receivers to enhance the bass sound. This is accomplished by separating the bass from the treble and mid-range tones and reproducing and improving it so that it sounds more realistic. TV receivers with wave resonance try to emulate large, full sound within the limited, available space of their cabinets.

Waveform. A graphic picturization of an electronic signal. All signals such as video, blanking and sync have a distinct waveform which can be displayed on either an OSCILLOSCOPE or waveform monitor. Some video cameras display a waveform in the electronic viewfinder as part of the information needed for proper adjustment. Waveforms are most often alluded to in VCR repair manuals as a point of reference. Each waveform should be adjusted to match its drawing in the manual.

Waveform/Color Picture Monitor. A professional/industrial instrument designed chiefly to be used in the field. The unit provides such features as on/off switchable IRE filter, on/off gain boost, sweep and the ability to accurately master pedestal adjustments. See OSCILLOSCOPE, VECTORSCOPE.

Waveform Digitizer. An electronic component designed to connect to an oscilloscope, converting the scope to a digital storage unit. Using its own memory to produce waveform storage without a storage CATHODE RAY TUBE, the digitizer generates its own CRT read-outs for video sweep rates, deflection factors and other functions. See OSCILLOSCOPE, WAVEFORM.

Waveform Generator. A professional device designed to be used with a standard OSCILLOSCOPE. The waveform generator converts composite color video signals to standard waveform monitor signals. Typical generators show luminance, chrominance, direct picture information and other related data.

Waveform Monitor. See OSCILLOSCOPE, WAVEFORM/COLOR PICTURE MONITOR.

Waveform Sampler. An electronic unit that permits the measurement of the composite video signal so that video equipment can be adjusted for optimum signal quality. In addition, the sampler checks such video functions as amplitude and pedestal level. See COMPOSITE VIDEO SIGNAL, PEDESTAL LEVEL CONTROL.

Wavelength. The distance between two cycles, at which point there is the same phase.

White. In video, the mixture of red, green and blue in color television.

White Balance. Refers to the amount of color that can be seen on a neutral object when the white balance controls on a video camera are adjusted for optimum. White balance is measured in IRE: the lower the IRE number, the better the color purity. A perfect white balance would measure zero IRE. Adjustments for white balance are necessary when using a color video camera so that all color and light values will register as true as possible. Some cameras have automatic controls and meters. Others have flashing warning lights and a hue control dial that is rotated until the appropriate white balance has been reached. See WHITE BALANCE CONTROL.

White Balance Control. A feature on a video camera to help set or define colors. It does this by matching the white balance of the camera exactly to prevailing light conditions. There are different kinds of controls for this purpose. Red/blue controls are convenient but ineffective with green. Independent red and blue controls can affect green by their being turned all the way up (reducing green) or down (increasing green). Some cameras provide two tint dials: one for red and blue and one to balance green and magenta. These controls operate in one of several ways, such as automatically, by meters, by indicator lights or by click-stop positions. There are various methods to control white balance. One approach is to pre-set the red/green/blue signals at the factory so that the camera is set for tungsten light and using COLOR CONVERSION FILTERS behind the camera lens. A switch then determines indoor/outdoor position (one filter) or, in some cases, indoor/hazy/sunny (two filters). Another technique for white balance control is to pre-set the camera at the factory for average lighting conditions such as tungsten, hazy light and bright sunlight. The proper level is then set electronically by a switch. For more flexibility, a red/blue knob is added to change the relative balance of red and blue in the picture. The knob controls red at one end of its turn and blue at the other. The effects can be witnessed on a color monitor or on cameras with a meter. Other cameras feature AUTOMATIC WHITE BALANCE while still

White Balance Hold

others provide three controls: a switch to choose a filter or an electronic setting; then auto white balance is set; finally, the manual red/blue knob is used to check or override the white balance. See MANUAL WHITE BALANCE.

White Balance Hold. A video camera feature that, when engaged, "sees" a white object and locks the unit into this position to maintain proper color as long as lighting conditions remain the same. See WHITE BALANCE CONTROL.

White Clip Level Extension. Part of the HQ (high quality) CIRCUITRY of particular VCRs. The white clip level phase of the circuit is designed to provide sharper edges and a more distinct contrast between light and dark portions of the picture. See WHITE PEAKING CIRCUIT.

White Level. In video, the carrier-signal level that matches the maximum picture brightness in television.

White Peak Carrier. Part of a carrier wavelength that holds the luminance signal. When the white peak carrier is increased, the amount of space available for recording video detail on tape is expanded. This contributes to the number of horizontal lines of resolution, which, in some cases, may be raised to 400 lines or more. See CARRIER FREQUENCY.

White Peaking Circuit. Advanced circuitry used on some TV monitor/receivers designed to produce a purer white on screen. This is accomplished by reducing the red beam of the electron gun and amplifying the blue beam.

Wideband Video Amplifier. A special electronic circuit, often found in TV monitor/receivers, designed to capture and reproduce the entire range of an incoming video signal. The wideband video amplifier boosts the video signal by raising its frequency response, which affects the horizontal resolution. TV monitor/receivers equipped with this active circuitry often attain 500 lines or more of horizontal resolution.

Wind Noise Switch. A camcorder feature that helps to minimize some types of unwanted extraneous audio interference that may reach the built-in microphone.

Wind Screen. A heavy foam rubber cover for a microphone, used outdoors to decrease wind noise and other similar types of audio interference.

Wipe. A special editing effect in which one image replaces another by means of a predetermined pattern. In video, hundreds of different wipes can be produced by using a professional device called a MIX/EFFECTS SWITCHER. Two of the more popular wipe effects are the ROTARY WIPE and MATRIX WIPE.

Wireless Cable Association. An organization of cable operators who use microwave transmitters, as opposed to coaxial cable, to broadcast cable programming to subscribers. By 1990, about 300,000 subscribers received wireless cable service, which requires a special microwave antenna and an addressable receiver/descrambler. See MULTIPOINT DISTRIBUTION SERVICE.

Wireless Infrared. See REMOTE CONTROL.

Wireless Remote Control. See REMOTE CONTROL.

Wise, Howard (1904–). Video art enthusiast, distributor and promoter; art gallery owner. He held the first video art survey in his New York gallery in 1969, entitled "TV As a Creative Medium." He closed his Howard Wise Gallery in 1970 and created the nonprofit Electronic Arts Intermix, an organization dedicated to working in video as art. With financial aid from the New York State Council of the Arts, he helped establish in 1973 an electronics equipment facility for video artists. By 1974, the EAI took on the job of distributing works of video artists. Howard Wise has produced a video catalog of excerpts from approximately 160 works of video art by 60 artists. See ELECTRONIC ARTS INTERMIX, VIDEO ART, VIDEO ARTIST.

Wollaston Prism. In a laser-type videodisc player, an optical component through which the laser passes before reaching the disc surface. It then returns in polarized form.

Word Graphics. The titles, credits, announcements or other word messages that appear superimposed on the TV screen. Word graphics may crawl up or down on the screen or move horizontally across the bottom, announcing news flashes, election returns and so on without interrupting program content. Word graphics are usually white or light-colored against a dark background, often created by means of a technique called "keying." See KEY.

Word Register. See DIGITAL IMAGE SUPERIMPOSER.

Wow. Refers to tape speed variations that result in the distortion of the audio signal.

Wrap-Around Theater Sound. See SURROUND SOUND.

Writing Speed. The effective speed at which videotape moves past the recording heads in relation to the tape travel speed. On both VHS and Beta VCRs, the video heads rotate at 1800 rpm while the linear tape speed for each differs (3.34 centimeters or 1 5/16 inches per second in SP mode for VHS format, 2 cm/sec for Beta II). Therefore, the effective writing speed for each format in its most popular mode equals 230 inches per second for VHS and 170 ips for Beta.

WTBS. A SUPERSTATION transmitted by satellite, offering baseball, basketball, professional wrestling, movies, TV reruns and some original shows. WTBS, owned by TED TURNER, was the first superstation in the United States. It began operations in 1976 and offers programming 24 hours a day. The third most popular cable TV advertiser-supported network, the superstation as of 1990 had more than 52 million subscribers.

X-Y-Z

X-1. The original Beta speed, providing 60 minutes of recording and playing time on an L-500 videocassette. Today, Beta I, as it is now known, is reserved for professional and industrial machines. Beta home VCRs usually offer Beta II and Beta III speeds with some machines providing Beta I only in playback. On the same L-500 cassette Beta II has a play time of two hours whereas Beta III extends the play to three hours.

Y Adapter. A connecting audio cable that is used to join two lines into a single input or output. The Y adapter can be applied to many tasks, such as copying a stereo tape onto a mono VCR. Using the stereo recorder as the playback machine, the owner connects the Y adapter from the two audio channel outputs of the stereo to the single audio input of the mono VCR. The adapter, sometimes called a Y connector or Y splitter, comes with various phono plugs or jacks.

Y/C Connector. A multipin input that helps to eliminate several types of video interference by processing the brightness (Y) and color (C) portions of the signal separately. Previously, the luminance and chrominance signals were mixed and had to be separated by the TV receiver. By avoiding this intermediate step, the Y/C connector eliminates such interference problems as crosstalk. Y/C connectors, usually installed on relatively more recent TV monitor/receivers and other similar equipment, accept connections from Super-VHS and ED-Beta VCRs and some models of laserdisc players. These inputs are sometimes known as S-video inputs or S-connectors.

Yield Strength. Refers to the degree of force necessary to produce a 5-percent elongation in a videotape. If a tape is stretched beyond this point, it may affect the overall quality whereby the tape is unwatchable. See VIDEOTAPE.

Zone Blanking. A procedure for turning off the cathode ray tube during part of the sweep of an antenna.

Zone Satellite. See SATELLITE FOCUS.

Zoom. To increase or reduce the size of a television image, usually in a gradual way. Zooming can be accomplished by means of electronics or optics. See DIGITAL ZOOM.

Zoom Lens. A lens that is capable of changing its focal length or range of view. Most video cameras come equipped with a standard 6:1 zoom lens (relationship of the longest focal length to its shortest) while some consumer models offer an 8×, 10× or even 12× variable power zoom lens. The zoom may go from wide angle through normal viewing to close-up or vice versa. This feature is usually adjusted by a ZOOM RING, a ZOOM RING LEVER or a POWER ZOOM which works electronically. The focal lengths for video lenses are different from those used in still photography. For example, a 12–75mm range in video would be equivalent to a 45–350mm lens on a 35mm camera. There are three types of zoom lenses: manual, power zoom and two-speed power zoom. The manual zoom is less costly and provides more control over the speed of the zoom, but is not as smooth as the other methods. Power zoom is simple to operate and convenient, but costs more. In addition, its zoom speed cannot be controlled. The two-speed zoom seems to overcome the disadvantages of the manual and the simple zoom. See MULTI-FUNCTION ZOOM LENS.

Zoom Light. An accessory normally used indoors with a video camera. Some zoom lights can be connected to wall outlets while other models, if used in the field, can be plugged into

Figure 49. A zoom light designed to be connected to the cigarette-lighter socket of an automobile. (Courtesy Vivitar Corp.)

the cigarette-lighter socket of an automobile.

Zoom Ratio. The telescoping range of a lens. The ratio depends on how close or large the subject appears in the finder compared to its original distance or size. The typical zoom ratio is 6:1 although some cameras feature an 8:1 ratio. EXTENDER LENSES are available; these extensions increase the zoom ratio. For example, a particular extender lens may increase an 8:1 ratio to a 12:1 telephoto zoom.

Zoom Ring. A control encircling the zoom lens. Its back-and-forth movement permits the mechanical adjustment of the focal length. When the zoom ring is rotated, the focus is changed. Other zoom controls are the ZOOM RING LEVER and the POWER ZOOM.

Zoom Ring Lever. A control on the zoom lens with an extended handle to help facilitate changing the focal

Zoom TV Control

length. An intermediate control, it is easier to operate than the basic ZOOM RING, but not as sophisticated as the POWER ZOOM.

Zoom TV Control. On a TV set, a feature that enlarges the center of the image approximately 50 percent. One of Zenith's 1979 models introduced this feature which operated by pressing a button on the remote control. The zoom tended to increase the snow and grain along with the image.

Zworykin, Vladimir (1889–1982). Considered by many as the "father of television." While working at Westinghouse in 1923, he patented the ICONOSCOPE camera tube. In addition, he invented the KINESCOPE picture tube in 1929. The former led to an eventual working model of a video camera. The two inventions together composed the first electronic television system. When his firm, Westinghouse, showed a lack of enthusiasm for his television invention, he went to work for DAVID SARNOFF at RCA.